Reactive Oxygen Species

Mohammad Faizan • Shamsul Hayat •
S. Maqbool Ahmed
Editors

Reactive Oxygen Species

Prospects in Plant Metabolism

Springer

Editors
Mohammad Faizan
Botany Section, School of Sciences
Maulana Azad National Urdu University
Hyderabad, India

Shamsul Hayat
Faculty of Life Sciences, Department of Botany
Aligarh Muslim University
Aligrah, India

S. Maqbool Ahmed
Botany Section, School of Sciences
Maulana Azad National Urdu University
Hyderabad, India

ISBN 978-981-19-9793-8 ISBN 978-981-19-9794-5 (eBook)
https://doi.org/10.1007/978-981-19-9794-5

This Springer imprint is published by the registered company Springer Nature Singapore Pte Ltd.
The registered company address is: 152 Beach Road, #21-01/04 Gateway East, Singapore 189721, Singapore

Preface

Reactive oxygen species (ROS) are key signalling molecules involved in the redox equilibrium and biological processes. In plants, ROS play an important role in biotic and abiotic stress sensing, integration of diverse environmental signals, and commencement of stress-response networks, thus contributing to the establishment of defense mechanisms and plant resilience. Critical amount of ROS is required by plants for regular operation of vital physiological mechanisms. Furthermore, if present at higher than critical concentration, ROS may be toxic and disturbing complete physiological performance of the plant. Thus, plants keep a threshold level of ROS by inherent antioxidant defense system. This book is dedicated to presenting the latest developments of the role of ROS in plants (positive and negative) to understand the importance and impacts on crop productivity.

This book comprises 15 chapters. Chapter 1 of the book discusses the latest update on ROS synthesis and its potential application. Chapter 2 deals with the mechanism of ROS regulation in plants. This chapter also describes the organelle-specific ROS regulation through different enzymatic defense system. Biomolecules targeted by ROS are covered in Chap. 3. Chapter 4 describes the functions of ROS in improving agriculture and future crop safety. Chapter 5 highlights the ecology of ROS signalling in plants. Chapter 6 addresses the physiological impact of ROS on leaf. Chapter 7 deals with ROS role in senescence and signal transduction in plants. In Chap. 8, hazardous phytotoxic nature of ROS in agriculture is discussed briefly. Chapter 9 describes the hormonal response in plants influenced by ROS. In Chap. 10, the dual role of ROS as signals that influence plant stress tolerance and programmed cell death is discussed. Chapter 11 deals with the insight into the interaction of strigolactones, abscisic acid, and ROS signals. In Chap. 12, the role of hydrogen peroxide as a regulator of plant development and abiotic stress response is presented. Chapter 13 demonstrates the content towards sustainable agriculture, its strategies involving phyto-protectants against ROS. In Chap. 14, signalling pathway of ROS in crop plants under abiotic stress is discussed. Chapter 15 covers adverse impact of ROS on nutrient accumulation and distribution in plants. This chapter also discusses the role of ROS in the life cycle of plants.

This book is not an encyclopedia of reviews but rather a compendium of newly composed, integrated, and illustrated contributions describing our knowledge of ROS in plants, signalling pathway under stress, ecology of ROS, and biomolecules damaged by the strike of ROS. The production sites are also explained with their hazardous effects. In addition, ROS effects on nature, nutrient uptake, seed physiology, hormonal uptake, senescence, and the crosstalk with other signalling molecules/ pathways information is also included in the book. Chapters incorporate both theoretical and practical aspects of ROS impacts on nature, plants, and tolerance against it and may serve as baseline information for future research through which significant development can be made. The book will be useful to researchers, instructors, and students both in universities and research institutes, especially in relation to biological and agricultural sciences.

With great pleasure, we extend our sincere thanks to all the contributors for their timely response, their excellent and up-to-date contributions, and their consistent support and cooperation. We are thankful to all who have helped us in any capacity during the preparation of this volume. We are extremely thankful to Springer Publishing for their expeditious acceptance of our proposal and completion of the review process. The subsequent cooperation and understanding by their staff are gratefully acknowledged. We express our sincere thanks to our family members for all the support they provided and the neglect and loss they suffered during the preparation of this book.

Finally, we are thankful to the Almighty who provided and guided all the channels to work in cohesion on the concept to the development of the final version of this treatise, *Reactive Oxygen Species—Prospects in Plant Metabolism.*

Hyderabad, India — Mohammad Faizan
Aligarh, India — Shamsul Hayat
Hyderabad, India — S. Maqbool Ahmed

Contents

Editors and Contributors

About the Editors

Mohammad Faizan is working as Assistant Professor, in the Botany Section, Maulana Azad National Urdu University, Hyderabad, India. He completed his postdoc in 2021, from the Department of Silviculture, College of Forestry, Nanjing Forestry University, Nanjing, China. In 2018, he received his Ph.D. degree in Botany from Aligarh Muslim University, Aligarh, India and his M.Sc. in 2014 at Chhatrapati Shahu Ji Maharaj University, Kanpur, India. His ongoing research is based on abiotic stress tolerance mediated by nanoparticles, and investigating the bioaccumulation, bio-transformations, uptake, translocation, and effects of bulk- and nano-forms of metals on plant physiology, morphology, anatomy, the ultrastructure of cellular and subcellular organelles, and modifications. With long experience and experimental work, he comprehensively detailed the state of research in environmental science in regard to "how nanoparticles/heavy metals interact with plants, soil, microbial community, and the larger environment as well as possible remediation technology using nanoparticles." He has published (total of 47 scientific publications) 42 peer-reviewed full-length articles, 4 books (3 in Springer), 15 chapters (Scopus indexed), and achieved h-index: 18 by Scopus with 1410 citations, till March, 2023. He is an internationally recognized reviewer, reviewed lot of manuscripts. Dr. Faizan is an editor and also an editorial board member of various high-impact journals.

Shamsul Hayat is Professor, in the Department of Botany, Aligarh Muslim University, Aligarh, India. He received his Ph.D. degree in Botany from Aligarh Muslim University, Aligarh, India. Before joining the Department as faculty, he has worked as Research Associate and Young Scientist in the same department. He has also worked as Associate Professor in King Saud University, Riyadh, Saudi Arabia, as a BOYSCAST Fellow at National Institute of Agrobiological Sciences, Tsukuba, Japan and as visiting scientist through INSA-Bilateral exchange program at Faculty of Biology and Chemistry, Institute of Biology, Department of Plant Biochemistry and Toxicology, University of Bialystok, Poland. The major area of research includes plant hormone, nanoscience, and abiotic stress in plants. It has been reported from his group that phytohormones such as brassinosteroids and salicylic acid play an important role in increasing the photosynthetic efficiency of the plant and regulate the antioxidant system even under abiotic stress. He is also studying the protein profiling in hormone-treated plants under abiotic stress. Dr. Hayat has been awarded Prof. Hira Lal Chakravorty Award by the Indian Science Congress Association, Kolkata, India, Associate of National Academy of Agricultural Sciences, New Delhi, India, BOYSCAST fellow by Department of Science & Technology, Government of India, New Delhi and young scientist by Association of the Advancement of Science, Aligarh, India. Dr. P.S. Khankhoje gold medal has also been awarded to him by Dr. Punjab Rao Deshmukh Krishi Vidyapeeth, Akola. He has been the Principal Investigator of the various projects sanctioned by different agencies and guided seven students for the award of Ph.D. degree and two students for the award of M. Phil degree besides a number of M.Sc. students. Dr. Hayat has published more than 160 research papers in leading journals of the world such as *Protoplasma, Photosynthetica, and Acta Physiologiae Plantarum* with high impact factor and also published 12 books by Springer and other publishers. Besides this 25 book chapters have also been published. Dr. Hayat has presented his work at several National and International conferences in Japan, Brazil, Spain, China, Poland, and Saudi Arabia. He is a regular reviewer and on the panel of editorial boards of National and International journals, Dr. Hayat is also the member of important National and International scientific societies. Recently, he has been appointed as Indian representative in Asia Association of Plant Scientist.

S. Maqbool Ahmed is presently working as a Professor of Botany, in Botany Section of School of Sciences of Maulana Azad National Urdu University, Hyderabad. He obtained his Master's and Ph.D. degree from Barkatullah University, Bhopal. He has teaching experience of 25 years and before joining MANUU he has worked as Research Fellow in Regional College of Education (NCERT) Bhopal in one project sanctioned by CSIR-NEERI. He has also worked as a Lecturer in Botany in Saifia PG College of Science and in Regional College of Education (NCERT), Bhopal. He was holding different administrative positions in the University like OSD-Exams, Overseas Coordinator, Nodal Officer of B.Voc programs, Coordinator of UGC Schemes, and so on. He had published more than 25 Research Articles in National and International journals of repute and also presented more than 30 papers in National and International conferences. He has also published two books and several book chapters. His research area is Allelopathy, Plant Physiology, and Environmental Science.

Contributors

Bilal Ahmad Department of Botany, Aligarh Muslim University, Aligarh, India

Khan Bilal Mukhtar Ahmed Department of Botany, Aligarh Muslim University, Aligarh, India

S. Maqbool Ahmed Botany Section, School of Sciences, Maulana Azad National Urdu University, Hyderabad, India

Arshiya Akeel Plant Physiology and Biochemistry Section, Department of Botany, Aligarh Muslim University, Aligarh, UP, India

Iqra Akhtar Department of Botany, Lahore College for Women University, Lahore, Pakistan

Abid Ali Laboratory of Germplasm Innovation and Molecular Breeding, College of Agriculture and Biotechnology, Zhejiang University, Hangzhou, China

Mohd Soban Ali Department of Botany, Faculty of Life Sciences, Aligarh Muslim University, Aligarh, India

Yamshi Arif Faculty of Life Sciences, Department of Botany, Aligarh Muslim University, Aligarh, India

Huma Arshad Department of Plant Sciences, Quaid-i-Azam University, Islamabad, Pakistan

Ritu Bala Department of Biotechnology, School of Bioengineering and Biosciences, Lovely Professional University, Phagwara, Punjab, India

Gausiya Bashri Department of Botany, Faculty of Life Sciences, Aligarh Muslim University, Aligarh, India

Divya Bhagat Department of Biotechnology, School of Bioengineering and Biosciences, Lovely Professional University, Phagwara, Punjab, India

Chen Chen School of Landscape and Horticulture, Yangzhou Polytechnic College, Yangzhou, Jiangsu, China

Priyanka Devi Department of Agronomy, School of Agriculture, Lovely Professional University, Phagwara, India

Abhijit Dey Department of Life Sciences, Presidency University, Kolkata, India

Shipa Rani Dey Department of Agronomy, School of Agriculture, I, Lovely Professional University, Phagwara, India

Padmanabh Dwivedi Department of Plant Physiology, Institute of Agricultural Sciences, Banaras Hindu University, Varanasi, India

Mohammad Faizan Botany Section, School of Sciences, Maulana Azad National Urdu University, Hyderabad, India

Ahmad Faraz School of Life Sciences, Glocal University, Saharanpur, UP, India

Mohd Gulfishan School of Agricultural Sciences, Global University, Saharanpur, India

Asif Hussain Hajam Department of Botany, Faculty of Life Sciences, Aligarh Muslim University, Aligarh, India

Hanan A. Hashem Botany Department, Faculty of Science, Ain Shams University, Cairo, Egypt

Shamsul Hayat Faculty of Life Sciences, Department of Botany, Aligarh Muslim University, Aligarh, India

Anjuman Hussain Plant Physiology Lab, Department of Botany, Faculty of Life Sciences, Aligarh Muslim University, Aligarh, India

Noshin Ilyas PMAS Arid Agriculture University, Rawalpindi, Pakistan

Sumera Iqbal Department of Botany, Lahore College for Women University, Lahore, Pakistan

Khajista Jabeen Department of Botany, Lahore College for Women University, Lahore, Pakistan

Ajmat Jahan Department of Botany, Aligarh Muslim University, Aligarh, India

Summera Jahan Department of Botany, University of Gujrat, Gujrat, Pakistan

Hassan Jaleel Department of Botany, GDC Billawar, Billawar, Jammu and Kashmir, India

Sumera Javad Department of Botany, Lahore College for Women University, Lahore, Pakistan

Radwan Khalil Botany Department, Faculty of Science, Benha University, Benha, Egypt

M. Masroor A. Khan Department of Botany, Aligarh Muslim University, Aligarh, India

Khusbhoo Department of Biotechnology, School of Bioengineering and Biosciences, Lovely Professional University, Phagwara, Punjab, India

Nimra Kousar Department of Botany, College for Women University, Lahore, Pakistan

Mukesh Kumar Department of Biotechnology, School of Bioengineering and Biosciences, Lovely Professional University, Phagwara, Punjab, India

Prasann Kumar Department of Agronomy, School of Agriculture, Lovely Professional University, Phagwara, India

Anayat Rasool Mir Faculty of Life Sciences, Department of Botany, Aligarh Muslim University, Aligarh, India

Muhammad Mohiuddin Department of Environmental Sciences, Kohsar University, Murree, Pakistan
Faculty of Biosciences, Kohsar University, Murree, Pakistan

Sidra tul Muntha Faculty of Biosciences, Kohsar University, Murree, Pakistan

Laboratory of Germplasm Innovation and Molecular Breeding, College of Agriculture and Biotechnology, Zhejiang University, Hangzhou, China

Ghazala Mustafa Department of Plant Sciences, Faculty of Biological Sciences, Quaid-i-Azam University, Islamabad, Pakistan

Komal Nawaz Department of Botany, University of Education, Lahore, Pakistan

Ramish Nida Department of Botany, Lahore College for Women University, Lahore, Pakistan

Ram Prakash Pandey Department of Biotechnology, Chandigarh University, Chandigarh, India

Sonam Panigrahi School of Life Sciences, Sambalpur University, Sambalpur, India

Sheo Mohan Prasad Ranjan Plant Physiology and Biochemistry Laboratory, Department of Botany, University of Allahabad, Prayagraj, India

Rohan Samir Kumar Sachan Department of Microbiology, School of Bioengineering and Biosciences, Lovely Professional University, Phagwara, Punjab, India

Lalit Saini Department of Agronomy, School of Agriculture, Lovely Professional University, Phagwara, India

Samrana Samrana Department of Botany, University of Swabi, Swabi, Pakistan

Anis Ali Shah Department of Botany, University of Education, Lahore, Pakistan

Shareen Environmental Biotechnology Lab, College of Biology and Environment, Nanjing Forestry University, Nanjing, China

Khushbu Sharma Department of Agronomy, School of Agriculture, Lovely Professional University, Phagwara, India

Junaid Shehzad Department of Plant Sciences, Faculty of Biological Sciences, Quaid-i-Azam University, Islamabad, Pakistan

Manpreet Kaur Somal Department of Biotechnology, School of Bioengineering and Biosciences, Lovely Professional University, Phagwara, Punjab, India

Mohammad Suhel Ranjan Plant Physiology and Biochemistry Laboratory, Department of Botany, University of Allahabad, Prayagraj, India

Amina Tariq Department of Botany, Lahore College for Women University, Lahore, Pakistan

Fangyuan Yu Collaborative Innovation Center of Sustainable Forestry in Southern China, College of Forest Science, Nanjing Forestry University, Nanjing, China

An Update on Reactive Oxygen Species Synthesis and Its Potential Application

1

Manpreet Kaur Somal, Rohan Samir Kumar Sachan, Divya Bhagat, Khusbhoo, Ritu Bala, and Mukesh Kumar

Abstract

Reactive oxygen species (ROS) are by-products of aerobic electron transport pathways produced in organelles like mitochondria, apoplast, chloroplast, and peroxisomes. These are also generated during plant–microbe interaction and are secondary messengers during plant developmental pathways. ROS plays a vital role in the dormancy and germination of seeds by inhibiting transcription and translation process. This results in carbonylation of proteins, gene damage causing loss of germination and viability. ROS is also involved in the perception and transmission of environmental factors that control germination during seed inhibition that favors germination. The levels of ROS are kept to a proportion that causes cellular damage called apoptosis by the release of cytochrome c through an opening of mitochondrial permeability transition pores. The present chapter will focus on the production of ROS in different compartments, its signaling pathways, its role in seed germination, and how it protects the seeds from pathogens. The study of the synthesis of ROS and its application will provide a basis for future research in the maintenance of plant quality, viability, and factors that may help increase plant yield.

Keywords

Reactive oxygen species · Endosperm weakening · Antioxidants · Apoptosis

M. K. Somal · D. Bhagat · Khusbhoo · R. Bala · M. Kumar
Department of Biotechnology, School of Bioengineering and Biosciences, Lovely Professional University, Phagwara, Punjab, India

R. S. K. Sachan (✉)
Department of Microbiology, School of Bioengineering and Biosciences, Lovely Professional University, Phagwara, Punjab, India

M. Faizan et al. (eds.), *Reactive Oxygen Species*,
https://doi.org/10.1007/978-981-19-9794-5_1

1.1 Introduction

Reactive oxygen species are the by-product of biochemical reactions and have become essential for plants. Oxygen reduction and activation produce ROS that can be superoxide(O_2–), singlet oxygen ($1O_2$), hydroxyl radical (HO.), and hydrogen peroxide (H_2O_2) (Sharma et al. 2012). ROS synthesis is triggered by different abiotic stresses such as light, heavy metal, temperature, salinity, and pathogen attack. It is produced in separate cellular compartments such as chloroplast, apoplast, mitochondria, and peroxisomes. It is also linked to plant hormones such as salicylic acid, jasmonic acid, ethylene, etc. (Mhamdi and Van Breusegem 2018). Leakage of electrons onto O_2 from the electron transport activities of chloroplasts, plasma membranes, mitochondria, or as a side product of many reactions in a different compartment of cells leads to ROS formation (Poór 2020). A cell is in oxidative stress when the ROS level exceeds the defense mechanism. The antioxidative defense protects against the damage caused due to oxidative stress in plants (Sachdev et al. 2021). Signaling pathways and redox sensing regulate the spatiotemporal titer of ROS by modifying the scavenging mechanisms and synthesis of ROS.

During photosynthesis, environmental stress inhibits the Calvin cycle's operation. It results in slowed photosynthetic electron transport and a reduction in the components of stromal electron acceptors and the electron transport chain of chloroplasts (Foyer and Shigeoka 2011). As a result, the plant develops oxygen species, which is followed by the generation of ROS such as superoxide radicals and hydroxyl radical (OH.), hydrogen peroxide (H_2O_2), as well as singlet oxygen ($1O_2$). Excessive ROS slows down electron transportation in the photosynthetic ETC and results in the activation of pseudo-cyclic electron transport and photorespiration (Foyer and Noctor 2003). $1O_2$ is synthesized initially, followed by H_2O_2 being synthesized either enzymatically catalyzed by SOD or nonenzymatically in superoxide diffusion with a low reaction yield.

Hydrogen peroxide is converted to water with the help of ascorbate peroxidase and ascorbate. At the expense of NADH, ascorbate is oxidized and then regenerated by reduced glutathione (Kreslavski et al. 2012) (Table 1.1).

Several enzymes, including NADPH oxidases, SOD, catalase, ascorbate peroxidase, glutathione reductase, dehydroascorbate reductase, GOPX, and glutathione-*S*-transferase, have been found associated with the synthesis of ROS (Tewari et al. 2021). NADPH is an integral protein involved in the oxidation of NADPH in cells and in reducing superoxide radicals. It also plays a vital role in the defense development of signaling pathways.

Table 1.1 ROS, their properties, enzymatic and non-enzymatic oxidants, and scavenging systems

ROS	Source	Action	Migration capacity	Scavenging	Enzymatic oxidants	Non-Enzymatic oxidants	References
Hydrogen peroxide (H_2O_2)	The plasma membrane, chloroplasts, the endoplasmic reticulum, ETC mitochondria, and peroxisomes	Reacts with protein (cysteine), reacts with $O2^-$, and produces $OH^{\cdot}$	1 μm	Catalases, flavonoids, and peroxidases	CAT, APX, GPX, SOD,	GSH, A. A., flavonoids	Karuppanapandian et al. (2011), Bhattacharjee et al. (2010), and Dat et al. (2000)
Hydroxyl radical ($OH^{\cdot}$)	Chloroplast, membranes, mitochondria, decomposition of O3 in apoplastic space	Reacts with lipids, macromolecules, DNA, and proteins	1 nm	Prevention of $OH^{\cdot}$ formation by sequencing Fe and flavonoids	–	Flavonoids	Karuppanapandian et al. (2011) and Halliwell and Gutteridge (1989)
Superoxide radical ($O_2^{\cdot -}$)	Chloroplasts (Mehler reaction), plasma membrane, ETC mitochondria, glyoxysomal photorespiration, and peroxisomes	Reacts with compounds such as iron-sulfur, nitric oxide to form peroxynitrite	30 nm	Superoxide dismutase	SOD	–	Karuppanapandian et al. (2011) and Miller et al. (2009)
Singlet oxygen ($1O_2$)	Chloroplast, mitochondria, membranes	Oxidizes polyunsaturated fatty acids and react with protein (Trp, His, Tyr, Met, Cys)	30 nm	Protects membrane lipids by detoxification of lipid peroxidation, and quenching 1O2		TOCs, CARs	Karuppanapandian et al. (2011)

Abbreviations: A.A.: ascorbic acid; APX: ascorbate peroxidase; CARs: carotenoids; CAT: catalase; GPX: guaiacol peroxidase; GSH: glutathione; SOD: superoxide dismutase; TOCs: tocopherols

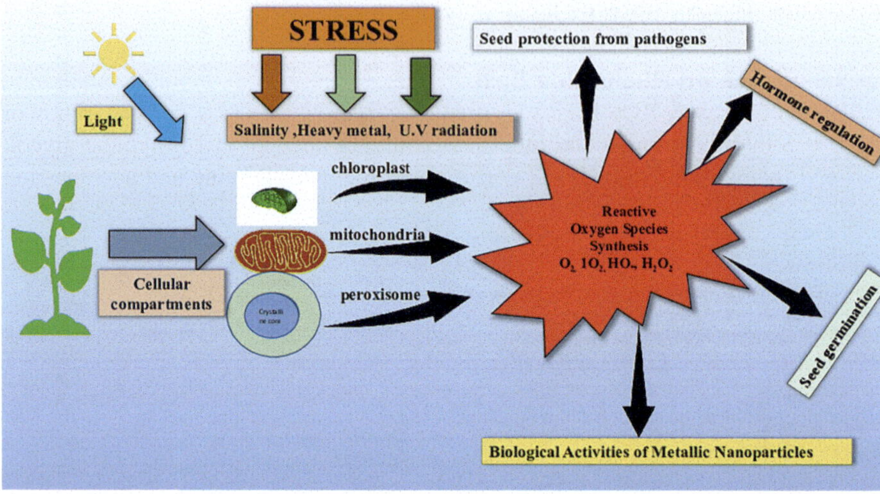

A diagrammatic illustration shows the factors that trigger the synthesis of reactive oxygen species in cellular compartments, mitochondria, chloroplast, and peroxisomes of plants. ROS generated plays a vital role in seed aging, seed dormancy, germination, and protection against pathogens.

ROS plays a vital role in plant development, such as pollen tube growth, root hair growth, etc. (Choudhary et al. 2020). ROS plays a role in endosperm weakening, mobilization of seed reserves, pathogen defense, and programmed cell death in seeds. ROS can aid in the mobilization of molecules during germination by breaking down polysaccharides, DNA, RNA, and fatty acids, as well as protein carbonylation (Adetunji et al. 2021). ROS are secondary messengers, and they carry information to the nucleus via redox processes. They act as a signal transduction molecule and regulate different pathways during the acclimation of the plant under environmental stress. ROS are essential for the success of various natural processes, such as the differentiation and proliferation of cells (Hasanuzzaman et al. 2020). The present chapter will focus on ROS synthesis in chloroplast, mitochondria, peroxisomes, and apoplast. The application part will focus on the pathways related to the role of ROS in seed germination, plant–pathogen interaction specifying defense mechanisms, hormones, ROS interaction, and biological activity of metallic nanoparticles.

1.2 Chloroplast ROS Production

Chloroplasts generate O_2 in three ways under suboptimal conditions, such as sudden shifts from low to high light and vice versa, UV radiation, drought, osmotic, temperature stress, or pathogen attack.

1. The first process involves the electron donor side of photosystem II, where incomplete water oxidation, one of the fundamental processes of photosynthesis, produces H_2O_2.
2. The second method occurs inside the electron transport chain when plastohydroquinone (PQH2) is reduced to plastosemiquinone, which interacts with O_2 to produce O_2^- (Pospíšil 2016).
3. The third process occurs when the electron transport chain associated with photosystem I is overloaded. The electron from the photo-excited photosystem I reaction center is used by ferredoxin to reduce O_2 and produce O_2^- instead of being transferred to $NADP^+$. The Mehler reaction is part of the photosynthetic water–water cycle and is responsible for ROS scavenging (Asada 2000).

Increased ROS generation is closely linked to light-dependent photosynthetic reactions. It acts as a signal of altering internal or external factors that necessitate metabolic acclimation or adaptation. Within thylakoid membranes, the singlet state of oxygen is formed primarily by energy transfer from the triplet state of photosystem IIs (P680) to ground state molecular oxygen ($3O_2$) (Fischer et al. 2013). There have been no direct enzymatic scavengers of $1O_2$. Rather, $1O_2$ is scavenged predominantly through interactions with other molecules, including carotenoids, tocopherols, and membrane lipids (Chan et al. 2016).

Recent research has added to our understanding of 1O2-induced carotene cleavage, revealing that cyclocitral and other breakdown products are essential in $1O_2$-induced chloroplast retrograde signaling. Superoxide anions form due to various enzyme reactions and processes characterized by high rates of electron flow, such as mitochondrial and chloroplastic electron transport chains. Hydrogen peroxide is produced through the enzymatic and spontaneous dismutation of superoxide anions and the activity of glycolate oxidases (Waszczak et al. 2018) (Fig. 1.1).

In the Mehler reaction, PSI reduces molecular oxygen, resulting in the synthesis of ATP without NADPH. The thylakoid copper/zinc SODs convert superoxide anion radicals to H_2O_2. Oxygen reduction provides an alternative electron sink and generates superoxide anion radicals, transformed to H_2O_2. The chloroplast APXS and PRXs can then convert H_2O_2 to water (Awad et al. 2015). Because it has negligible reactivity to most biological compounds and is promptly transformed to H_2O_2 by the action of thylakoid and stromal SODs, superoxide has received little study concerning chloroplast functions. Superoxide reacts with nitric oxide (NO) in chloroplasts and is reduced to H_2O_2 by ascorbic acid. Iron-SODs (FeSODs) and copper/zinc SODs (Cu/ZnSODs) are two types of chloroplast SODs that are essential for chloroplast function and development. Ascorbate peroxidases (APXS), glutathione peroxidase-like (GPXLs) enzymes, and peroxiredoxins (Prxrs) detoxify H_2O_2 in the chloroplast stroma (Foyer 2018).

In ch1 mutant and wild-type (W.T.) Arabidopsis plants, excessive light-driven 1O2 synthesis causes oxidation and cleavage of b-carotene in the PSII RC in the grana core, resulting in different carbonyl compounds, especially apocarotenoids. b-cyclocitral (bCC) and dihydroactinidiolide (DhA) are two of these products demonstrated to be biologically active volatile chemicals that facilitate ORS.

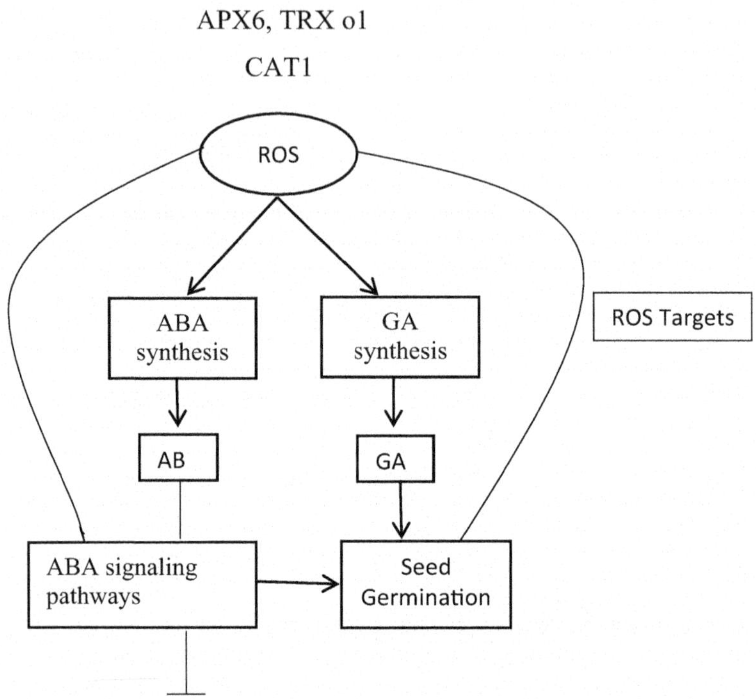

Fig. 1.1 Synthesis and scavenging of ROS

Nuclear transcriptome generated by bCC and DhA promotes plant tolerance to numerous abiotic stresses.

PSII generates $1O_2$, and $1O_2$ generates volatile signaling chemicals by the oxidation of b-carotene, implying that PSII functions as a sensory mechanism that converts critical values of $1O_2$ into volatile signals (Li and Kim 2021).

1.3 Mitochondria ROS Production

Any atom or molecule with an unpaired electron in its outermost shell is known as a free radical. The free radicals can lead to molecular damage as they are associated with high indiscriminate reactivity (Harman 1992). Any oxygen-containing molecule either radical or non-radical which is capable of initiating the deleterious reaction is termed a reactive oxygen species. ATP production in the cells or the synthesis of ATP energy is generated from the mitochondria which come across the inner membrane of mitochondria through the electrochemical proton gradient. The pH component and the electrical component and the enzymes from the electron transport chain are utilized as they pump protons to the intermembrane space to the mitochondrial membrane to generate this gradient or also known as the proton motive force. With a high potential of energy, the electron passes from carriers in

Complex I to Complex III which carries in Complex IV with less potential energy to generate the energy for the proton pumping (Dröge 2002). At Complex I, Complex II, and Complex IV the pair of electrons entering the chain from the reduced substrates reduces oxygen to water by dehydrogenases. There are certain sites within the chain such as FeS centers from where the electron passes singly to carriers in which they take the electron one at a time or to the stable one-electron reduced intermediates such as quinones and flavin. Instead of passing these electrons down to the respiratory chain to Complex IV these single electrons occasionally react directly with the oxygen which results in the formation of superoxide (Lebovitz et al. 1996).

The ROS production mechanism has been obtained in isolated mitochondria due to their high experimental accessibility. The change in characteristics such as absorbance, fluorescence, paramagnetism, and the luminescence is due to the reaction of superoxide with some compound on which the superoxide detection system relies. The compounds such as acylated cytochrome c, epinephrine, DMPO, lucigenin, and coelenterazine are known as detector compounds. The direct detection of superoxide in intact mitochondria is very problematic which generates in the matrix by using the probes. The superoxide cannot diffuse the mitochondria because it cannot cross the inner membrane of the mitochondria and due to this the probe must be delivered into the matrix where several non-specific reactions must be taken place (Gardner 2002). Firstly, the production of superoxide by the probe is a particular concern as some probes undergo many redox cycling. Secondly, by reaction with the superoxide if the probe becomes charged or was charged then according to the potential of the membrane which can change the physicochemical properties, it may distribute across the inner membrane of the mitochondria (Miwa and Brand 2005). The rate of superoxide production with the direct quantification is as difficult as standard superoxide which cannot generate the standard curve as it cannot be added to the standard curve. To measure the production of superoxide, there are two indirect alternative methods. The first method in which the matrix enzyme of the mitochondria contains the labile iron that is sensitive to superoxide in its iron-sulfur is known as inactivation of aconitase. The aconitase activity of mitochondria depends on inactivation by superoxide or with other ROS and the rate of reactivation by the iron replacement or reduction into its cluster (Belov et al. 1998).

$$\text{Citrate} \rightarrow \text{Isocitrate}.$$

This reaction is based on the aconitase assay where citrate to isocitrate is catalyzed by aconitase.

$$\text{Isocitrate} + \text{NADP}^{+} \rightarrow \alpha - \text{ketoglutarate} + \text{NADPH} + \text{H}^{+}.$$

This reaction is catalyzed by isocitrate dehydrogenase and monitored by the increase in fluorescence and absorbance.

Under a specific amount of time and conditions basically, mitochondria are incubated and then lysed to determine the aconitase activity. But the rate of the total production of superoxide is not possible which is one drawback of this method. For the production of superoxide, the most commonly used method is by measuring the hydrogen peroxide indirectly which diffuses out readily of the mitochondria and is cooperative with the simple detection protocols. This method relies on the fact that the production of superoxide is mostly by the mitochondria which are converted by the endogenous superoxide dismutase into hydrogen peroxide (St-Pierre et al. 2002). The hydrogen peroxide removal process into the matrix does not change under any experimental conditions which are assumed by this method. Some common methods used for the production of hydrogen peroxide in mitochondria are based on the reduced oxidation detector compound combined with the HRP (horseradish peroxide) to the hydrogen peroxide enzymatic reduction.

$$\mathrm{HRP} + \mathrm{H_2O_2} \rightarrow \mathrm{HRP{-}H_2O_2}.$$

Or

$$\mathrm{HRP} - \mathrm{H_2O_2} + 2\mathrm{AH} \rightarrow \mathrm{HRP} + 2\mathrm{H_2O} + \mathrm{A_2}.$$

AH = p-hydroxyphenyl acetic acid or homovanillic acid.

The quantification is easy and achievable by obtaining a standard curve of the known amount of hydrogen peroxide (H_2O_2) in this system. This technique is an advantageous method.

1.4 Apoplastic ROS Production

The apoplast fills in as a point of interaction for trading supplements and signals between plant cells and the climate. In several cases, plants reacted to both ecological and endogenous boosts which include the amassing of ROS inside this compartment. There are myriad of subcellular compartments such as chloroplast, mitochondria, cell wall, plasma membrane, and peroxisomes/oxosomes by which reactive oxygen species (ROS) generation takes place under normal and stress conditions. The production of ROS during the plant–pathogen interaction is majorly through the apoplast. Enzymes like NADPH oxidases, apoplastic peroxidases, and polyamine oxidases are responsible for the production of apoplastic ROS (Qi et al. 2017). The job of apoplastic peroxidases as ROS markers was at first shown pharmacologically and later by hushing or producing stable mutant lines. The polyamine oxidases produce the H_2O_2 under the biotic as well as abiotic conditions and also help in the development of the plants. The NADPH oxidases are considered valuable among all the ROS-producing enzymes that transfer the electrons over the plasma membrane which further produces the O_2- (Waszczak et al. 2018).

A recent study demonstrated that both the NADPH-dependent respiratory burst oxidases (RBOH) and apoplastic peroxidases mainly the class III peroxidases (POX) are responsible for the production of apoplastic ROS in the *Arabidopsis thaliana* plants treated with high ammonium nitrogen nutrition. Moreover, the rise in ROS burst could be considered a sensory mechanism for such types of metabolic stress conditions (Podgórska et al. 2021). One of the previous studies concluded the overexpression of *OsPRX38,* a class III peroxidase enzyme of *Oryza sativa L.* in the *Arabidopsis thaliana enhances their arsenic tolerance conditions by apoplastic lignification. OsPRX38 enzyme is limited on the apoplast and found to upregulate to both* arsenate (AsV) and arsenite (AsIII) stresses (Kidwai et al. 2019). A newly identified mechanism has been concluded that melatonin rises the abiotic stress tolerance conditions by the regulation of the H2O2 signaling in the plants. A novel signal transduction pathway has been identified in which RBOH activity and H_2O_2 signaling are the key factors involved in melatonin-induced stress tolerance in tomato plants. The lower level of NO and SNOs by melatonin results in higher RBOH activity and then the formation of H_2O_2 takes place. A similar function also happened in the case of watermelon which concludes that H_2O_2 is a commonly known signal transduction pathway of melatonin in plant stress tolerance (Gong et al. 2017).

1.5 Peroxisomes ROS Production

In eukaryotic cells, peroxisomes are ubiquitous organelles that do not contain DNA and are bounded by a single membrane. Researchers have identified and proposed the functional term known as peroxisomes which are based on the presence of the oxidase and catalase containing H_2O_2. The granular matrix and the crystalline and amorphous inclusions which are composed of catalase are present in plant peroxisomes. Due to the action of catalase and the production of H_2O_2 in different organelles by oxidase peroxisomes were considered as cell garbage depots that were removed from the organelles. Now according to the researchers, they are well known due to their function metabolically and dynamically which participate in the development of cell response and stress as well as morphogenesis in different cellular processes, H_2O_2 detoxification, and oxidation of fatty acids. There are several important characteristics of peroxisomes which are given below

1. Oxidative metabolism of peroxisomes which contains the important number of oxidases to produce H_2O_2.
2. Presence of different superoxide radicals' sources in the organelles.
3. The H_2O_2 regulation and the accumulation of superoxide radicals in peroxisomes are due to the presence of antioxidant defense complex battery and avoid the toxicity level.
4. Due to the abiotic factors such as xenobiotics, heavy metals, and the H_2O_2 production in alteration under the stress conditions, leading to severe oxidative damage in peroxisomes. Because of this reactive oxygen species as well as H_2O_2 acts as signaling molecules and helps in the activation of the mechanism of the stress response.

The role of peroxisomal ROS metabolism causes and alters the human disease which imposes the considerable risk. The reduction of oxidative species which includes the superoxide free radicals, peroxyl, hydroxyl, and alkoxyl groups refers to the term known as reactive oxygen species. There are some non-radical compounds which also include in the reactive oxygen species term such as H_2O_2 (hydrogen peroxide), singlet oxygen from the excited oxygen species, hypochlorous acid as well as ozone and peroxynitrite. In different metabolic pathways, the aerobic metabolism and the antioxidative complex mechanism with their accumulation continue as a by-product of this metabolism Each ROS carries a different pathway as some of them are stronger oxidizing species with a very short lifetime such as OH which also carries the capacity to react with all types of biomolecules and generate disturbances in the metabolism of cells especially fatty acids, proteins as well as DNA and the chloroplast, mitochondria membrane, nuclei, and the plasma membrane as well as other compartments of plant cells in which ROS takes place.

1.6 Role of ROS in Seed Germination

In most of the seeds, rupturing of the testa and rupturing of endosperm are two independent occurrences in the germination process. Endosperm breakage involves the weakening of the endosperm layer's cell walls. The tool called puncture force measurement is a good way to figure out how much the endosperm is weakening. Endosperm weakening has been linked to several processes. Cell wall weakening is required for endosperm weakening (Wang et al. 2020). It involves the breakdown of cell wall polymers or the loosening of polymer bonds and linkages. Endosperm weakening has been linked to several molecular processes. The breakdown of cell wall polymers in the endosperm by reactive oxygen species, or more specifically, apoplastic hydroxyl radicals (OH) generated when superoxide (O2) and hydrogen peroxide (H2O2) conduct a Fenton reaction in the presence of peroxidases, is the most prominent among them. The interaction of hormones such as Gibberellin (GA), Abscisic Acid (ABA), and ethylene further complicates seed germination regulation. Furthermore, the role of ROS in the signaling of hormones for this type of control is yet unknown. In Vigna radiata seed germination, ROS was found to have a positive interacting impact with GA, ABA, and ethylene. In vascular tissue, ROS generation is required for lignification and cross-linking of cell wall polymers. H2O2 enhances seed germination by oxidizing the germination inhibitor(s) present in the pericarp. Seed germination was decreased in a dose-dependent manner by antioxidants that are derivatives of well-known germination inhibitors. An oxidant such as H2O2 should be used to break down the germination inhibitors to start seed germination (Wang et al. 2020).

Plant embryos and the surrounding endosperm have relatively restricted metabolic activity in dry and dormant seeds; hence, ROS generation is thought to be quite minimal (Bailly 2019). However, metabolism quickly returns after seed imbibition and throughout germination and this rapid metabolic resumption appears to be linked to enhanced ROS generation via multiple paths and at multiple subcellular

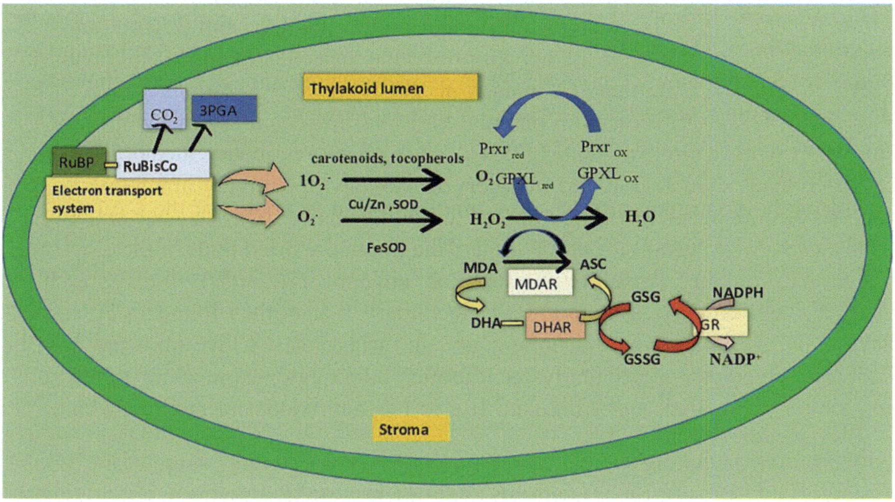

Fig. 1.2 During seed germination, ROS interacts with the ABA and GA pathways. The breaking of dormancy and the beginning of germination are influenced positively by the buildup of ROS (which can be generated by pharmacological or genetic techniques). At the top, metabolites and enzymes that may play a role in keeping ROS levels in developing seeds under control are displayed. ROS activities are mostly dependent on interaction with the ABA and GA signaling pathways. However, there are some more direct effects (shown by curved arrows) as well

locations. This comprises NADPH oxidase synthesis, mitochondrial respiration, lipid catabolism, and lipid-oxidation in glyoxysomes (Ishibashi et al. 2017). Experiments in which exogenously applied oxidants, such as H2O2 (El-Maarouf-Bouteau et al. 2015), and a pharmacologically or genetically induced reduction in catalase or other antioxidant activities, corroborated the spatiotemporal correlation of increased ROS production and accumulation during the onset of germination were shown to positively influence the release of dormancy and the onset of germination (Fig. 1.2). On the other hand, overexpression of CAT in barley (Hordeum vulgare) seeds has been found to prevent precocious germination (Ishibashi et al. 2017). As a result, elevated ROS levels are essential for successful germination and serve as favorable indicators of dormancy breaking.

After seed imbibition, the increased ROS level indicates that germination is started. Furthermore, when ROS levels are above a particular threshold, they are either too low to enable germination or too high to affect viability of embryo which leads in preventing or delaying the germination. Thus, during germination, ROS homeostasis must be strictly managed, resulting in a “oxidative window” for germination that limits competent seedling growth within particular boundaries of increasing levels of ROS. Several abnormalities are consistently reported in mutants with disrupted antioxidant equilibrium. For example, knocking off cytosolic APX6, which has high transcript levels in dry seeds and reduces germination rates due to increase in the protein carbonylation (Chen et al. 2016). Stress and ABA sensitivity are also elevated in these apx6 mutants, which is induced by disrupted signaling of

ABA and auxin. It shows that various signaling pathways are interlinked and that the accumulation of ABA and auxin, as well as the activation of ROS and redox signals, is needed. In contrast, increased H2O2 levels as well as faster germination have been shown in the mitochondrial thioredoxin O1 (trxo1) mutants (Ortiz-Espín et al. 2017). In Arabidopsis thaliana, the transcription factor ABI5 helps to regulate H2O2 homeostasis in addition to its core role in ABA-dependent signaling. ABI5 also facilitates the germination process by binding to the promoter of the CATALASE 1 gene and controlling its expression along with the H2O2 levels (Bi et al. 2017).

Endosperm weakening, cell wall loosening, and radicle elongation all lead to increase in the ROS levels. According to treatment of pea (Pisum sativum) seeds with H2O2 enhances the germination of seed and growth of seedling. In Arabidopsis, ABA inhibits ROS-mediated germination, which can be counterbalanced by the activity of GA. In barley, H2O2 is necessary to break dormancy, and this is accomplished by GA deposition and the regulation of GA synthesis and signaling genes, rather through ABA signaling suppression (Gong et al. 2021). Upregulation of synthesis genes increases ABA levels in the ascorbate-deficient Arabidopsis mutant vtc1 and ascorbate-deficient vtc2 and vtc5 mutants exhibit seedling-lethal symptoms that can be recovered by the treatment with ascorbate or its precursor galactose. Similarly, the ascorbate pool in apx6 mutants shows minimal changes.

Collectively, these data support the idea that ROS activity during seed germination is strongly reliant on interactions with ABA and GA, the two primary phytohormones that play antagonistic roles in seed germination regulation (Fig. 1.2). A better understanding of the molecular mechanisms underlying ROS function in seed physiology will undoubtedly open up new ways for improving the seed quality and pathogen tolerance, as well as provide new directions for engineering germination-recalcitrant species.

1.7 ROS in Seed Protection from Pathogen

ROS have been shown to perform important signaling functions in abiotic and biotic stressors in a number of investigations. Seeds may be subjected to biotic stress during their growth and longevity. The buildup of reactive oxygen species (ROS) would protect seedlings against disease infection. ROS is well established for its ability to influence infections directly or indirectly by inducing the hypersensitive reaction that leads to PCD. •O2 produced by plasma membrane NADPH oxidase activation is transformed to H2O2 either by spontaneous dismutation or by the activity of superoxide dismutase (SOD). This mechanism produces H2O2, which kills local infected cells and so prevents the invasion of pathogen. Furthermore, it has a main significance in cell infection resistance, implying a substrate involvement in cell wall peroxidation (Foyer et al. 2020). Additionally, it has been proven that the production of reactive oxygen species (ROS) from germinating seeds has an inhibitory effect on infections. Much more study is needed to confirm the notion of pathogen-induced protection via ROS.

One of the primary causes of crop loss has been recognized as seed pathogen infection (Bi et al. 2017). In a previous research *Pseudomonas syringae pv.* tomato DC300 (Pst DC3000) was inoculated into Arabidopsis thaliana seedlings in order to explore seed–pathogen interactions under mild heat stress at 30 °C. These parameters were ideal for stress combinations, such as heat stress and pathogen infection. Following translocation study of bacterial effector proteins, Pst DC3000, the results of this experiment revealed that heat promoted bacterial secretion (Strange and Scott 2005). Under modest heat stress, pathogen susceptibility was raised by around a two- to fourfold difference. SA mediates seed defense against pathogens or bacterial pathogen effectors (Xin and He 2013). The salicylic acid (SA) pathway was examined in wild-type and ics1 mutant seedlings under a combination of moderate stress and pathogenic infection due to enhanced pathogen susceptibility. The Arabidopsis mutant ics1 suppresses the biosynthesis of salicylic acid. Under mild heat stress and pathogenic infection, ics1 showed significant inhibition of SA activity, along with a two-fold difference. The temperature difference in wild-type plant seedlings had no effect on activity of SA pathway (Tsuda et al. 2008). The researchers examined the pathogen susceptibility in seedlings under mild heat stress; it would be fascinating to study into the same circumstances in seedlings under more severe heat stress. Seed immune defenses may be lowered in both wild-type and isc1 mutants when temperatures rise, making them more vulnerable to pathogenic infection.

1.8 Conclusion

Multiple cellular processes generate ROS or even as a response to an external stimulus. Many plant organelles like chloroplast, mitochondria, and plasma membrane generate ROS as a by-product during electron transport activities. The ROS application has been applied for study of biotic and abiotic stress in plants, seed germination, development of root, shoot, and flower. Still certain gray areas are left to be explored like effect of short half-life of ROS on plant growth. Future need for progress in genomics, proteomics will clear the understanding of plant cellular response to oxidative stress and production of ROS. There is also a further need for a biotechnological approach (like transformation or gene manipulation) to produce an in-built tolerance power toward ROS in plants.

References

Adetunji AE, Adetunji TL, Varghese B, Pammenter NW (2021) Oxidative stress, ageing and methods of seed invigoration: an overview and perspectives. Agronomy 11(12):2369

Asada K (2000) The water–water cycle as alternative photon and electron sinks. Philos Trans R Soc Lond B Biol Sci 355(1402):1419–1431

Awad J, Stotz HU, Fekete A, Krischke M, Engert C, Havaux M et al (2015) 2-cysteine peroxiredoxins and thylakoid ascorbate peroxidase create a water-water cycle that is essential to protect the photosynthetic apparatus under high light stress conditions. Plant Physiol 167(4): 1592–1603

Bailly C (2019) The signalling role of ROS in the regulation of seed germination and dormancy. Biochem J 476(20):3019–3032

Belov L, Sztejnberg L, Fridovich I (1998) Critical evaluation of the use of hydroethidine as a measure of superoxide anion radical. Free Radic Biol Med 25(7):826–831

Bhattacharjee S (2010) Sites of generation and physicochemical basis of formation of reactive oxygen species in plant cell, in Reactive oxygen species and antioxidants in higher plants. S. Dutta

Bi C et al (2017) Arabidopsis ABI5 plays a role in regulating ROS homeostasis by activating CATALASE 1 transcription in seed germination. Plant Mol Biol 94(1):197–213

Chan KX, Phua SY, Crisp P, McQuinn R, Pogson BJ (2016) Learning the languages of the chloroplast: retrograde signaling and beyond. Annu Rev Plant Biol 67:25–53

Chen D, Li Y, Fang T, Shi X, Chen X (2016) Specific roles of tocopherols and tocotrienols in seed longevity and germination tolerance to abiotic stress in transgenic rice. Plant Sci 244:31–39

Choudhary A, Kumar A, Kaur N (2020) ROS and oxidative burst: roots in plant development. Plant Divers 42(1):33–43

Dat J, Vandenabeele S, Vranov'a E, van Montagu M, Inz'e D, van Breusegem F (2000) Dual action of the active oxygen species during plant stress responses. Cell Mol Life Sci 57(5):779–795

Dröge W (2002) Free radicals in the physiological control of cell function. Physiol Rev 82:47–95

El-Maarouf-Bouteau H, Sajjad Y, Bazin J, Langlade N, Cristescu SM, Balzergue S, Baudouin E, Bailly C (2015) Reactive oxygen species, abscisic acid and ethylene interact to regulate sunflower seed germination. Plant Cell Environ 38:364–374. https://doi.org/10.1111/pce.12371

Fischer BB, Hideg E, Krieger-Liszkay A (2013) Production, detection, and signaling of singlet oxygen in photosynthetic organisms. Antioxid Redox Signal 18(16):2145–2162

Foyer CH (2018) Reactive oxygen species, oxidative signaling and the regulation of photosynthesis. Environ Exp Bot 154:134–142

Foyer CH, Noctor G (2003) Redox sensing and signaling associated with reactive oxygen in chloroplast, peroxisome and mitochondria. Physiol Plant 119:255–264

Foyer CH, Shigeoka S (2011) Understanding oxidative stress and antioxidant functions to enhance photosynthesis. Plant Physiol 155:93–100

Foyer CH, Kyndt T, Hancock RD (2020) Vitamin C in plants: novel concepts, new perspectives, and outstanding issues. Antioxid Redox Signal 32(7):463–485

Gardner PR (2002) Aconitase: sensitive target and measure of superoxide. Methods Enzymol 349: 9–23

Gong B, Yan Y, Wen D, Shi Q (2017) Hydrogen peroxide produced by NADPH oxidase: a novel downstream signaling pathway in melatonin-induced stress tolerance in Solanum Lycopersicum. Physiol Plant 160(4):396–409

Gong F et al (2021) H2O2 response gene 1/2 are novel sensors or responders of H2O2 and involve in maintaining embryonic root meristem activity in Arabidopsis thaliana. Plant Sci 310:110981

Halliwell B, Gutteridge JMC (1989) Free radicals in biology and medicine. Clarendon Press, Oxford. Foyer, C.H. and Noctor, G.D., Redox Regulation in Photosynthetic Organisms: Signaling, Acclimation, and Practical Implications, Antiox. Redox Signal., 2009, vol. 11, pp. 861–905

Harman D (1992) Free radical theory of ageing. Mutat Res/DNAging 275(3-6):257–266

Hasanuzzaman M, Bhuyan MHM, Zulfiqar F, Raza A, Mohsin SM, Mahmud JA et al (2020) Reactive oxygen species and antioxidant defense in plants under abiotic stress: revisiting the crucial role of a universal defense regulator. Antioxidants 9(8):681

Ishibashi Y, Aoki N, Kasa S, Sakamoto M, Kai K, Tomokiyo R (2017) The interrelationship between abscisic acid and reactive oxygen species plays a key role in barley seed dormancy and germination. Front Plant Sci 8:275

Karuppanapandian T, Moon JC, Kim C, Manoharan K, Kim W (2011) Reactive oxygen species in plants: their generation, signal transduction, and scavenging mechanisms. Aust J Crop Sci 5(6): 709–725

Kidwai M, Dhar YV, Gautam N, Tiwari M, Ahmad IZ, Asif MH, Chakrabarty D (2019) Oryza sativa class III peroxidase (OsPRX38) overexpression in Arabidopsis thaliana reduces arsenic accumulation due to apoplastic lignification. J Hazard Mater 362:383–393
Kreslavski VD, Los DA, Allakhverdiev SI, Kuznetsov VV (2012) Signaling role of reactive oxygen species in plants under stress. Russ J Plant Physiol 59(2):141–154
Lebovitz RM, Zhang H, Vogel H, Cartwright J, Dionne L, Lu N, Huang S, Matzuk MM (1996) Neurodegeneration, myocardial injury, and perinatal death in mitochondrial superoxide dismutase-deficient mice. Proc Natl Acad Sci 93(18):9782–9787
Li M, Kim C (2021) Chloroplast ROS and stress signaling. Plant Commun 3:100264
Mhamdi A, Van Breusegem F (2018) Reactive oxygen species in plant development. Development 145(15):dev164376
Miller G, Schlauch K, Tam R et al (2009) The plant NADPH oxidase RBOHD mediates rapid systemic signaling in response to diverse stimuli. Sci Signal 2(84):45–49
Miwa S, Brand MD (2005) The topology of superoxide production by complex III and glycerol 3-phosphate dehydrogenase in Drosophila mitochondria. Biochim Biophys Acta Bioenerg 1709(3):214–219
Ortiz-Espín A et al (2017) Mitochondrial AtTrxo1 is transcriptionally regulated by AtbZIP9 and AtAZF2 and affects seed germination under saline conditions. J Exp Bot 68(5):1025–1038
Podgórska A, Burian M, Dobrzyńska K, Rasmusson AG, Szal B (2021) Respiratory burst oxidases and apoplastic peroxidases facilitate ammonium syndrome development in Arabidopsis. Environ Exp Bot 181:104279
Poór P (2020) Effects of salicylic acid on the metabolism of mitochondrial reactive oxygen species in plants. Biomol Ther 10(2):341
Pospíšil P (2016) Production of reactive oxygen species by photosystem II as a response to light and temperature stress. Front Plant Sci 7:1950
Qi J, Wang J, Gong Z, Zhou JM (2017) Apoplastic ROS signaling in plant immunity. Curr Opin Plant Biol 38:92–100
Sachdev S, Ansari SA, Ansari MI, Fujita M, Hasanuzzaman M (2021) Abiotic stress and reactive oxygen species: generation, signaling, and defense mechanisms. Antioxidants 10(2):277
Sharma P, Jha AB, Dubey RS, Pessarakli M (2012) Reactive oxygen species, oxidative damage, and antioxidative defense mechanism in plants under stressful conditions. J Bot 2012:217037
St-Pierre J, Buckingham JA, Roebuck SJ, Brand MD (2002) Topology of superoxide production from different sites in the mitochondrial electron transport chain. J Biol Chem 277(47): 44784–44790
Strange RN, Scott PR (2005) Plant disease: a threat to global food security. Annu Rev Phytopathol 43:83–116
Tewari RK, Yadav N, Gupta R, Kumar P (2021) Oxidative stress under micronutrient deficiency in plants. J Soil Sci Plant Nutr 21(1):832–859
Tsuda K, Sato M, Glazebrook J, Cohen JD, Katagiri F (2008) Interplay between MAMP-triggered and SA-mediated defense responses. Plant J 53:763–775
Wang H et al (2020) Factors affecting seed germination and emergence of Aegilops tauschii. Weed Res 60(3):171–181
Waszczak C, Carmody M, Kangasjärvi J (2018) Reactive oxygen species in plant signaling. Annu Rev Plant Biol 69:209–236
Xin XF, He SY (2013) Pseudomonas syringae pv. tomato DC3000: a model pathogen for probing disease susceptibility and hormone signaling in plants. Annu Rev Phytopathol 51:473–498

Mechanism of Reactive Oxygen Species Regulation in Plants

2

Junaid Shehzad and Ghazala Mustafa

Abstract

Reactive oxygen species (ROS) are the products of physiological metabolism in various cellular compartments such as mitochondria, peroxisomes, and chloroplasts. Under biotic and abiotic stress, ROS are significantly accumulated and can progressively induce oxidative damage and ultimately lead to cell death. Moreover, ROS generation is a fundamental process in plants to transmit and transport the signal toward the nucleus to increase tolerance against diverse conditions. The defense system against ROS not only consists of a scavenging system but is also comprised of an enzymatic and non-enzymatic defense system against various environmental issues. The enzymatic defense system against ROS includes various enzymes such as peroxidase (POD), superoxide dismutase (SOD), polyphenol oxidase (PPO), ascorbate peroxidase (APX), glutathione peroxidase (GPX), and catalase (CAT). This chapter covers a detailed study of the organelle-specific generation of ROS and existing enzymatic systems to balance the redox state. Moreover, the role of non-enzymatic and low-molecular-weight antioxidants in ROS detoxification and retrograde signaling will also be discussed. Plants have also evolved several interconnected signaling pathways to control the expression of different transcriptional factors and stress-responsive genes for producing various classes of proteins that result in stress regulation. More importantly, the relationship between ROS and epigenetic modifications to regulate gene expression will be precisely discussed in this chapter. Excess ROS accumulation can lead to the activation of cell death processes such as apoptosis which is crucial for plant development and survival. The role of apoptosis in

J. Shehzad · G. Mustafa (✉)
Department of Plant Sciences, Faculty of Biological Sciences, Quaid-i-Azam University, Islamabad, Pakistan
e-mail: mghazala@qau.edu.pk

M. Faizan et al. (eds.), *Reactive Oxygen Species*,
https://doi.org/10.1007/978-981-19-9794-5_2

maintaining normal cellular homeostasis under stress conditions is discussed in detail. In this chapter, the different mechanisms involved in ROS regulation that will offer a new platform for the improvement of abiotic stress tolerance in crops are summarized.

Keywords

Apoptosis · Enzymatic antioxidants · Genes · Proteins · ROS

2.1 Introduction

Global climate change and associated adverse environmental conditions negatively influence plant growth and development (Hasanuzzaman et al. 2020). Under various environmental stresses, the plant produces oxygen radicals and their derivatives, which are called ROS (Zainab et al. 2021). ROS comprise free radicals (hydroxyl radical $OH^{\bullet}$, alkoxy radical $RO^{\bullet}$, hydroperoxyl radical $HO_2^{\bullet}$, and superoxide anion $O_2^{\bullet-}$) and non-radical molecules such as hydrogen peroxide H_2O_2 and singlet oxygen $^1O_2$ (Mehla et al. 2017; Hasanuzzaman et al. 2019a). Different cellular compartments such as apoplast, plasma membranes, mitochondria, chloroplasts, and peroxisomes are the main sites of ROS production in plants (Janků et al. 2019; Singh et al. 2019). An increase in ROS levels causes a serious threat to plants by inducing lipid peroxidation, nucleic acids damage, enzymes inhibition through protein oxidation, and stimulation of the programmed cell death (PCD) pathway (Raja et al. 2017; Tripathi et al. 2020; Zainab et al. 2021). Apart from ROS damaging effects, they act as secondary messengers and transport the signal to the nucleus by utilizing mitogen-activated protein kinase pathway that increases tolerance against adverse environmental conditions (Singh et al. 2019). For instance, ROS also actively participates in the development of root and shoot apical meristem, leaves, lateral roots, root hair cells, pollen tubes, and flowering (Noctor et al. 2018; Mittler 2017). Therefore, ROS are considered toxic molecules and products of various metabolic processes which are important for cellular mechanisms such as proliferation and differentiation.

The most critical impact of ROS is the disruption of the balance between the ROS generation and the activity of plant defense systems to reduce the ROS accumulation. Notably, the enzymatic and nonenzymatic defense systems maintained the equilibrium between ROS generation and detoxification under various environmental stresses (Hasanuzzaman et al. 2020). The enzymatic defense system comprises various antioxidant enzymes, like CAT, SOD, GPX, POD, PPO, thioredoxins (TRXs), peroxiredoxins (PRXs), and enzymes of ascorbate glutathione cycle, such as monodehydroascorbate reductase (MDHAR), glutathione reductase (GR), ascorbate peroxidase (APX), (Nath et al. 2018; Laxa et al. 2019), while the non-enzymatic system includes low-molecular-weight components like carotenoids, tocopherol, alkaloids, flavonoids, phenolic compounds, and non-protein amino acids, like glutathione (GSH) and AsA (Gill and Tuteja 2010; Hasanuzzaman et al. 2012; Kaur et al.

2019). The enzymatic and non-enzymatic antioxidants are linked with each other to maintain the stability between ROS production and detoxification. Furthermore, the genes expression and protein abundance change, which activates many stress-responsive transcription factors for the downstream responses against oxidative stress (Wang et al. 2016; Zhu 2016). An equilibrium is needed between ROS generation and safe detoxification for the survival of plants under different challenging conditions. This chapter provides an in-depth look at recent findings related to ROS generation and regulation by various antioxidant and nonantioxidant mechanisms in plants under various environmental stresses.

2.2 ROS Generation in Plants

Different kinds of environmental issues such as heavy metals, drought, nutrient deficiency, salinity, temperature, light, and pathogen attacks triggered ROS generation (Tripathy and Oelmüller 2012). The unavoidable leakage of electrons onto O_2 in both stressed and unstressed cells generates ROS at various cellular locations including mitochondria, chloroplasts, cell walls, peroxisomes, apoplast, and plasma membranes. Out of total O_2 utilized by plants in various subcellular compartments, nearly 1% is diverted to produce ROS (Sharma et al. 2012). Below we will analyze the generation of ROS in different cellular organelles and its regulation by different antioxidant and nonantioxidant defense systems present in these organelles.

Chloroplast is the pigment that is more susceptible to being attacked by ROS due to its diversity and abundance (Tripathy and Oelmüller 2012). The photosystem I and II are the main sites for $O_2^{\bullet -}$ anion production in the chloroplast. During the Mehler reaction the low concentrations of $NADP^+$, ferredoxin reduction, iron-sulfur protein, and thioredoxin presence are responsible for ROS production in photosystem I (Janků et al. 2019), while the leakage of O_2 on the photosystem II acceptor side also produces $O_2^{\bullet -}$ radical (Pospíšil 2016). During photosynthesis, ribulose-1,5-bisphosphate oxygenase activity produces phosphoglycerate, and phosphoglycolate is another source of chloroplastic ROS generation (Mueller-Cajar and Whitney 2008). Triplet state $^1O_2$ production also takes place in the photosystem II antenna complex and the reaction center (Triantaphylidès and Havaux 2009; Roach and Krieger-Liszkay 2014). Lipoxygenase activity under light stress conditions also boosted $^1O_2$ production in plant chloroplast (Dogra et al. 2018).

Mitochondria are the main sites for cellular ROS production due to the occurrence of an electron transport chain (ETC) at the inner mitochondrial membrane which is responsible for aerobic respiration (Corpas et al. 2015). Out of total O_2 consumed by mitochondria, about 2–5% is derived toward the formation of $O_2^{\bullet -}$ radical (Gupta and Igamberdiev 2015). Under the normal respiratory chain operation $O_2^{\bullet -}$ is also produced but the production rate is extremely enhanced under decelerated respiratory rates, e.g., limited ADP availability, and respiratory chain inhibition which results in a reduction of mitochondrial ETC. Mitochondrial matrix SOD strongly accelerates the $O_2^{\bullet -}$ disproportionation to H_2O_2 and O_2 (Morgan et al. 2008). The plant mitochondria-generated ROS has a huge impact on the respiratory process and cellular functions, including retrograde signaling, PCD, plant hormone action, and defense against pathogens (Huang et al. 2016).

Single membrane-bounded ubiquitous cellular organelle peroxisomes are another main site of ROS generation because of an oxidative type of environment (Del Río and López-Huertas 2016). Different types of ROS, including $O_2^{\bullet -}$, H_2O_2, and $^1O_2$ have been reported to produce in the peroxisome. The spontaneous or enzymatic dismutation of $O_2^{\bullet -}$ radicals, fatty acid b-oxidation enzymes, flavin oxidases, enzymatic reactions, and photorespiratory glycolate oxidase (GOX) are the main H_2O_2 producing metabolic processes in peroxisome (Foyer et al. 2009; Fransen et al. 2013). The $O_2^{\bullet -}$ generation also occurs in the peroxisomal membrane, due to ETC's presence. The re-oxidation of NADH by the peroxisomal ETC to regenerate NAD^+ for peroxisomal metabolic processes might be the main reason for $O_2^{\bullet -}$ production in membranes (Sandalio et al. 2008). Similarly, the generation of $^1O_2$ via a Haber–Weiss mechanism in peroxisome might be the result of $O_2^{\bullet -}$ reaction with H_2O_2 (Mor et al. 2014). Peroxisomes also play a vital role in cellular signaling and redox homeostasis because of the production of different types of ROS.

Under different stress stimuli and physiological conditions, cell walls, apoplast, and cytoplasmic membrane are also involved in ROS production (Qi et al. 2017). Intensive ROS production occurs after the recognition of stress stimuli by plasma membrane receptors which activates the signaling pathways (Zipfel 2014). The plasma membrane NADPH-dependent oxidase inhibitors also induce ROS production (Chen and Yang 2020). The endoplasmic reticulum is also an important cell compartment where hydroxylation, deamination, and oxidation of xenobiotics and cellular components occur (Neve and Ingelman-Sundberg 2010). Overall, the light-driven reactions in active green tissues and mitochondria in non-green tissues or dark are the major source of ROS production.

2.3 ROS Regulation Mechanism in Plants

It is well-acquainted fact that various redox and aerobic reactions in plant cellular organelles produce a considerable amount of ROS. However, under favorable conditions plant counteracts ROS production with various strategies to maintain redox homeostasis (Kumari et al. 2021). The lower ROS level within the cells can also act as secondary messengers that can participate in cell maintenance, division, differentiation, and organogenesis (Zeng et al. 2017; Fujita et al. 2006; Dvořák et al. 2021).

2.3.1 Organelle-Specific ROS Regulation Through Different Enzymatic Defense Systems

Under abnormal conditions, excess ROS are generated in different cellular compartments that disturb different metabolic processes (Hirayama and Shinozaki 2010). In this situation, plants maintain ROS homeostasis and redox signaling by a

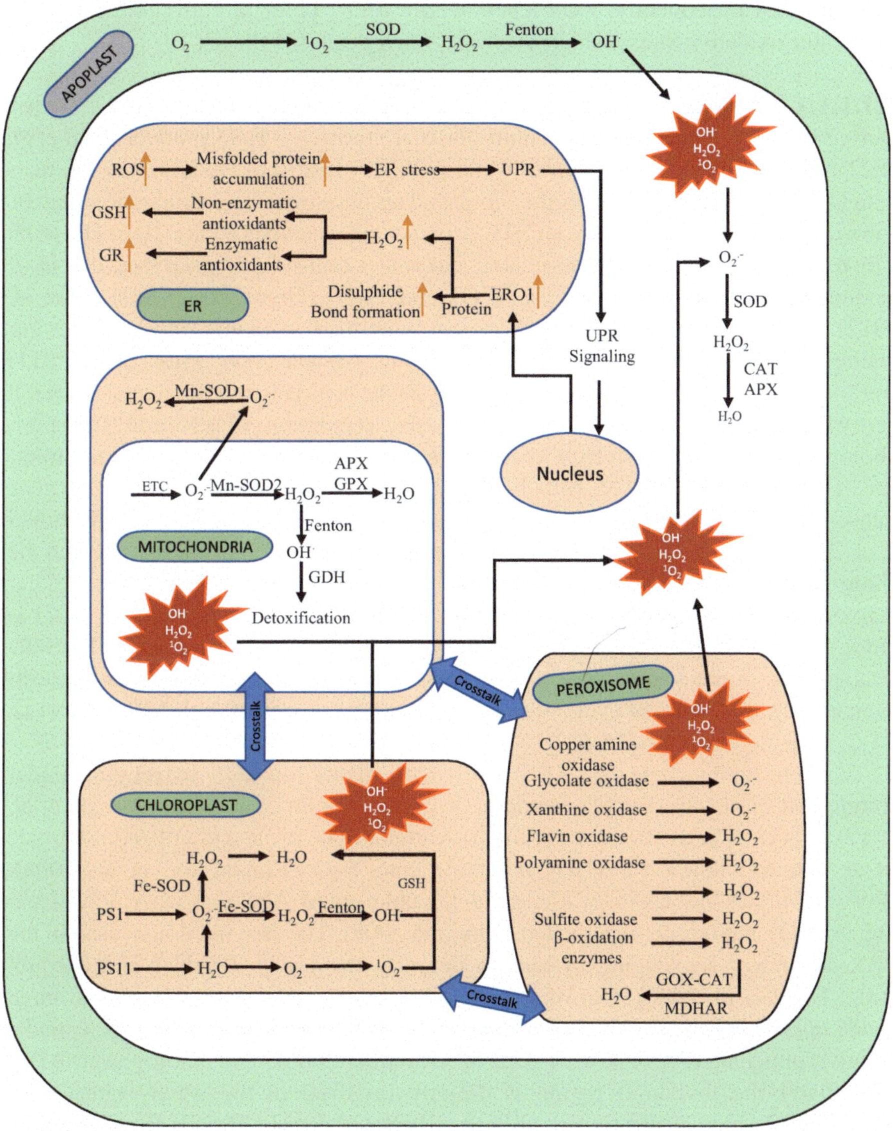

Fig. 2.1 The processes for the generation and detoxification of ROS in different plant cell organelles. The OH^- H_2O_2 $O_2^{·-}$ formation occurs in chloroplast, mitochondria, peroxisome, and ER. After formation the antioxidant enzyme changes the ROS to less toxic form. However, under extreme stress conditions ROS production increases the rate of detoxification and as a result ROS leakage occurs from different organelles. Abbreviations: H_2O_2, hydrogen peroxide; $O_2^{•-}$, superoxide anion; $^1O_2$, singlet oxygen; $^•OH$, hydroxyl radical; SOD, superoxide dismutase; CAT, Catalase

peculiar ROS scavenging system (Fig. 2.1). The antioxidant defense pathway AsA-GSH requires energy in the form of NADPH, and once this energy is depleted, this pathway is incapable of avoiding ROS toxicity (Choudhury et al. 2017; Mittler et al. 2004). Consequently, ROS levels within cells exceed the threshold level and cells enter oxidative stress which may cause cell death (Mittler 2017).

2.3.1.1 Chloroplast

During photosynthesis in plants, chloroplasts are the major ROS producers (Li et al. 2022). The electron generated from water at PSII is lost from the ETC and interacts with O_2 to form the $O_2^{\bullet-}$. Under the biotic and abiotic stress $O_2^{\bullet-}$ production may be enhanced at any location where an ETC is present (Leonowicz et al. 2018). The SOD which is the principal defense-related enzyme against ROS catalyzes the $O_2^{\bullet-}$ dismutation to less toxic forms, **i.e.,** O_2 and H_2O_2 (Grene 2002; Sharma et al. 2012). Out of three different SOD classes identified in plants, the Fe-SOD has mainly been considered a chloroplastic enzyme (Asada 2006; Pilon et al. 2011) and found in the stroma (Masuda and Fujita 2008; Amirjani and Sundqvist 2004). It performed several functions in chloroplasts: regulation of electron transport, photoprotection, protection from oxidative stress, abiotic stress signaling, and tolerance (Plöscher et al. 2011). Mostly the changes in the expression of SOD gene transcript are indicators of the level of oxidative stress (Ślesak et al. 2006; Foyer et al. 2009; Ślesak et al. 2008; Huseynova et al. 2014). Under oxidative stress in the chloroplast, the SOD works as part of a ROS scavenging system with either peroxidase or peroxiredoxin. Scavenging would be more efficient if the SOD is close to the ROS production site and enzyme that can consume H_2O_2. The SOD arrangement in the stroma near to thylakoid membrane surface increases the efficiency of the water–water cycle because there it works together with APX (Wu et al. 2012; Cruz de Carvalho 2008).

The hem-binding enzyme APX is used by plants for ROS scavenging under abiotic and biotic stresses (Caverzan et al. 2012; Anjum et al. 2016; Pandey et al. 2017). It is used in the first step of the AsA-glutathione cycle and utilized ascorbate as an electron donor for H_2O_2 reduction into H_2O (Kangasjärvi et al. 2008). However, under low levels of ascorbate, the chloroplast APX was extremely sensitive to H_2O_2 and rapidly inactivated (Kitajima 2008). The enzyme is also used in the PRX-dependent scavenging system to detoxify the H_2O_2 (Rouhier and Jacquot 2002; Dietz et al. 2006). Out of several APX isoenzymes present in plants, three reside in chloroplasts: A 33 kDa stromal APX, 38 kDa thylakoid bound APX, and a putative lumenal APX (Chew et al. 2003; Kitajima 2008). The chloroplastic APX, particularly the thylakoid region is directly involved in the water–water cycle because of its high susceptibility to H_2O_2 (Davletova et al. 2005; Hua et al. 2012). In the mutant line of wheat, a reduction in thylakoid APX activity decreases the photosynthetic carbon assimilation which resulted in reduced seed germination and growth rate (Danna et al. 2003). On the other hand, *Nicotiana tabacum* and transgenic *Arabidopsis thaliana* overexpressing thylakoid APX exhibit improved tolerance to photo-oxidative stress (Murgia et al. 2004). Previous findings have also supported that APX is responsible for biosynthesis of secondary cell walls during cotton fiber development stages (Tao et al. 2018).

The CAT is the principal ROS scavenging enzyme in the chloroplast, and it was noted that no plant or animal exists without that enzyme (Kirkman and Gaetani

2007; Mhamdi et al. 2010). Under ROS stress in chloroplasts CAT prevents the photosynthetic pigments degradation and improved the photosynthetic ability. It was also reported that CAT inhibits the chloroplast clustering under salt stress in plants. Another key component of the ROS detoxification system is the GR, a flavoprotein oxidoreductase which is mainly present in chloroplasts. It reduces the oxidized form of glutathione (GSSG) to GSH by using NADPH, which detoxifies the ROS species such as OH• and 1O2 (Gill et al. 2013). Furthermore, the heat-acclimated turfgrass maintained the low ROS levels by enhancing the GSH and AA biosynthesis under elevated temperatures (Kirkman and Gaetani 2007; Hasanuzzaman et al. 2018). Overall, these experiments suggest the role of antioxidant enzymes in heat stress tolerance.

2.3.1.2 Mitochondria

Mitochondria are the main ROS production site in non-green parts of the plant, especially in roots, where $O_2^{\bullet-}$ is produced due to electron leakage from both complexes I and which is later changed into H_2O_2 by CuZn-SOD and Mn-SOD (Bose et al. 2014; Singh et al. 2019). Therefore, the mitochondrial respiratory chain is a major ROS generation site in non-green tissues. The production sites of $O_2^{\bullet-}$ or H_2O_2 are located on the inner face of the inner membrane facing the matrix and mostly released their product in the mitochondrial matrix (Larosa and Remacle 2018). Mitochondrial enzymes like GSH, GR, ascorbate, and DHAR are present in the matrix, while the APX and MDHAR reductase are bound to the mitochondrial membranes (Teixeira and Glaser 2013). The occurrence of these enzymes in the vicinity of the ROS production site is beneficial for their detoxification.

Plant mitochondrial MnSOD is a tetramer enzyme located in the matrix space of mitochondria. The $O_2^{\bullet-}$ directly generated by cytosolic cytochrome P450 or oxidases or escaping from mitochondria is converted into H_2O_2 by the cytosolic CuZnSOD, which contains zinc and copper instead of Mn. Plant MnSODs have approximately 65% sequence similarity to one another (Grene 2002). Under oxidative stress, the MnSOD level decreased and caused specific perturbations in mitochondrial redox status. In salt-tolerant cultivars, mitochondrial SOD is overexpressed under salt-induced stress (Rubio et al. 2009; Kaminaka et al. 1999; Dai et al. 2009; Hernández and Almansa 2002). The tomato, poplar, rice, and transgenic *A. thaliana* plants also exhibited improved salt tolerance due to MnSOD gene transcript overexpression (Tanaka et al. 2002; Wang et al. 2004, 2010).

The ascorbate biosynthesis occurs in mitochondrial intermembrane space, and it gets oxidized before transportation toward the mitochondrial matrix where it enters the ascorbate–GSH cycle. In the mitochondrial matrix, APX reduces the H_2O_2 into H_2O by using ascorbate. These findings also suggest that the electron donor and acceptor sites of this protein are on the internal side of the mitochondrial membrane. As compared to chloroplasts and peroxisomal APX the activity of mitochondrial APX resulted in at least two isoenzymes that have different substrate sensibility and specificity (Jiménez et al. 1998). In mitochondrial APX expression remained unchanged in *Oryza sativa* when exposed to salinity stress (Teixeira et al. 2006), while Yamane et al. (2010) described the activation of the enzyme in rice.

The main mitochondrial enzyme system that deals with H_2O_2 produced by mitochondrial MnSOD are the GPx and GSH system. The enzymes, GPX, and peroxiredoxin (Prx) metabolize most of the H_2O_2 in the mitochondria. The activities of these enzymes to reduce H_2O_2 depend on the thiol groups of the cysteine residues (Cox et al. 2010). GPx is also involved in the prevention of lipid peroxidation by the reduction of organic hydroperoxides. When ROS production is high, the conversion of the oxidized forms of GPx or Prx to their reduced forms cannot meet the conversion rate from reduced to an oxidized form. Hence, a portion of the GPx or Prx remains in their oxidized form. Hence, an increase in the activity of other antioxidants would decrease the concentration of H_2O_2 and as a result the concentration of the reduced form of Prxs and Gpxs also increased. Overaccumulation of **GPx** transcripts has been reported in several plant species in response to salt stress, heavy metal toxicity, PCD, and exogenous hormone treatments (Chen and Dickman 2004; Sreenivasulu et al. 2004; Miao et al. 2006).

GSH biosynthesis only occurs in cytosol and plastids and is then transported to mitochondria by different transporter proteins. Since GSH synthesis is restricted to the plastid and cytosol, its subcellular distribution indicates that GSH transport likely takes place through all cellular endomembranes (Dorion et al. 2021). Several GSH transporters have been identified in the tonoplast, plastid inner envelope, mitochondrial inner membrane, and plasma membrane (Zhang et al. 2016; Oestreicher and Morgan 2019). Under normal conditions, GSH is present in reduced form, but oxidative stress and ROS detoxification can result in its oxidation and change the cellular redox state (Vanacker et al. 2006; Jiménez et al. 1998; Noctor et al. 2012). Zaffagnini et al. (2012) reported that GSH plays a significant role in the protection and signaling through PTMs and the regulation of intracellular redox potential.

2.3.1.3 Peroxisome

Peroxisomes with an extremely oxidative type of metabolic reaction are the main sites of cellular ROS production. Apart from fatty acid β-oxidation, it also controls a wide range of cellular processes within the plant cells (Terlecky and Titorenko 2009). Fatty acid β-oxidation enzymes, acyl-CoA oxidase, photorespiratory glycolate oxidase reaction, and the dismutation of $O_2^{\bullet -}$ radicals are the key metabolic processes involved in peroxisomal ROS generation (Foyer et al. 2009). In response to ROS presence of CAT and metalloenzyme SOD was reported in the peroxisome.

SODs are mainly distributed in different cellular locations such as cytoplasm, apoplasts, chloroplasts, mitochondria, and nuclei. The SOD's occurrence in peroxisomes has been also reported and confirmed by immunogold electron microscopy in 10 distinct species of plants (Rodríguez-Serrano et al. 2007; Mateos et al. 2003; Corpas et al. 2006). The $O_2^{\bullet -}$ generated by peroxisomal oxidases (xanthine oxidase) is scavenged by MnSOD and CuZnSOD (Schrader and Fahimi 2006). The role of peroxisomal MnSOD was reported in pea for the removal of $O_2^{\bullet -}$ formed by the action of xanthine oxidase. Similarly, Mittova et al. (2003) investigated that CuZn-SOD and Mn-SOD activities were increased (by 350 and 70%, respectively) in *Lycopersicon pennellii* under salt stress. In another experiment, Rodríguez-Serrano et al. (2007) reported the MnSOD presence in peroxisomal membranes which suggests its role in cell signaling events. The product of this reaction, H_2O_2, is an important signal transduction molecule involved in different physiological pathways and stress signaling in the plant cell (Apel and Hirt 2004; Neill et al. 2002; Del Río et al. 2006).

The most dynamic antioxidant enzyme APX is present in the peroxisomal membrane and has a relatively higher affinity for H_2O_2 and converts it to H_2O (Bunkelmann and Trelease 1996). Peroxisomal APX plays a critical role in a wide range of cellular processes such as seedling growth, leaf senescence, and apoptosis (Corpas and Trelease 1998; Ribeiro et al. 2017). The APX and MDAR also maintain a constant supply of NAD^+ by reoxidizing endogenous NADH in the leaf peroxisomal membranes for the metabolism process. Apart from this, APX and MDAR modulate H_2O_2, which is leaked from peroxisomes, or continuously formed by $O_2^{\bullet-}$ dismutation in the peroxisomal membrane by NAD(P)H-dependent electron transport system (Del Río et al. 2006). The overexpression of the peroxisomal APX gene (*SbpAPX*) in *Salicornia brachiata* and transgenic *Arachis hypogaea* confers salt stress tolerance in extreme halophytic conditions (Singh et al. 2014).

The major antioxidant enzyme CAT is present in the peroxisome matrix (Palma et al. 2020; Mhamdi et al. 2012; Mhamdi et al. 2010). Peroxisomal GOX in photosynthetic tissues is involved in 70% of H_2O_2 generated during photorespiration (Noctor et al. 2002). Furthermore, Zhang et al. (2016) have reported an efficient mechanism for GOX-CAT association–dissociation that regulates peroxisomal H_2O_2 levels in rice. The GOX and CAT association kept H_2O_2 level under control; however, under stress conditions, this complex GOX-CAT dissociation inhibits the CAT activity, leading to H_2O_2 accumulation (Zhang et al. 2016; Kohli et al. 2019). In another experiment, Yamauchi et al. (2019) reported that GOX and CAT association is essential for stomatal movement. In contrast, the peroxisomal Hsp17.6CII is present in *A. thaliana* and enhanced the CAT activity under stress conditions (Li et al. 2017). Similarly, the peroxisomal CAT activity upregulated in *Solanum lycopersicum* roots and leaves resulted in ROS regulation under stress conditions (Mittova et al. 2003).

2.3.1.4 Endoplasmic Reticulum

Proteins after synthesis in ER must undergo proper modification and fold before their transportation to their final destinations via the Golgi apparatus (Urade 2007; Lai et al. 2007). The unique oxidative environment in ER favors protein folding and modification (Schröder and Kaufman 2005). However, under the onset of adverse environmental conditions, the demand for protein folding increases than ER folding capacity which led to the misfolded protein accumulation (Howell 2013; Iwata and Koizumi 2012). The buildup of misfolded proteins induces ER stress which activates the unfolded protein response, and it is a significantly adaptive signaling pathway to prevent the accumulation of misfolded proteins (Angelos and Brandizzi 2018). The presence of antioxidant enzymes within ER regulates ROS production. ER peroxidases, such as glutathione peroxidases (GPx7, GPx8) and peroxiredoxin

4 (Prx4), scavenge the luminal H_2O_2 and prevent H_2O_2 leakage from the ER. As an alternative pathway, Prx4 can convert H_2O_2 to disulfide bonds on nascent polypeptides (Yoboue et al. 2018). Ozgur et al. (2014) findings revealed that ER under stress activates the enzymatic and nonenzymatic antioxidants to deal with stress conditions. GSH present in the lumen of ER breaks or rearranges the incorrect disulfide bonds present in misfolded proteins (Tu and Weissman 2004). Under higher ROS production GSH depletion occurs and as a result, GR activity is enhanced in ER.

2.3.2 ROS Regulation Through Nonenzymatic Defense Systems

During stressful conditions, the non-enzymatic and low molecular substances, such as tocopherol, ascorbic acid (AA), glutathione (GSH), phenolics, carotenoids, proline, and flavonoids, maintain redox equilibrium in plants (Miller et al. 2010; Gill et al. 2011; Gill and Tuteja 2010). Nonenzymatic components engage in an extensive range of cellular functions, which affect the development and yield of plants. In addition to playing key roles in plant defense, they can also regulate different processes ranging from mitosis to cell elongation and from aging and apoptosis (De Pinto and De Gara 2004). Stress hypersensitivity has been found in mutants with low nonenzymatic levels of antioxidants (Semchuk et al. 2009; Gao and Zhang 2008). The significance of ROS detoxification for cellular viability stems from the pervasiveness of both components of the oxidative mechanism (Gill et al. 2011).

2.3.2.1 Carotenoids

Carotenoids are a type of lipophilic antioxidant found in both photosynthetic and non-photosynthetic plant tissue. They can be identified in microorganisms as well as plants (Mezzomo and Ferreira 2016). They are part of the family of antennae molecules that collect light in the range of 450–570 nm and transmit it to the chlorophyll molecule (Zehra et al. 2015; Hasanuzzaman et al. 2019b). Under stress photosynthetic machinery is protected by carotenoids in four different ways: reacting with LPO products to end chain reactions, inhibiting the formation of $^1O_2$ through reacting with $^3Chl^*$ as well as energized chlorophyll (Chl*), scavenging $^1O_2$, producing heat as a by-product, and releasing excess excitation energy through the xanthophyll phase (Li et al. 2008). Carotenoids are made up of a polyene backbone, which is made up of several C=C bonds. This property is mainly responsible for pigmentation and the capability to quench ROS (Young and Lowe 2018).

2.3.2.2 Flavonoid

Flavonoids can be found in a wide variety of plant parts, including floral organs, leaves, and pollen grains. Flavones, flavonols, anthocyanins, and isoflavones are the four types of flavonoids based on their structure (Liu et al. 2014). They provide a variety of roles in pollen germination, pathogen defense, and plant fertility, including giving pigmentation to flowers, seeds, and fruits (Løvdal et al. 2010). Flavonoids have been proposed as an alternative scavenging system of ROS in plants whose photosynthetic apparatus has been harmed by excessive energy excitation (Fini et al. 2011). The process of $^1O_2$ recycling and the repair of the outer envelope of the chloroplast membrane also needed flavonoids (Agati et al. 2012). During stress, genes involved in flavonoid production are significantly upregulated, triggering defense systems via raising flavonoid contents. Flavonoids have antioxidant properties as well as the ability to modulate auxin transport in vivo and provide photoprotection. Furthermore, flavonoids work as sunscreens and protect plants from UV light harm by trapping ultraviolet rays (Sisa et al. 2010). Although light is required for the biosynthetic pathway, UV radiation causes plants to produce more flavonoids, which then aids in the elimination of ROS.

2.3.2.3 α-Tocopherol

The tocopherol is a member of lipophilic antioxidant, which is an effective scavenger of lipid radicals and ROS, manufacturing them as key defenders and fundamental parts of biological membranes (Kiffin et al. 2006; Holländer-Czytko et al. 2005). It is associated with the cell reinforcement system synergistically with glutathione (GSH) and ascorbate (AsA) under abiotic stress to keep up with ROS homeostasis. This triad is accounted to be linked with plants' tolerance mechanism under abiotic stresses (Foyer and Noctor 2005). Among the four isomers, α-tocopherol has the most noteworthy antioxidant capacity. The tocopherols are exclusively synthesized by photosynthetic organisms and hence they are only present in green tissues (Hasanuzzaman et al. 2014). They are known for their capacity to safeguard lipids and other chloroplast constituents by responding to O_2 and reducing its overabundance of energy, consequently safeguarding the PSII, both functionally and structurally. During AsA-GSH cycle, these two soluble antioxidants and low molecular weight collaborate. Tocopherol is a significant component of the ascorbate-glutathione-α-tocopherol chord (Szarka et al. 2012). It has been reported that tocopherol is involved in reducing singlet molecular oxygen ($^1O_2$) and a single α-tocopherol scavenges all 120 particles of $^1O_2$ (Mishra et al. 2019). In the ascorbate-glutathione-α-tocopherol chord pathways, tocopherols reduced the lipid per-oxy radical and changed over into tocopherol radicals which were further decreased by AsA (Szarka et al. 2012). Tocopherol likewise fills in as a compelling free radicle setup by stopping the chain-producing step of the LPO cycle. It responds with the lipid radicals $ROO^•$, RO*, $RO^•$, and at the water–membrane interface, where α-tocopherol diminishes them and itself gets changed over into $TOH^•$ (Igamberdiev et al. 2004). The $TOH^•$ radicle goes through reprocessing to its reduced structure by linking with AA and GSH.

2.3.2.4 Proline

Proline, an amino acid is likewise considered a strong antioxidant. Usually, it is involved across the various process and neutralizes the harmful impacts of ROS (Szabados and Savouré 2010). Proline is synthesized from pyrroline 5-carboxylate intermediate, which is generated by using glutamic acid as a substrate. Two compounds pyrroline-5-carboxylate reductase and ð1-pyrroline-5-carboxylate synthetase catalyzed this pathway in plants (Saibi et al. 2015). Proline metabolism is straightforwardly associated with the NAD(P)H/NAD(P) + redox balance, making it possible that proline goes about as a redox transport (Giberti et al. 2014; Sharma et al. 2011). Enhancing synthesis of proline under salt or drought stress is a mechanism to keep up with redox potential at values fit for metabolism. The biosynthesis of the rubisco enzyme creates one CO_2 particle and two NADPH particles. Under stress, carbon reduction continues because of the regeneration of $CO_{2,}$ and NAD(P)H is used in the synthesis of proline and to avoid photoinhibition in the plastids (Verslues and Sharma 2010). Proline is also a proficient scavenger of $OH^{\bullet}$ and $^1O_2$ and can restrain the harm because of LPO. Different environmental stresses enormously increased the proline contents in the cell due to enhanced biosynthesis or reduced degradation of proline (Verbruggen and Hermans 2008). In this manner, proline metabolism has a double role as a pro-oxidant and ROS scavenger.

2.3.2.5 Glutathione (GSH)

The thiol tripeptide GSH is richly observed in nearly all cellular parts like mitochondria, peroxisomes, chloroplasts, ER, vacuoles, cytosol, and the apoplast. GSH controls an array of processes like cell division/growth, senescence, xenobiotics detoxification, enzymatic activity regulation, conjugation of metabolites, phytochelatins formation, and regulation of sulfate transport, proteins, and nucleotides synthesis, and genes expression (Mullineaux and Rausch 2005). This adaptability of GSH is all because of its high reductive potential. The nucleophilic cysteine residue, in the center of GSH, is the main source of its reducing power. In presence of free organic radicals, GSH safeguards various biomolecules by reducing them and scavenges $O_2^{\bullet -}$ $^1O_2$, $OH^{\bullet}$, and H_2O_2 producing GSSG as a by-product or by forming adducts. It also enhances the GSSG production by stimulating AA. The GSSG thus produced is transformed back to GSH, both by GR and de novo synthesis. This eventually refills the cellular GSH level. Under heavy metal stress, GSH also stimulates phytochelatin synthase for phytochelatins synthesis (Choudhury et al. 2012a), which acts as a chelating agent (Choudhury et al. 2012b). Hence, the precise stability between GSSG and GSH is essential for sustaining the cell redox state.

2.4 ROS-Mediated Epigenetic Regulation of Gene

The elevated ROS acts as signaling molecules and enhances the expression level of different stress-related genes. Under drought conditions, the overexpression of CAT OsCATB prevents H_2O_2 accumulation and protects the plant from oxidative damage of ROS (Ye et al. 2011). It was reported that histone modifications such as H3k27me3, H3K4me2/3, H3K9me2, and H3K79me3 increased under harsh environmental conditions (Niu et al. 2015). The reprogramming of the genes such as RD22 and RD29B enhances the drought tolerance in *Coffea arabica* (Menezes-Silva et al. 2017).

The chromatin structure also plays a dynamic role in gene expression and epigenetic regulation in response to ROS (Struhl and Segal 2013). The accumulation of H3K4 di- and trimethylation is needed in *A. thaliana* for thermotolerance maintenance under heat stress. The H3K4 methylation depends on the HSFA2 TF (Lämke et al. 2016). Furthermore, the oxidative stress under the cold treatment upregulates the ADH1 gene expression in the *Chorispora bungeana*. The epigenetic modifications especially H3K4 trimethylation and H3K9 acetylation regulate the ADH1 gene expression (Liu et al. 2017). In *A. thaliana*, miR408 expression is not only involved in cold and salinity stress tolerance but also improved the sensitivity to drought treatment (Ma et al. 2015). Salt stress significantly enhanced the expression of miR397, miR156, miR393, miR319, miR159, and miR169 in *A. thaliana* however, the expression of miR398 is predominantly decreased (Liu et al. 2008). Similarly, the expression of miR156 significantly decreased under salinity treatment in maize (Ding et al. 2009).

The plant ROS generating systems, such as the respiratory burst oxidase homologues, belong to the plasma membrane enzymes family and have NADPH oxidase activity. The RBOH is activated under biotic and abiotic stress and resulted in O_2^- production which is dismutated to oxygen and H_2O_2 inside the apoplast (Drerup et al. 2013). The double mutants *A. thaliana* defective in RBOHD2-RBOHF2 and RBOHD1-RBOHF1exhibit reduced ROS production and reduced salt tolerance in comparison with wild-type (Ma et al. 2012). The precise control of OsCPK17 gene expression is vital for rice responses against cold stress (Almadanim et al. 2017). Yu et al. (2012) claim that any disturbance in the biosynthesis of thylakoid proteins occurring in rps1 distorts the membrane stability. This event not also disturbs the ROS production but also affects the redox-dependent retrograde signaling.

2.5 ROS Scavenging or Detoxification Proteins

Regarding ROS scavenging proteins, the authors hypothesize that the increased number of photosynthetic proteins and higher APX activity in plants confer to expand ROS scavenging capability, increase the photosynthetic rate, and reduce lipid peroxidation in subsequent stress response (Wang et al. 2014). The utterance of HSPs and HSFs is also produced by other stressful situations, like salinity and anoxia, which put forward these proteins' role in stress-tolerance systems (Fu et al. 2016). HSPs are produced by drought and heat stress in wheat, with the outcome of improving plant tolerance to extreme-heat stress (Zhang et al. 2016). Constantly, upregulation of certain heat shock proteins consults to a wide array of biotic and abiotic stresses tolerance. In chrysanthemum, constituent expression of HSP70 presents improved stress tolerance to heat, salt, and drought (Song et al. 2014). In transgenic rice plants overexpressed OsHsp23.7 & OsHsp17.0 proteins exhibit advanced salt and drought stress tolerance in comparison with the control plant (Zou et al. 2012). Furthermore, *A. thaliana* overexpressed HsfA1a exhibited tolerance to distinct stresses by upregulation of these HSPs (Qian et al. 2014). Plant responses to stress also include the induction of JmjC proteins by the stress-associated genes, which formed a complex defensive network and reduce the stress-generated ROS level (Shen et al. 2016). Though, it's not obvious that ROS directly altered the activity of JmjC proteins. Further, ROS features the activation of trimeric G proteins signaling pathway and mitogen-activated protein kinases (MAPK) which helps to remove the stressed and damaged plant cells (Jalmi and Sinha 2015). To protect plants against fungal pathogens, they established the pattern-recognition receptor for chitin sensitivity, which activates the intracellular stimulation of MAPK for instant ROS production (Kawasaki et al. 2017). The presence of cutin constituents in Alfalfa–Colletotrichum species interaction induces the gene expressions which encode for LIPK (lipid-induced protein kinase) and protein kinase C (PKC). These encoded results either prime the suppression of disease development or in the production of more infection structures (Dickman et al. 2003). In rice, an inhibiting apoptosis protein (BCL-2) can control (VPE) vacuolar processing enzymes by adjusting ion fluxes, which play a vital role in many PCD activities. Overexpressed BCL-2 lessens NaCl-induced K^+ efflux, suppresses the VPEs expression, and thus lessens PCD signs (Kim et al. 2014). Another mode of reducing stress and PCD responses is through cystatins. For example, overexpressed two cystatins—AtCYSa and AtCYSb in *A. thaliana* enhance the tolerance to all abiotic stresses, e.g., salt, oxidation, drought, and cold (Zhang et al. 2008).

2.6 ROS-Mediated Programmed Cell Death

When the concentration of a disadvantageous factor is high, the resistance program activated by plants is the initiation of PCD, a genetically meticulous process began to remove or isolate injured tissues which ensure the organism's survival. Different stress encouraged PCD death is not only an important biological process but also has a significant interest in the agricultural procedure to impact the yield of crops (Petrov et al. 2015). ROS have not only lethal effects on aerobic metabolic rate with regulated cellular levels, but they also function as controlling many biological activities like producing signaling agents and pleiotropic effects. Over the last period, ROS is known as an essential modulator of a plant's PCD. This mechanism plays a central role such as the formation and maturation of many cell tissues and types, plant adaptation to different environmental conditions, and embryo development. Molecular genetic approaches using transcriptome studies and plant mutants associated with ROS-mediated PCD have contributed to an elaborate redox signaling network and a wide array of plant specialized cell death regulators (Gadjev et al. 2008). Several endogenous and exogenous aspects include ROS production and plant cell homeostasis through PCD in plants. Endogenously generated ROS may impact extremely controlled signal transduction systems that deliver either negative or positive feedback control processes. To retain cells in homeostatic conditions, the antioxidants system is actively involved in inhibiting the increase of ROS levels in the cell. As ROS different grade levels decide the fate of plant cells, the optimum level of ROS possesses cells in homeostatic conditions for normal development and plant growth. A reduced amount of ROS primes weakened physiological function causing defects in the plant development. Increased ROS level leads to specific signaling paths for aging and diseases, autophagy, necrosis, PCD, and accidental cellular damage (Nath and Lu 2015).

Drought stress is accompanied by ROS accretion, mostly due to enhanced electron drip to triplet oxygen, and decreased CO_2 fixation which might ultimately lead to PCD (Gechev et al. 2012). Thus, one of the primary methods to inhibit ROS proliferation and avoid redundant PCD under drought stress is to decrease ROS development by downregulating the chloroplast components involved in photosynthetic and chlorophyll synthesis (Farrant 2007). Mature-green bananas exposed to abiotic stress at 1 °C for 7 days revealed critical chilling damage established the peel of the banana, a necrotic manner of ROS-refereed PCD, and an augmented deficit to sustain oxidoreduction homeostasis. Results showed distinct variation in metabolic rate due to oxidative destruction (Ramírez-Sánchez et al. 2022).

Polybrominated diphenyl ethers (PBDEs) exist as a sequence of extremely constant biological pollutants universally dispersed in marine ecosystems. To examine the toxic mechanism of PBDEs on microalgae cells, the process of PCD and its mediating mechanism was studied in *Thalassiosira pseudonana*. This indicated the *T. pseudonana* cell death happened because of BDE-47 stress. Furthermore, chloroplast and the cell membrane were recognized as the major sites for ROS production and signaling molecules to enable the process of the PCD (Zhao et al. 2020). The localization and type of ROS species may affect the capability of the PCD that observes. The exogenous product of the general antioxidant's glutathione and ascorbate to high temperature-treated *A. thaliana* cells prefers the induction of PCD, perhaps as uncontrolled necrotic death is prevented, and cellular stress levels are reduced. In contrast, dealing together with CAT, which is particular for H_2O_2, briefly represses PCD. Hence concluded that H_2O_2 functions as the PCD-generating mobile signal, though the other ROS forms and their findings serve as negative or positive managers of the elements of the PCD system (Doyle and McCabe 2010).

2.7 Conclusion

ROS are regarded as the by-products of metabolic reactions and under normal conditions; their production is low in different cellular compartments. However, various environmental stresses elevate the intracellular ROS level and induce oxidative stress. The elevated level of ROS not only causes irreversible DNA damage but also poses a serious threat to plant cells by inducing lipid peroxidation, enzyme inhibition through protein oxidation, and reduction in metabolic process. Apart from cellular damages ROS also function as important signaling molecules that activate many stress-responsive genes through a complex signal transduction network. As a result, the activated enzymatic and nonenzymatic antioxidant defense systems maintain equilibrium between ROS generation and detoxification. An imbalance between

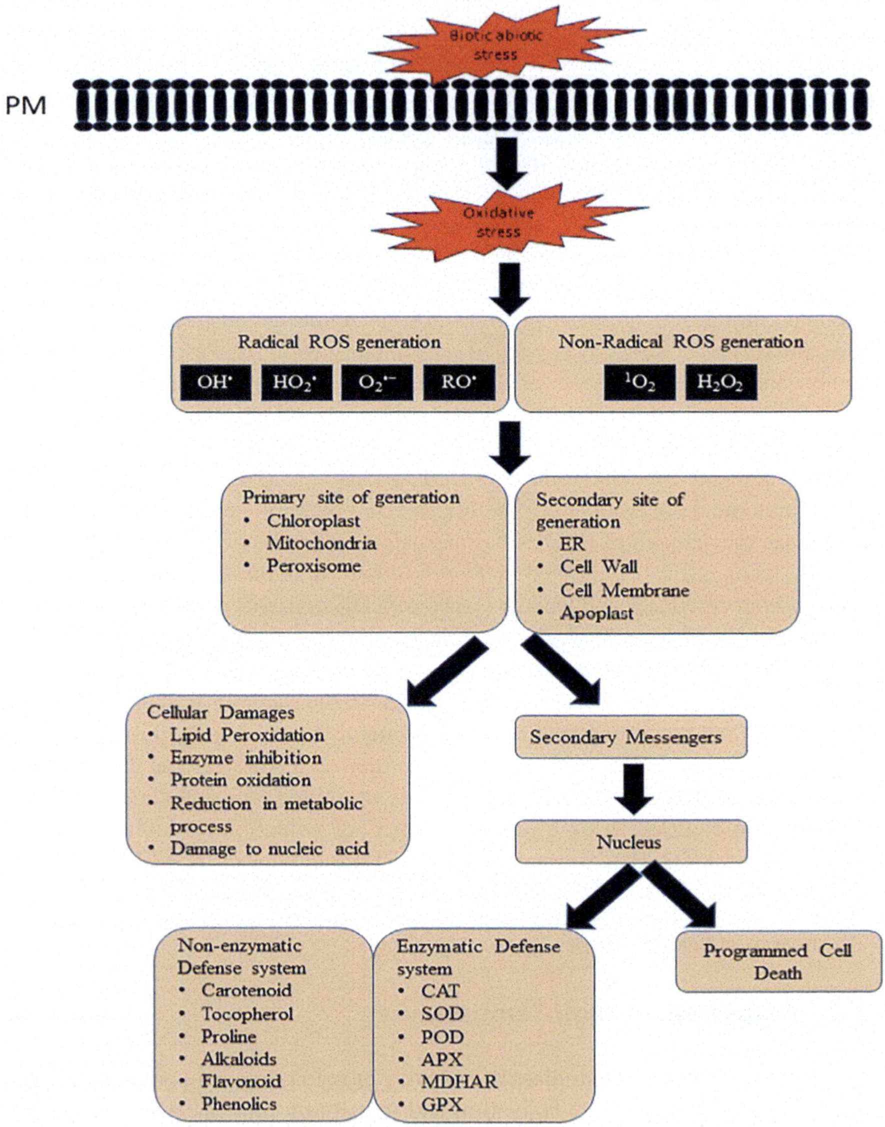

Fig. 2.2 Oxidative stress in plants and its consequences. The role of ROS in signal transduction pathway and programmed cell death

ROS production and the antioxidant defense system can lead to apoptosis or PCD (Fig. 2.2).

References

Agati G, Azzarello E, Pollastri S, Tattini M (2012) Flavonoids as antioxidants in plants: location and functional significance. Plant Sci 196:67–76

Almadanim MC, Alexandre BM, Rosa MT, Sapeta H, Leitão AE, Ramalho JC, Oliveira MM (2017) Rice calcium-dependent protein kinase OsCPK17 targets plasma membrane intrinsic protein and sucrose-phosphate synthase and is required for a proper cold stress response. Plant Cell Environ 40(7):1197–1213

Amirjani MR, Sundqvist C (2004) Regeneration of protochlorophyllide in green and greening leaves of plants with varying proportions of protochlorophyllide forms in darkness. Physiol Plant 121(3):377–390

Angelos E, Brandizzi F (2018) NADPH oxidase activity is required for ER stress survival in plants. Plant J 96(6):1106–1120

Anjum NA, Sharma P, Gill SS, Hasanuzzaman M, Khan EA, Kachhap K, Tuteja N (2016) Catalase and ascorbate peroxidase—representative H2O2-detoxifying heme enzymes in plants. Environ Sci Pollut Res 23(19):19002–19029

Apel K, Hirt H (2004) Reactive oxygen species: metabolism, oxidative stress, and signal transduction. Annu Rev Plant Biol 55:373–399

Asada K (2006) Production and scavenging of reactive oxygen species in chloroplasts and their functions. Plant Physiol 141(2):391–396

Bose J, Rodrigo-Moreno A, Shabala S (2014) ROS homeostasis in halophytes in the context of salinity stress tolerance. J Exp Bot 65(5):1241–1257

Bunkelmann JR, Trelease RN (1996) Ascorbate peroxidase (a prominent membrane protein in oilseed glyoxysomes). Plant Physiol 110(2):589–598

Caverzan A, Passaia G, Rosa SB, Ribeiro CW, Lazzarotto F, Margis-Pinheiro M (2012) Plant responses to stresses: role of ascorbate peroxidase in the antioxidant protection. Genet Mol Biol 35(4):1011–1019

Chen S, Dickman MB (2004) Bcl-2 family members localize to tobacco chloroplasts and inhibit programmed cell death induced by chloroplast-targeted herbicides. J Exp Bot 55(408): 2617–2623

Chen Q, Yang G (2020) Signal function studies of ROS, especially RBOH-dependent ROS, in plant growth, development and environmental stress. J Plant Growth Regul 39(1):157–171

Chew O, Whelan J, Millar AH (2003) Molecular definition of the ascorbate-glutathione cycle in Arabidopsis mitochondria reveals dual targeting of antioxidant defenses in plants. J Biol Chem 278(47):46869–46877

Choudhury AR, Das K, Ghosh S, Mukherjee RN, Banerjee R (2012b) Transgenic plants: benefits and controversies. J Bot Soc Bengal 66:29–35

Choudhury FK, Rivero RM, Blumwald E, Mittler R (2017) Reactive oxygen species, abiotic stress and stress combination. Plant J 90(5):856–867

Corpas FJ, Trelease RN (1998) Differential expression of ascorbate peroxidase and a putative molecular chaperone in the boundary membrane of differentiating cucumber seedling peroxisomes. J Plant Physiol 153(3–4):332–338

Corpas FJ, Fernandez-Ocana A, Carreras A, Valderrama R, Luque F, Esteban FJ, Barroso JB (2006) The expression of different superoxide dismutase forms is cell-type dependent in olive (Olea europaea L.) leaves. Plant Cell Physiol 47(7):984–994

Corpas FJ, Gupta DK, Palma JM (2015) Production sites of reactive oxygen species (ROS) in organelles from plant cells. In: Reactive oxygen species and oxidative damage in plants under stress. Springer, Cham, pp 1–22

Cox AG, Winterbourn CC, Hampton MB (2010) Mitochondrial peroxiredoxin involvement in antioxidant defence and redox signalling. Biochem J 425(2):313–325

Cruz de Carvalho MH (2008) Drought stress and reactive oxygen species: production, scavenging and signaling. Plant Signal Behav 3(3):156–165

Dai J, Huff DR, Schlossberg MJ (2009) Salinity effects on seed germination and vegetative growth of greens-type Poa annua relative to other cool-season turfgrass species. Crop Sci 49(2): 696–703

Danna CH, Bartoli CG, Sacco F, Ingala LR, Santa-María GE, Guiamet JJ, Ugalde RA (2003) Thylakoid-bound ascorbate peroxidase mutant exhibits impaired electron transport and photosynthetic activity. Plant Physiol 132(4):2116–2125

Davletova S, Rizhsky L, Liang H, Shengqiang Z, Oliver DJ, Coutu J, Mittler R (2005) Cytosolic ascorbate peroxidase 1 is a central component of the reactive oxygen gene network of Arabidopsis. Plant Cell 17(1):268–281

De Pinto MC, De Gara L (2004) Changes in the ascorbate metabolism of apoplastic and symplastic spaces are associated with cell differentiation. J Exp Bot 55(408):2559–2569

Del Río LA, López-Huertas E (2016) ROS generation in peroxisomes and its role in cell signaling. Plant Cell Physiol 57(7):1364–1376

Del Río LA, Sandalio LM, Corpas FJ, Palma JM, Barroso JB (2006) Reactive oxygen species and reactive nitrogen species in peroxisomes. Production, scavenging, and role in cell signaling. Plant Physiol 141(2):330–335

Dickman MB, Ha YS, Yang Z, Adams B, Huang C (2003) A protein kinase from Colletotrichum trifolii is indued by plant cutin and is required for appressorium formation. Mol Plant-Microbe Interact 16:411–421

Dietz KJ, Jacob S, Oelze ML, Laxa M, Tognetti V, de Miranda SMN, Finkemeier I (2006) The function of peroxiredoxins in plant organelle redox metabolism. J Exp Bot 57(8):1697–1709

Ding D, Zhang L, Wang H, Liu Z, Zhang Z, Zheng Y (2009) Differential expression of miRNAs in response to salt stress in maize roots. Ann Bot 103(1):29–38

Dogra V, Rochaix JD, Kim C (2018) Singlet oxygen-triggered chloroplast-to-nucleus retrograde signalling pathways: an emerging perspective. Plant Cell Environ 41(8):1727–1738

Dorion S, Ouellet JC, Rivoal J (2021) Glutathione metabolism in plants under stress: beyond reactive oxygen species detoxification. Meta 11(9):641

Doyle SM, McCabe PF (2010) Type and cellular location of reactive oxygen species determine activation or suppression of programmed cell death in Arabidopsis suspension cultures. Plant Signal Behav 5:467–468

Drerup MM, Schlücking K, Hashimoto K, Manishankar P, Steinhorst L, Kuchitsu K, Kudla J (2013) The calcineurin B-like calcium sensors CBL1 and CBL9 together with their interacting protein kinase CIPK26 regulate the Arabidopsis NADPH oxidase RBOHF. Mol Plant 6(2): 559–569

Dvořák P, Krasylenko Y, Zeiner A, Šamaj J, Takáč T (2021) Signaling toward reactive oxygen species-scavenging enzymes in plants. Front Plant Sci 11:2178

Farrant JM (2007) Mechanisms of desiccation tolerance in angiosperm resurrection plants. Plant Desiccat Tolerance 72:51–90

Fini A, Brunetti C, Di Ferdinando M, Ferrini F, Tattini M (2011) Stress-induced flavonoid biosynthesis and the antioxidant machinery of plants. Plant Signal Behav 6(5):709–711

Foyer CH, Noctor G (2005) Redox homeostasis and antioxidant signaling: a metabolic interface between stress perception and physiological responses. Plant Cell 17(7):1866–1875

Foyer CH, Bloom AJ, Queval G, Noctor G (2009) Photorespiratory metabolism: genes, mutants, energetics, and redox signaling. Annu Rev Plant Biol 60:455–484

Fransen M, Nordgren M, Wang B, Apanasets O, Van Veldhoven PP, del Río LA (2013) Peroxisomes and their key role in cellular signaling and metabolism. Springer, Dordrecht

Fu X, Wang J, Feng X, Wang D, Chen W, Song C, Zeng S (2016) Early Jurassic carbon-isotope excursion in the Qiangtang Basin (Tibet), the eastern Tethys: implications for the Toarcian oceanic anoxic event. Chem Geol 442:62–72

Fujita M, Fujita Y, Noutoshi Y, Takahashi F, Narusaka Y, Yamaguchi-Shinozaki K, Shinozaki K (2006) Crosstalk between abiotic and biotic stress responses: a current view from the points of convergence in the stress signaling networks. Curr Opin Plant Biol 9(4):436–442

Gadjev I, Stone JM, Gechev TS (2008) Programmed cell death in plants: new insights into redox regulation and the role of hydrogen peroxide. Int Rev Cell Mol Biol 270:87–144

Gao Q, Zhang L (2008) Ultraviolet-B-induced oxidative stress and antioxidant defense system responses in ascorbate-deficient vtc1 mutants of Arabidopsis thaliana. J Plant Physiol 165(2): 138–148

Gechev TS, Dinakar C, Benina M, Toneva V, Bartels D (2012) Molecular mechanisms of desiccation tolerance in resurrection plants. Cell Mol Life Sci 69(19):3175–3186

Giberti S, Funck D, Forlani G (2014) Δ1-pyrroline-5-carboxylate reductase from Arabidopsis thaliana: stimulation or inhibition by chloride ions and feedback regulation by proline depend on whether NADPH or NADH acts as co-substrate. New Phytol 202(3):911–919

Gill SS, Tuteja N (2010) Reactive oxygen species and antioxidant machinery in abiotic stress tolerance in crop plants. Plant Physiol Biochem 48(12):909–930

Gill SS, Anjum NA, Hasanuzzaman M, Gill R, Trivedi DK, Ahmad I, Tuteja N (2013) Glutathione and glutathione reductase: a boon in disguise for plant abiotic stress defense operations. Plant Physiol Biochem 70:204–212

Gill SS, Khan NA, Anjum NA, Tuteja N (2011) Amelioration of cadmium stress in crop plants by nutrients management: morphological, physiological and biochemical aspects. Plant Stress 5(1): 1–23

Grene R (2002) Oxidative stress and acclimation mechanisms in plants. In: The arabidopsis book, vol 1. American Society of Plant Biologists, Rockville, MD

Gupta KJ, Igamberdiev AU (eds) (2015) Reactive oxygen and nitrogen species signaling and communication in plants. Springer International Publishing, Cham

Hasanuzzaman M, Alam M, Rahman A, Hasanuzzaman M, Nahar K, Fujita M (2014) Exogenous proline and glycine betaine mediated upregulation of antioxidant defense and glyoxalase systems provides better protection against salt-induced oxidative stress in two rice (Oryza sativa L.) varieties. BioMed Res Int 2014:757219

Hasanuzzaman M, Bhuyan MHM, Anee TI, Parvin K, Nahar K, Mahmud JA, Fujita M (2019a) Regulation of ascorbate-glutathione pathway in mitigating oxidative damage in plants under abiotic stress. Antioxidants 8(9):384

Hasanuzzaman M, Bhuyan MHM, Zulfiqar F, Raza A, Mohsin SM, Mahmud JA, Fotopoulos V (2020) Reactive oxygen species and antioxidant defense in plants under abiotic stress: revisiting the crucial role of a universal defense regulator. Antioxidants 9(8):681

Hasanuzzaman M, Fotopoulos V, Nahar K, Fujita M, (Eds.). (2019b) Reactive oxygen, nitrogen and Sulfur species in plants: production, metabolism, Signaling and Defense mechanisms. John Wiley & Sons, Newark

Hasanuzzaman M, Hossain MA, Silva JA, Fujita M (2012) Plant response and tolerance to abiotic oxidative stress: antioxidant defense is a key factor. In: Crop stress and its management: perspectives and strategies. Springer, Dordrecht, pp 261–315

Hasanuzzaman M, Oku H, Nahar K, Bhuyan MHM, Mahmud JA, Baluska F, Fujita M (2018) Nitric oxide-induced salt stress tolerance in plants: ROS metabolism, signaling, and molecular interactions. Plant Biotechnol Rep 12(2):77–92

Hernández JA, Almansa MS (2002) Short-term effects of salt stress on antioxidant systems and leaf water relations of pea leaves. Physiol Plant 115(2):251–257

Hirayama T, Shinozaki K (2010) Research on plant abiotic stress responses in the post-genome era: past, present and future. Plant J 61(6):1041–1052

Holländer-Czytko H, Grabowski J, Sandorf I, Weckermann K, Weiler EW (2005) Tocopherol content and activities of tyrosine aminotransferase and cystine lyase in Arabidopsis under stress conditions. J Plant Physiol 162(7):767–770

Howell SH (2013) Endoplasmic reticulum stress responses in plants. Annu Rev Plant Biol 64:477–499

Hua D, Wang C, He J, Liao H, Duan Y, Zhu Z, Gong Z (2012) A plasma membrane receptor kinase, GHR1, mediates abscisic acid-and hydrogen peroxide-regulated stomatal movement in Arabidopsis. Plant Cell 24(6):2546–2561

Huang S, Van Aken O, Schwarzländer M, Belt K, Millar AH (2016) The roles of mitochondrial reactive oxygen species in cellular signaling and stress response in plants. Plant Physiol 171(3): 1551–1559

Huseynova IM, Aliyeva DR, Aliyev JA (2014) Subcellular localization and responses of superoxide dismutase isoforms in local wheat varieties subjected to continuous soil drought. Plant Physiol Biochem 81:54–60

Igamberdiev AU, Seregelyes C, Hill RD (2004) NADH-dependent metabolism of nitric oxide in alfalfa root cultures expressing barley hemoglobin. Planta 219(1):95–102

Iwata Y, Koizumi N (2012) Plant transducers of the endoplasmic reticulum unfolded protein response. Trends Plant Sci 17(12):720–727

Jalmi SK, Sinha AK (2015) ROS mediated MAPK signaling in abiotic and biotic stress- striking similarities and differences. Front Plant Sci 6:769

Janků M, Luhová L, Petřivalský M (2019) On the origin and fate of reactive oxygen species in plant cell compartments. Antioxidants 8(4):105

Jiménez A, Hernández JA, Pastori G, del Rıo LA, Sevilla F (1998) Role of the ascorbate-glutathione cycle of mitochondria and peroxisomes in the senescence of pea leaves. Plant Physiol 118(4): 1327–1335

Kaminaka H, Morita S, Tokumoto M, Masumura T, Tanaka K (1999) Differential gene expressions of rice superoxide dismutase isoforms to oxidative and environmental stresses. Free Radic Res 31(sup1):219–225

Kangasjärvi S, Lepistö A, Hännikäinen K, Piippo M, Luomala EM, Aro EM, Rintamäki E (2008) Diverse roles for chloroplast stromal and thylakoid-bound ascorbate peroxidases in plant stress responses. Biochem J 412(2):275–285

Kaur N, Kaur J, Grewal SK, Singh I (2019) Effect of heat stress on Antioxidative defense system and its amelioration by heat acclimation and salicylic acid pre-treatments in three pigeonpea genotypes. Indian J Agricu Biochem 32(1):106–110

Kawasaki T, Yamada K, Yoshimura S, Yamaguchi K (2017) Chitin receptor-mediated activation of MAP kinases and ROS production in rice and Arabidopsis. Plant Signal Behav 12:e1361076

Kiffin R, Bandyopadhyay U, Cuervo AM (2006) Oxidative stress and autophagy. Antioxid Redox Signal 8(1–2):152–162

Kim Y, Wang M, Bai Y, Zeng Z, Guo F, Han N et al (2014) Bcl-2 suppresses activation of VPEs by inhibiting cytosolic Ca2+ level with elevated K+ efflux in NaCl-induced PCD in rice. Plant Physiol Biochem 80:168–175

Kirkman HN, Gaetani GF (2007) Mammalian catalase: a venerable enzyme with new mysteries. Trends Biochem Sci 32(1):44–50

Kitajima S (2008) Hydrogen peroxide-mediated inactivation of two chloroplastic peroxidases, ascorbate peroxidase and 2-Cys peroxiredoxin. Photochem Photobiol 84(6):1404–1409

Kohli SK, Khanna K, Bhardwaj R, Abd Allah EF, Ahmad P, Corpas FJ (2019) Assessment of subcellular ROS and NO metabolism in higher plants: multifunctional signaling molecules. Antioxidants 8(12):641

Kumari A, Mahendar SB, Sachin S, Manoj KC, Ashim D, Chandan M, Manoj P, Binny S (2021) ROS regulation mechanism for mitigation of abiotic stress in plants. In: Ahmad R (ed) Reactive oxygen species. IntechOpen, London

Lai E, Teodoro T, Volchuk A (2007) Endoplasmic reticulum stress: signaling the unfolded protein response. Physiology 22(3):193–201

Lämke J, Brzezinka K, Altmann S, Bäurle I (2016) A hit-and-run heat shock factor governs sustained histone methylation and transcriptional stress memory. EMBO J 35(2):162–175

Larosa V, Remacle C (2018) Insights into the respiratory chain and oxidative stress. Biosci Rep 38(5):BSR20171492

Laxa M, Liebthal M, Telman W, Chibani K, Dietz KJ (2019) The role of the plant antioxidant system in drought tolerance. Antioxidants 8(4):94

Leonowicz G, Trzebuniak KF, Zimak-Piekarczyk P, Ślesak I, Mysliwa-Kurdziel B (2018) The activity of superoxide dismutases (SODs) at the early stages of wheat deetiolation. PLoS One 13(3):e0194678

Li C, Li J, Du X, Zhang J, Zou Y, Liu Y, Lu H (2022) Chloroplast Thylakoidal ascorbate peroxidase, PtotAPX, has enhanced resistance to oxidative stress in Populus tomentosa. Int J Mol Sci 23(6):3340

Li F, Vallabhaneni R, Yu J, Rocheford T, Wurtzel ET (2008) The maize phytoene synthase gene family: overlapping roles for carotenogenesis in endosperm, photomorphogenesis, and thermal stress tolerance. Plant Physiol 147(3):1334–1346

Li G, Li J, Hao R, Guo Y (2017) Activation of catalase activity by a peroxisome-localized small heat shock protein Hsp17. 6CII. J Genet Genomics 44(8):395–404

Liu KL, Shen L, Sheng JP (2008) Improvement in cadmium tolerance of tomato seedlings with an antisense DNA for 1-aminocyclopropane-1-carboxylate synthase. J Plant Nutr 31:809–827

Liu L, Song Y, Xu J, Li D, Li G, An L (2017) Differential expression by chromatin modifications of alcohol dehydrogenase 1 of Chorispora bungeana in cold stress. Gene 636:1–16

Liu S, Ju J, Xia G (2014) Identification of the flavonoid 3′-hydroxylase and flavonoid 3′, 5′--hydroxylase genes from Antarctic moss and their regulation during abiotic stress. Gene 543(1): 145–152

Løvdal T, Olsen KM, Slimestad R, Verheul M, Lillo C (2010) Synergetic effects of nitrogen depletion, temperature, and light on the content of phenolic compounds and gene expression in leaves of tomato. Phytochemistry 71(5–6):605–613

Ma CS, Deenick EK, Batten M, Tangye SG (2012) The origins, function, and regulation of T follicular helper cells. J Exp Med 209(7):1241–1253

Ma C, Burd S, Lers A (2015) mi R 408 is involved in abiotic stress responses in a rabidopsis. Plant J 84(1):169–187

Masuda T, Fujita Y (2008) Regulation and evolution of chlorophyll metabolism. Photochem Photobiol Sci 7(10):1131–1149

Mateos RM, León AM, Sandalio LM, Gómez M, Luis A, Palma JM (2003) Peroxisomes from pepper fruits (Capsicum annuum L.): purification, characterisation and antioxidant activity. J Plant Physiol 160(12):1507–1516

Mehla N, Sindhi V, Josula D, Bisht P, Wani SH (2017) An introduction to antioxidants and their roles in plant stress tolerance. In: Reactive oxygen species and antioxidant Systems in Plants: role and regulation under abiotic stress. Springer, Singapore, pp 1–23

Menezes-Silva PE, Sanglard LM, Ávila RT, Morais LE, Martins SC, Nobres P, DaMatta FM (2017) Photosynthetic and metabolic acclimation to repeated drought events play key roles in drought tolerance in coffee. J Exp Bot 68(15):4309–4322

Mezzomo N, Ferreira SR (2016) Carotenoids functionality, sources, and processing by supercritical technology: a review. J Chem 2016:3164312

Mhamdi A, Noctor G, Baker A (2012) Plant catalases: peroxisomal redox guardians. Arch Biochem Biophys 525(2):181–194

Mhamdi A, Queval G, Chaouch S, Vanderauwera S, Van Breusegem F, Noctor G (2010) Catalase function in plants: a focus on Arabidopsis mutants as stress-mimic models. J Exp Bot 61(15): 4197–4220

Miao Y, Lv D, Wang P, Wang XC, Chen J, Miao C, Song CP (2006) An Arabidopsis glutathione peroxidase functions as both a redox transducer and a scavenger in abscisic acid and drought stress responses. Plant Cell 18(10):2749–2766

Miller GAD, Suzuki N, Ciftci-Yilmaz SULTAN, Mittler RON (2010) Reactive oxygen species homeostasis and signalling during drought and salinity stresses. Plant Cell Environ 33(4): 453–467

Mishra B, Chand S, Sangwan NS (2019) ROS management is mediated by ascorbate-glutathione-α-tocopherol triad in co-ordination with secondary metabolic pathway under cadmium stress in Withania somnifera. Plant Physiol Biochem 139:620–629

Mittler R (2017) ROS are good. Trends Plant Sci 22(1):11–19

Mittler R, Vanderauwera S, Gollery M, Van Breusegem F (2004) Reactive oxygen gene network of plants. Trends Plant Sci 9(10):490–498

Mittova V, Tal M, Volokita M, Guy M (2003) Up-regulation of the leaf mitochondrial and peroxisomal antioxidative systems in response to salt-induced oxidative stress in the wild salt-tolerant tomato species Lycopersicon pennellii. Plant Cell Environ 26(6):845–856

Mor A, Koh E, Weiner L, Rosenwasser S, Sibony-Benyamini H, Fluhr R (2014) Singlet oxygen signatures are detected independent of light or chloroplasts in response to multiple stresses. Plant Physiol 165(1):249–261

Morgan MJ, Lehmann M, Schwarzlander M, Baxter CJ, Sienkiewicz-Porzucek A, Williams TC et al (2008) Decrease in manganese superoxide dismutase leads to reduced root growth and affects tricarboxylic acid cycle flux and mitochondrial redox homeostasis. Plant Physiol 147(1): 101–114

Mueller-Cajar O, Whitney SM (2008) Directing the evolution of rubisco and rubisco activase: first impressions of a new tool for photosynthesis research. Photosynth Res 98(1):667–675

Mullineaux PM, Rausch T (2005) Glutathione, photosynthesis and the redox regulation of stress-responsive gene expression. Photosynth Res 86(3):459–474

Murgia I, Tarantino D, Vannini C, Bracale M, Carravieri S, Soave C (2004) Arabidopsis thaliana plants overexpressing thylakoidal ascorbate peroxidase show increased resistance to Paraquat-induced photooxidative stress and to nitric oxide-induced cell death. Plant J 38(6):940–953

Nath K, Lu Y (2015) A paradigm of reactive oxygen species and programmed cell death in plants. Journal of Cell Science and Therapy 6(2):1–2

Nath M, Bhatt D, Bhatt MD, Prasad R, Tuteja N (2018) Microbe-mediated enhancement of nitrogen and phosphorus content for crop improvement. In: Crop improvement through microbial biotechnology. Elsevier, pp 293–304

Neill SJ, Desikan R, Clarke A, Hurst RD, Hancock JT (2002) Hydrogen peroxide and nitric oxide as signalling molecules in plants. J Exp Bot 53(372):1237–1247

Neve EP, Ingelman-Sundberg M (2010) Cytochrome P450 proteins: retention and distribution from the endoplasmic reticulum. Curr Opin Drug Discov Devel 13(1):78–85

Niu Y, DesMarais TL, Tong Z, Yao Y, Costa M (2015) Oxidative stress alters global histone modification and DNA methylation. Free Radic Biol Med 82:22–28

Noctor G, Mhamdi A, Chaouch S, Han YI, Neukermans J, Marquez-Garcia BELEN, Foyer CH (2012) Glutathione in plants: an integrated overview. Plant Cell Environ 35(2):454–484

Noctor G, Gomez L, Vanacker H, Foyer CH (2002) Interactions between biosynthesis, compartmentation and transport in the control of glutathione homeostasis and signalling. J Exp Bot 53(372):1283–1304

Noctor G, Reichheld JP, Foyer CH (2018) ROS-related redox regulation and signaling in plants. In: Seminars in Cell & Developmental Biology, vol 80. Academic Press, pp 3–12

Oestreicher J, Morgan B (2019) Glutathione: subcellular distribution and membrane transport. Biochem Cell Biol 97(3):270–289

Ozgur R, Turkan I, Uzilday B, Sekmen AH (2014) Endoplasmic reticulum stress triggers ROS signalling, changes the redox state, and regulates the antioxidant defence of Arabidopsis thaliana. J Exp Bot 65(5):1377–1390

Palma JM, Mateos RM, López-Jaramillo J, Rodríguez-Ruiz M, González-Gordo S, Lechuga-Sancho AM, Corpas FJ (2020) Plant catalases as NO and H2S targets. Redox Biol 34:101525

Pandey S, Fartyal D, Agarwal A, Shukla T, James D, Kaul T, Reddy MK (2017) Abiotic stress tolerance in plants: myriad roles of ascorbate peroxidase. Front Plant Sci 8:581

Petrov V, Hille J, Mueller-Roeber B, Gechev TS (2015) ROS-mediated abiotic stress-induced programmed cell death in plants. Front Plant Sci 6:69

Pilon M, Ravet K, Tapken W (2011) The biogenesis and physiological function of chloroplast superoxide dismutases. Biochim Biophys Acta (BBA) Bioenerg 1807(8):989–998

Plöscher M, Reisinger V, Eichacker LA (2011) Proteomic comparison of etioplast and chloroplast protein complexes. J Proteome 74(8):1256–1265

Pospíšil P (2016) Production of reactive oxygen species by photosystem II as a response to light and temperature stress. Front Plant Sci 7:1950

Qi J, Wang J, Gong Z, Zhou JM (2017) Apoplastic ROS signaling in plant immunity. Curr Opin Plant Biol 38:92–100

Qian J, Chen J, Liu YF, Yang LL, Li WP, Zhang LM (2014) Overexpression of Arabidopsis HsfA1a enhances diverse stress tolerance by promoting stress-induced Hsp expression. Genet Mol Res 13(1):1233–1243

Raja V, Majeed U, Kang H, Andrabi KI, John R (2017) Abiotic stress: interplay between ROS, hormones and MAPKs. Environ Exp Bot 137:142–157

Ramírez-Sánchez M, Huber DJ, Vallejos CE (2022) Abiotic stress triggers ROS-mediated programmed cell death in banana (Musa sp., AAA group, Cavendish sub-group) fruit. Sci Hortic 293:110748

Ribeiro CW, Korbes AP, Garighan JA, Jardim-Messeder D, Carvalho FE, Sousa RH et al (2017) Rice peroxisomal ascorbate peroxidase knockdown affects ROS signaling and triggers early leaf senescence. Plant Sci 263:55–65

Roach T, Krieger-Liszkay A (2014) Regulation of photosynthetic electron transport and photoinhibition. Curr Protein Peptide Sci 15(4):351–362

Rodríguez-Serrano M, Romero-Puertas MC, Pastori GM, Corpas FJ, Sandalio LM, del Río LA, Palma JM (2007) Peroxisomal membrane manganese superoxide dismutase: characterization of the isozyme from watermelon (Citrullus lanatus Schrad.) cotyledons. J Exp Bot 58(10): 2417–2427

Rouhier N, Jacquot JP (2002) Plant peroxiredoxins: alternative hydroperoxide scavenging enzymes. Photosynth Res 74(3):259–268

Choudhury A, Pradhan S, Chaudhuri B, Das K (2012a) 11 phytoremediation of toxic metals and the involvement of brassica species. In: Phytotechnologies: remediation of environmental contaminants. CRC Press, Taylor and Francis, Boca Raton, pp 219–252

Rubio MC, Bustos-Sanmamed P, Clemente MR, Becana M (2009) Effects of salt stress on the expression of antioxidant genes and proteins in the model legume Lotus japonicus. New Phytol 181(4):851–859

Saibi W, Feki K, Ben Mahmoud R, Brini F (2015) Durum wheat dehydrin (DHN-5) confers salinity tolerance to transgenic Arabidopsis plants through the regulation of proline metabolism and ROS scavenging system. Planta 242(5):1187–1194

Sandalio LM, Rodríguez-Serrano M, Romero-Puertas MC, Luis A (2008) Imaging of reactive oxygen species and nitric oxide in vivo in plant tissues. Methods Enzymol 440:397–409

Schrader M, Fahimi HD (2006) Peroxisomes and oxidative stress. Biochim Biophys Acta (BBA) Mol Cell Res 1763(12):1755–1766

Schröder M, Kaufman RJ (2005) ER stress and the unfolded protein response. Mutat Res 569(1–2): 29–63

Semchuk NM, Lushchak V, Falk J, Krupinska K, Lushchak VI (2009) Inactivation of genes, encoding tocopherol biosynthetic pathway enzymes, results in oxidative stress in outdoor grown Arabidopsis thaliana. Plant Physiol Biochem 47(5):384–390

Sharma P, Jha AB, Dubey RS, Pessarakli M (2012) Reactive oxygen species, oxidative damage, and antioxidative defense mechanism in plants under stressful conditions. J Bot 2012:217037

Sharma S, Villamor JG, Verslues PE (2011) Essential role of tissue-specific proline synthesis and catabolism in growth and redox balance at low water potential. Plant Physiol 157(1):292–304

Shen Q, Zhou W, Li H, Hu L, Mo H (2016) ROS involves the fungicidal actions of thymol against spores of aspergillus flavus via the induction of nitric oxide. PLoS One 11(5):e0155647

Singh A, Kumar A, Yadav S, Singh IK (2019) Reactive oxygen species-mediated signaling during abiotic stress. Plant Gene 18:100173

Singh N, Mishra A, Jha B (2014) Over-expression of the peroxisomal ascorbate peroxidase (SbpAPX) gene cloned from halophyte Salicornia brachiata confers salt and drought stress tolerance in transgenic tobacco. Mar Biotechnol 16(3):321–332

Sisa M, Bonnet SL, Ferreira D, Van der Westhuizen JH (2010) Photochemistry of flavonoids. Molecules 15(8):5196–5245

Ślesak I, Haldaś W, Ślesak H (2006) Influence of exogenous carbohydrates on superoxide dismutase activity in Trifolium repens L. explants cultured in vitro. Acta Biol Cracov Ser Bot 48(1):93–98

Ślesak I, Ślesak H, Libik M, Miszalski Z (2008) Antioxidant response system in the short-term post-wounding effect in Mesembryanthemum crystallinum leaves. J Plant Physiol 165(2):127–137

Song A, Zhu X, Chen F, Gao H, Jiang J, Chen S (2014) A chrysanthemum heat shock protein confers tolerance to abiotic stress. Int J Mol Sci 15(3):5063–5078

Sreenivasulu N, Miranda M, Prakash HS, Wobus U, Weschke W (2004) Transcriptome changes in foxtail millet genotypes at high salinity: identification and characterization of a PHGPX gene specifically up-regulated by NaCl in a salt-tolerant line. J Plant Physiol 161(4):467–477

Struhl K, Segal E (2013) Determinants of nucleosome positioning. Nat Struct Mol Biol 20(3): 267–273

Szabados L, Savouré A (2010) Proline: a multifunctional amino acid. Trends Plant Sci 15(2):89–97

Szarka A, Tomasskovics B, Bánhegyi G (2012) The ascorbate-glutathione-α-tocopherol triad in abiotic stress response. Int J Mol Sci 13(4):4458–4483

Tanaka H, Matsumura I, Ezoe S, Satoh Y, Sakamaki T, Albanese C, Kanakura Y (2002) E2F1 and c-Myc potentiate apoptosis through inhibition of NF-κB activity that facilitates MnSOD-mediated ROS elimination. Mol Cell 9(5):1017–1029

Tao C, Jin X, Zhu L, Xie Q, Wang X, Li H (2018) Genome-wide investigation and expression profiling of APX gene family in Gossypium hirsutum provide new insights in redox homeostasis maintenance during different fiber development stages. Mol Gen Genomics 293(3):685–697

Teixeira FK, Menezes-Benavente L, Galvão VC, Margis R, Margis Pinheiro M (2006) Rice ascorbate peroxidase gene family encodes functionally diverse isoforms localized in different subcellular compartments. Planta 224(2):300–314

Teixeira PF, Glaser E (2013) Processing peptidases in mitochondria and chloroplasts. Biochim Biophys Acta (BBA) Mol Cell Res 1833(2):360–370

Terlecky SR, Titorenko VI (2009) Emergent functions of the peroxisome. Research Signpost, Kerala

Triantaphylidès C, Havaux M (2009) Singlet oxygen in plants: production, detoxification and signaling. Trends Plant Sci 14(4):219–228

Tripathi DK, Singh VP, Lux A, Vaculik M (2020) Silicon in plant biology: from past to present, and future challenges. J Exp Bot 71(21):6699–6702

Tripathy BC, Oelmüller R (2012) Reactive oxygen species generation and signaling in plants. Plant Signal Behav 7(12):1621–1633

Tu BP, Weissman JS (2004) Oxidative protein folding in eukaryotes: mechanisms and consequences. J Cell Biol 164(3):341–346

Urade R (2007) Cellular response to unfolded proteins in the endoplasmic reticulum of plants. FEBS J 274(5):1152–1171

Vanacker H, Sandalio LM, Jiménez A, Palma JM, Corpas FJ, Meseguer V, Del Río LA (2006) Roles for redox regulation in leaf senescence of pea plants grown on different sources of nitrogen nutrition. J Exp Bot 57(8):1735–1745

Verbruggen N, Hermans C (2008) Proline accumulation in plants: a review. Amino Acids 35(4): 753–759

Verslues PE, Sharma S (2010) Proline metabolism and its implications for plant-environment interaction. In: The arabidopsis book, vol 8. American Society of Plant Biologists, Rockville, MD

Wang LI, Neuberg D, Christiani DC (2004) Asbestos exposure, manganese superoxide dismutase (MnSOD) genotype, and lung cancer risk. J Occup Environ Med 46:556–564

Wang P, Xie K, Wang C, Bi J (2014) Oxidative stress induced by lipid peroxidation is related with inflammation of demyelination and neurodegeneration in multiple sclerosis. Eur Neurol 72(3–4):249–254

Wang QJ, Sun H, Dong QL, Sun TY, Jin ZX, Hao YJ, Yao YX (2016) The enhancement of tolerance to salt and cold stresses by modifying the redox state and salicylic acid content via the cytosolic malate dehydrogenase gene in transgenic apple plants. Plant Biotechnol J 14(10): 1986–1997

Wang V, Chen SY, Chuang TC, Shan DE, Soong BW, Kao MC (2010) Val-9Ala and Ile+ 58Thr polymorphism of MnSOD in Parkinson's disease. Clin Biochem 43(12):979–982

Wu GQ, Zhang LN, Wang YY (2012) Response of growth and antioxidant enzymes to osmotic stress in two different wheat (Triticum aestivum L.) cultivars seedlings. Plant Soil Environ 58(12):534–539

Yamane K, Mitsuya S, Taniguchi M, Miyake H (2010) Transcription profiles of genes encoding catalase and ascorbate peroxidase in the rice leaf tissues under salinity. Plant Product Sci 13(2): 164–168

Yamauchi S, Mano S, Oikawa K, Hikino K, Teshima KM, Kimori Y, Takemiya A (2019) Autophagy controls reactive oxygen species homeostasis in guard cells that is essential for stomatal opening. Proc Natl Acad Sci 116(38):19187–19192

Ye N, Zhu G, Liu Y, Li Y, Zhang J (2011) ABA controls H2O2 accumulation through the induction of OsCATB in rice leaves under water stress. Plant Cell Physiol 52:689–698

Yoboue ED, Sitia R, Simmen T (2018) Redox crosstalk at endoplasmic reticulum (ER) membrane contact sites (MCS) uses toxic waste to deliver messages. Cell Death Dis 9(3):1–14

Young AJ, Lowe GL (2018) Carotenoids—antioxidant properties. Antioxidants 7(2):28

Yu HD, Yang XF, Chen ST, Wang YT, Li JK, Shen Q, Guo FQ (2012) Downregulation of chloroplast RPS1 negatively modulates nuclear heat-responsive expression of HsfA2 and its target genes in Arabidopsis. PLoS Genet 8(5):e1002669

Zaffagnini M, Bedhomme M, Marchand CH, Morisse S, Trost P, Lemaire SD (2012) Redox regulation in photosynthetic organisms: focus on glutathionylation. Antioxid Redox Signal 16(6):567–586

Zainab N, Khan AA, Azeem MA, Ali B, Wang T, Shi F et al (2021) PGPR-mediated plant growth attributes and metal extraction ability of Sesbania sesban L. Ind Contamin Soils Agron 11(9): 1820

Zehra A, Dubey MK, Tiwari A, Meena M, Kumari P, Singh VK, Upadhyay RS (2015) Fungal biomolecules and their implications. In: Fungal biomolecules: sources, applications and recent developments. Wiley, Hoboken, p 363

Zeng J, Dong Z, Wu H, Tian Z, Zhao Z (2017) Redox regulation of plant stem cell fate. EMBO J 36(19):2844–2855

Zhang X, Liu S, Takano T (2008) Two cysteine proteinase inhibitors from Arabidopsis thaliana, AtCYSa and AtCYSb, increasing the salt, drought, oxidation, and cold tolerance. Plant Mol Biol 68:131–143

Zhang Z, Xie Q, Jobe TO, Kau AR, Wang C, Li Y, Schroeder JI (2016) Identification of AtOPT4 as a plant glutathione transporter. Mol Plant 9(3):481–484

Zhao Y, Tang X, Qu F, Lv M, Liu Q, Li J et al (2020) ROS-mediated programmed cell death (PCD) of Thalassiosira pseudonana under the stress of BDE-47. Environ Pollut 262:114342

Zhu JK (2016) Abiotic stress signaling and responses in plants. Cell 167(2):313–324

Zou J, Liu C, Liu A, Zou D, Chen X (2012) Overexpression of OsHsp17. 0 and OsHsp23. 7 enhances drought and salt tolerance in rice. J Plant Physiol 169(6):628–635

Zipfel C (2014) Plant pattern-recognition receptors. Trends Immunol 35(7):345–351

Biomolecules Targeted by Reactive Oxygen Species

3

Arshiya Akeel and Hassan Jaleel

Abstract

Reactive oxygen species (ROS) are known to be produced as a by-product of plant cellular metabolism. The formation of ROS in plant cells has both negative and positive consequences. ROS plays a crucial role in signaling pathways that control plant acclimation and defense responses. High levels of ROS, on the other hand, can disrupt redox equilibrium, resulting in cell death and, as a result, a reduction in biomass and yield. ROS promptly inactivates enzymes, damages essential plant cellular organelles, and disintegrates membranes by triggering the destruction of pigments, proteins, lipids, and nucleic acids, resulting in cell death. In addition to destroying macromolecules, ROS acts as a stimulant in signal transduction pathways and a secondary messenger in various plant developmental pathways. In light of recent researches, ROS is known to engage in a variety of processes throughout the plant life cycle, including seed development and germination, root, shoot, and flower development.

Furthermore, ROS production rates are directly linked to soluble sugars through various metabolic reactions and regulations like mitochondrial respiration or photosynthesis regulation and carotenoid biosynthesis. So, it indicates that ROS interacts with different biomolecules to produce multiple plant responses. Therefore, the chapter deposits information on biomolecules targeted by ROS and an in-depth study of changes bought in targeted biomolecules structure and function is also canvassed.

A. Akeel (✉)
Plant Physiology and Biochemistry Section, Department of Botany, Aligarh Muslim University, Aligarh, UP, India

H. Jaleel
Department of Botany, GDC Billawar, Billawar, Jammu and Kashmir, India

M. Faizan et al. (eds.), *Reactive Oxygen Species*,
https://doi.org/10.1007/978-981-19-9794-5_3

3.1 What Are ROS?

Reactive oxygen species (ROS) are a class of free radicals, reactive molecules, and ions that are known to be emanated from O_2. ROS are known to be produced primarily as a byproduct of plant cellular metabolism (Asada and Takahashi 1987). Initially, reactive oxygen species (ROS) were thought to be toxic by-products of aerobic metabolism. ROS has been shown to play an important signaling role in plants, controlling processes such as growth, development, and especially response to biotic and abiotic environmental cues in recent times (Das and Roychoudhury 2014). Amidst their destructive nature, they are well-studied second messengers in a wide range of cellular processes, including the conferment of tolerance to various environmental stresses. The delicate balance between ROS generation and scavenging determines whether ROS function as signaling molecules or cause oxidative damage to tissues. A potent and effective scavenging of ROS, produced during various environmental stresses necessitates the response of several nonenzymatic and enzymatic antioxidants found in tissues (Sharma et al. 2012). The main members of the ROS family comprise free radicals like O^{-}_{2}, $OH^{\bullet}$ and non-radicals like H_2O_2 and $^1O_2$.

3.2 ROS: History and Evolution

Around 2.7 billion years ago, O_2-evolving photosynthetic organisms brought molecule oxygen into the early reducing atmosphere of the Earth, resulting in the emergence of reactive oxygen species (ROS) as undesirable by-products (Halliwell 2006).

3.3 Role in Plants

The ROS family plays a dual role in cellular homeostasis, acting as secondary messengers in a number of important physiological processes while also inducing oxidative damage in response to a variety of environmental stressors, such as salinity, drought, cold, heavy metals, and UV radiation, among others (Sharma et al. 2012). At large quantities, all ROS are extremely toxic to living things. The term "oxidative stress" refers to the condition in which a cell is in when the level of ROS exceeds the defense systems. By triggering peroxidation of lipids, oxidation of proteins, damage to nucleic acids, enzyme inhibition, activation of the programmed cell death (PCD) pathway, and eventually cell death, the increased generation of ROS amid environmental conditions can be dangerous to cells (Shah et al. 2001; Meriga et al. 2004; Maheshwari and Dubey 2009; Sharma et al. 2012).

3.4 Site of ROS Generation in Plants

The chloroplast, mitochondria, and peroxisomes are where plants produce most of their ROS. The endoplasmic reticulum, cell membrane, cell wall, and apoplast are examples of secondary sites (Das and Roychoudhury 2014; Noctor et al. 2018). Chloroplasts and peroxisomes are the primary sources of ROS generation in the presence of light, whereas the mitochondrion is the prime source of ROS in the absence of light (Choudhury et al. 2013).

3.5 Targeted Biomolecules

Cellular damage manifests itself in the degradation of biomolecules such as pigments, proteins, lipids, carbohydrates, and DNA, resulting in plant cellular death (McCord 2000; Khanna-Chopra 2012; Das and Roychoudhury 2014; Sachdev et al. 2021). The ROS mainly comprise $^1O_2$, H_2O_2, $O^{\bullet-}{}_2$, and $OH^{\bullet}$. These are very lethal and cause extensive damage to protein, DNA, and lipids and thereby affect normal cellular functioning (Apel and Hirt 2004; Foyer and Noctor 2005; Sachdev et al. 2021). Production and removal of ROS must be tightly regulated in order to prevent antioxidant stress. However, under a number of stressful situations, such as salinity, drought, bright light, toxicity owing to metals, viruses, and so forth, the equilibrium between the production and scavenging of ROS is disturbed. Increased ROS levels can harm biomolecules like lipids, proteins, and DNA (Fig. 3.1). These reactions can alter intrinsic membrane properties such as fluidity, ion transport, enzyme activity loss, protein cross-linking, protein synthesis inhibition, DNA damage, and so on, ultimately leading to cell death (Sharma et al. 2012; Abouzari and Fakheri 2015).

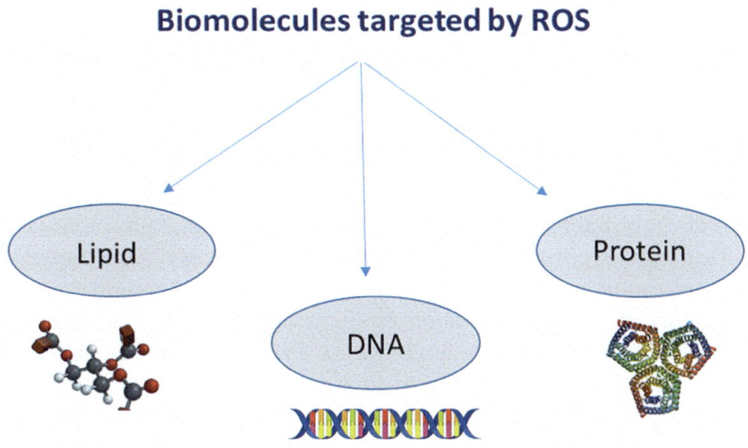

Fig. 3.1 Biomolecules targeted by ROS

3.5.1 Effect of ROS on Lipids

When ROS levels exceed a certain threshold, lipid peroxidation (LPO) increases in both cellular and organellar membranes, affecting normal cellular functioning (Van Breusegem and Dat 2006). And this peroxidation of membrane lipids occurs when ROS reacts with unsaturated fatty acids leading to leakage of cellular contents, rapid desiccation, and hence cell death (Ahmad et al. 2008). The harmful effect of ROS is due primarily to their ability to initiate a variety of auto-oxidative chain reactions on unsaturated fatty acids (Smirnoff 2000). The most harmful process known to take place in every living cell is the peroxidation of lipids (Repetto et al. 2012). The plasma membrane that surrounds the cell and aids in its ability to adapt to its dynamic environment is largely made up of lipids. The membrane's lipid peroxidation (LPO), however, becomes so harmful under stressed circumstances when the level of ROS rises above the threshold value that it is frequently used as the sole parameter to measure lipid degradation (Jambunathan 2010). It has been postulated by various researchers that an increase in lipid peroxidation in response to environmental stresses corresponds with an increase in ROS production (Oberschall et al. 2000; Shah et al. 2001; Ali and Alqurainy 2006; Sharma et al. 2012). Malondialdehyde (MDA) is one of the end products of peroxidation of unsaturated fatty acids in phospholipids and causes cell membrane damage (Esterbauer et al. 1991; Halliwell and Gutteridge 1999). Two common sites of ROS attack on phospholipid molecules are unsaturated (double) bonds between two carbon atoms and ester bonds between glycerol and fatty acids (Tiwari et al. 2017). Polyunsaturated fatty acids (PUFAs) present in membrane phospholipids are particularly vulnerable to attack by ROS. Since the reactions involved in this process are part of a cyclical chain reaction, one OH can cause peroxidation of many polyunsaturated fatty acids.

3.5.2 Effect of ROS on Proteins

ROS generated under stressful conditions leads to protein oxidation as well. Protein oxidation is generally described as the covalent modification of a protein caused by ROS or oxidative stress by-products. Excessive ROS production causes particular amino acid changes, disintegration of peptide chain, accumulation of cross-linked reaction products, and enhanced susceptibility to proteolysis (Mehta et al. 1992; Sharma et al. 2012). Alteration in protein due to ROS is broadly categorized in two categories: direct or indirect. Through direct modification, protein activity is altered as a result of various chemical modifications such as nitrosylation, carboxylation, disulfide bond formation, and glutathionylation. And during indirect modification proteins combine with the degradation products of fatty acid peroxidation, which causes its alteration (Yamauchi et al. 2008). As a result of these direct and indirect changes brought by an enhanced level of ROS, proteolysis takes place (Palma et al. 2002; Khanna-Chopra 2012). To assess the protein oxidation, carbonylation of protein is widely used as a marker because the tissues damaged by oxidative stress are found to have typically higher amounts of carbonylated proteins (Palma et al. 2002; Job et al. 2005; Møller et al. 2007).

3.5.2.1 Amino Acid Modifications

ROS-mediated changes in protein cause site-specific alterations in amino acids including Arg, Lys, Pro, Thr, and Trp and become more susceptible to proteolytic degradation (Rivett and Levine 1990; Dunlop et al. 2002; Møller et al. 2007). Although the vulnerability of the amino acids to ROS attack varies, the most susceptible amino acids are those with sulfur and thiol groups. It has also been observed that the reactive $^1O_2$ and $OH^\bullet$ cause harm to Cys and Met equally (Mittler 2017; Choudhary et al. 2020), while enzymes containing iron-sulfur centers are irreversibly inactivated when oxidized by $O^{\bullet -}{}_2$. Thus, oxidized proteins are better targets for proteolytic digestion by preparing them for proteasomal degradation via ubiquitination (Gardner and Fridovich 1991).

3.5.3 Effect of ROS on DNA

ROS have been found to play a significant role in DNA damage (Imlay and Linn 1988; Das and Roychoudhury 2014). ROS-mediated damage of DNA takes place in multiple stages. It incorporates deoxyribose sugar oxidation, nucleotide base modification, nucleotide alteration, DNA strand breakage, and DNA and protein cross-linkage (Sharma et al. 2012; Choudhary et al. 2020). Studies have revealed that ROS have the potential to cause oxidative damage to nuclear, mitochondrial, and chloroplastic DNA. However, the mitochondrial and chloroplastic DNA are the main victims of the ROS attack due to a lack of protective histones as well as closeness to ROS production machinery, on the other hand, nuclear DNA of plants is well protected by histones and associated proteins (Das and Roychoudhury 2014).

3.5.3.1 Mechanism of DNA Degradation

In ROS-mediated DNA damage both sugar and base moieties are equally susceptible to oxidation by ROS (Halliwell and Gutteridge 1999). Mechanism of DNA degradation by ROS mainly involves the oxidative attack by hydroxyl radical on bases, it reacts with double bonds of the purine and pyrimidine bases. On the other hand, it damages the sugar backbone by dehydrogenation of deoxyribose (Halliwell 2006).

3.5.3.2 Attack on Sugar Moiety

Studies revealed that ROS removes the C-4 H-atom from the deoxyribose sugar to create the deoxyribose radical which is further responsible for break in single strand of DNA (Dizdaroglu 1993; Evans et al. 2004).

3.5.3.3 Attack on Nucleotide Base

Oxidation of bases by ROS leads to formation of various products. Most common of which is 8-hydroxyquinine and other less common ones involves thymine glycol, hydroxyl methyl urea, dehydro-2′-deoxyguanosine, and thymine and adenine ring opened (Halliwell 2006; Das and Roychoudhury 2014). Studies have also revealed that hydroxyl radical begins pyrimidine damage by attacking its C5-C6 double bond to produce 5-hydroxy-6-yl and 6-hydroxy-5-yl radicals that continue to react and form various stable DNA lesions (Dizdaroglu 1999). Some of the significant oxidation products of pyrimidine are saturated derivatives of the ring, especially thymine glycol, 5,6-dihydro-thymine, and cytosine glycol (Roldán-Arjona and Ariza 2009). However, products obtained from oxidation of cytosine may transform into uracil derivatives; for example, cytosine glycol can be deaminated to produce uracil glycol or 5-hydroxyuracil (Zuo et al. 1995; Dizdaroglu 1999). Hydroxyl radicals are also known to react with purines by attacking on C4-, C5-, and C8-positions. Perhaps the most thoroughly researched purine oxidation product is 7-hydro-8-oxoguanine, which originates from the oxidation of the C8–OH adduct radical (Dizdaroglu and Jaruga 2012; Cadet and Wagner 2013). The imidazole ring-opened derivative 2,6-diamino-4-hydroxy-5-formamidopyrimidine is another significant guanine-derived lesion. And the oxidized derivatives of adenines, such as 8-oxoadenine and 4,6-diamino-5-formamidopyrimidine, have received less attention (Dizdaroglu 1999; Roldán-Arjona and Ariza 2009).

Hydroxyl ions also attack DNA binding proteins, which results in destructive cross-linking of the protein and DNA before transcription and replication (Das and Roychoudhury 2014). In particular, ROS directly damage DNA by disrupting G:C sites to induce mutations and indirectly by producing harmful macromolecules (Choudhary et al. 2020). Additionally, ROS can generate a variety of DNA lesions that result in deletions, mutations, and other detrimental genetic effects (Srivalli et al. 2003).

3.6 Conclusions

All biomolecules, including lipids, proteins, and DNA, are targets of ROS, which may lead to severe damage in plants, it compromises the integrity of the cell and ultimately results in cell death. To combat harsh climatic conditions, evolution has given plants a larger array of defense mechanisms, including morphological, metabolic, and genetic alterations.

Enhanced level of ROS causes oxidative damage to lipid, protein, and DNA leading to altered intrinsic membrane properties like fluidity, ion transport, loss of enzyme activity, protein cross-linking, inhibition of protein synthesis, DNA damage, ultimately resulting in cell death. In order to avoid the oxidative damage, higher plants possess a complex antioxidative defense system comprising of nonenzymatic and enzymatic components. From the various studies, it is now evident that an increased quantity of ROS results in oxidative damage to DNA, proteins, and lipids, altering intrinsic membrane features like fluidity, ion transport, enzyme function loss, protein cross-linking, and protein synthesis suppression, all of which ultimately cause cell death. But higher plants have a sophisticated antioxidative defense system with both enzymatic and nonenzymatic components to prevent oxidative damage.

Apart from the natural plant defense mechanism, researchers have also developed various techniques to mitigate the ROS damage. Although, another important point to keep in mind is that, despite the rapid advancement in the field of ROS research, there are still some significant limitations to it. Such as the numerous dubiety and discontinuity in our knowledge about ROS formation and their effects on plants, which are typically caused by the short half-life and high reactivity of ROS. So it can be concluded here that there is still a need of extensive and thorough study in this domain. A more comprehensive understanding of the function of ROS in plants will be possible with the use of improved analytical tools to study the generation and destiny of ROS.

References

Abouzari A, Fakheri BA (2015) Reactive oxygen species: generation, oxidative damage, and signal transduction. Int J Biol Sci 9:3–17

Ahmad P, Sarwat M, Sharma S (2008) Reactive oxygen species, antioxidants and signaling in plants. J Plant Biol 51:167–173

Ali AA, Alqurainy F (2006) Activities of antioxidants in plants under environmental stress. In: Motohashi N (ed) The lutein-prevention and treatment for diseases. Transworld Research Network, Trivandrum, pp 187–256

Apel K, Hirt H (2004) Reactive oxygen species: metabolism, oxidative stress, and signal transduction. Annu Rev Plant Biol 55:373–399

Asada K, Takahashi M (1987) Production and scavenging of active oxygen in photosynthesis. In: Kyle DJ, Osmond CB, Arntzen CJ (eds) Photoinhibition: topics of photosynthesis, 9th edn. Elsevier, Amsterdam, pp 227–287

Cadet J, Wagner JR (2013) DNA base damage by reactive oxygen species, oxidizing agents, and UV radiation. Cold Spring Harb Perspect Biol 5:a012559

Choudhary A, Kumar A, Kaur N (2020) ROS and oxidative burst: roots in plant development. Plant Divers 42:33–43

Choudhury S, Panda P, Sahoo L, Panda SK (2013) Reactive oxygen species signaling in plants under abiotic stress. Plant Signal Behav 8:e23681

Das K, Roychoudhury A (2014) A reactive oxygen species (ROS) and response of antioxidants as ROS-scavengers during environmental stress in plants. Front Environ Sci 2:53

Dizdaroglu M (1993) Chemistry of free radical damage to DNA and nucleoproteins. In: Halliwell B, Aruoma OI (eds) DNA and free radicals. Ellis Horwood, London, pp 19–39

Dizdaroglu M (1999) Mechanisms of oxidative DNA damage; lesions and their measurement. In: Dizdaroglu M, Karakaya AE (eds) Advances in DNA damage and repair: oxygen radical effects, cellular protection, and biological consequences, vol 302. Springer Science & Business Media, Boston, pp 67–87

Dizdaroglu M, Jaruga P (2012) Mechanisms of free radical-induced damage to DNA. Free Radic Res 46:382–419

Dunlop RA, Rodgers KJ, Dean RT (2002) Recent development in the intracellular degradation of oxidized proteins. Free Radic Biol Med 33:894–906

Esterbauer H, Schaur RJ, Zollner H (1991) Chemistry and biochemistry of 4-hydroxynonenal, malonaldehyde and related aldehydes. Free Radic Biol Med 11:81–128

Evans MD, Dizdaroglu M, Cooke MS (2004) Oxidative DNA damage and disease: induction, repair and significance. Mutat Res 567:1–61

Foyer CH, Noctor G (2005) Redox homeostasis and antioxidant signaling: a metabolic interface between stress perception and physiological responses. Plant Cell 17:1866–1875

Gardner PR, Fridovich I (1991) Superoxide sensitivity of the *Escherichia coli* 6-phosphogluconate dehydratase. J Biol Chem 266:1478–1483
Halliwell B (2006) Reactive species and antioxidants. Redox biology is a fundamental theme of aerobic life. Plant Physiol 141:312–322
Halliwell B, Gutteridge JMC (1999) Free radicals in biology and medicine, 3rd edn. Oxford University Press, Oxford
Imlay JA, Linn S (1988) DNA damage and oxygen radical toxicity. Science 240:1302–1309
Jambunathan N (2010) Determination and detection of reactive oxygen species (ROS), lipid peroxidation, and electrolyte leakage in plants. In: Sunkar R (ed) Plant stress tolerance, methods in molecular biology, vol 639. Humana press, New York, pp 291–297
Job C, Rajjou L, Lovigny Y, Belghazi M, Job D (2005) Patterns of protein oxidation in Arabidopsis seeds and during germination. Plant Physiol 138:790–802
Khanna-Chopra R (2012) Leaf senescence and abiotic stresses share reactive oxygen species-mediated chloroplast degradation. Protoplasma 249:469–481
Maheshwari R, Dubey RS (2009) Nickel-induced oxidative stress and the role of antioxidant defence in rice seedlings. Plant Growth Regul 59:37–49
McCord JM (2000) The evolution of free radicals and oxidative stress. Am J Med 108:652–659
Mehta RA, Fawcett TW, Porath D, Mattoo AK (1992) Oxidative stress causes rapid membrane translocation and in vivo degradation of ribulose-1,5-biphosphate carboxylase/oxygenase. J Biol Chem 267:2810–2816
Meriga B, Reddy BK, Rao KR, Reddy LA, Kishor PBK (2004) Aluminium-induced production of oxygen radicals, lipid peroxidation and DNA damage in seedlings of rice (*Oryza sativa*). J Plant Physiol 161:63–68
Mittler R (2017) ROS are good. Trends Plant Sci 22:11–19
Møller IM, Jensen PE, Hansson A (2007) Oxidative modifications to cellular components in plants. Annu Rev Plant Biol 58:459–481
Noctor G, Reichheld JP, Foyer CH (2018) ROS-related redox regulation and signaling in plants. Semin Cell Dev Biol 80:3–12
Oberschall A, Deák M, Török K, Sass L, Vass I, Kovács I, Fehér A, Dudits D, Horváth GV (2000) A novel aldose/aldehyde reductase protects transgenic plants against lipid peroxidation under chemical and drought stresses. Plant J 4:437–446
Palma JM, Sandalio LM, Corpas FJ, Romero-Puertas MC, McCarthy I, Luis A (2002) Plant proteases, protein degradation, and oxidative stress: role of peroxisomes. Plant Physiol Biochem 40:521–530
Repetto M, Semprine J, Boveris A (2012) Lipid peroxidation: chemical mechanism, biological implications and analytical determination. In: Lipid peroxidation, vol 1. IntechOpen, London, pp 3–30
Rivett AJ, Levine RL (1990) Metal-catalyzed oxidation of *Escherichia coli* glutamine synthetase: correlation of structural and functional changes. Arch Biochem Biophys 278:26–34
Roldán-Arjona T, Ariza RR (2009) Repair and tolerance of oxidative DNA damage in plants. Mutat Res 681:169–179
Sachdev S, Ansari SA, Ansari MI, Fujita M, Hasanuzzaman M (2021) Abiotic stress and reactive oxygen species: generation, signaling, and defense mechanisms. Antioxidants 10:277
Shah K, Kumar RG, Verma S, Dubey RS (2001) Effect of cadmium on lipid peroxidation, superoxide anion generation and activities of antioxidant enzymes in growing rice seedlings. Plant Sci 161:1135–1144
Sharma P, Jha AB, Dubey RS, Pessarakli M (2012) Reactive oxygen species, oxidative damage, and antioxidative defense mechanism in plants under stressful conditions. J Bot 2012:1–26
Smirnoff N (2000) The role of active oxygen in the response of plants to water deficit and desiccation. New Phytol 125:27–58
Srivalli B, Chinnusamy V, Khanna-Chopra R (2003) Antioxidant defense in response to abiotic stresses in plants. J Plant Biol 30:121–139

Tiwari S, Tiwari S, Singh M, Singh A, Prasad SM (2017) Generation mechanisms of reactive oxygen species in the plant cell: an overview. In: Singh V, Singh S, Tripathi D, Prasad S, Chauhan D (eds) Reactive oxygen species in plants: boon or bane-revisiting the role of ROS, vol 13. Wiley, Hoboken, pp 1–22

Van Breusegem F, Dat JF (2006) Reactive oxygen species in plant cell death. Plant Physiol 141: 384–390

Yamauchi Y, Furutera A, Seki K, Toyoda Y, Tanaka K, Sugimoto Y (2008) Malondialdehyde generated from peroxidized linolenic acid causes protein modification in heat-stressed plants. Plant Physiol Biochem 46:786–793

Zuo S, Boorstein RJ, Teebor GW (1995) Oxidative damage to 5-methylcytosine in DNA. Nucleic Acids Res 23:3239–3243

Functions of Reactive Oxygen Species in Improving Agriculture and Future Crop Safety

4

Anjuman Hussain, Chen Chen, Fangyuan Yu, S. Maqbool Ahmed, and Mohammad Faizan

Abstract

Each year, abiotic stress factors such as drought, heat, salinity, and cold, as well as their many combinations, take a toll on agricultural productivity all over the world. Because of the growing rate and intensity of global climate changes, the effects of these adverse conditions on plant productivity have become increasingly worrying in recent years. Improving crop tolerance to abiotic stress conditions necessitates a thorough understanding of plant responses to environmental changes. Early and late signal transduction events involving essential signalling molecules such as reactive oxygen species (ROS), various plant hormones, and other signalling molecules are required for this response. The orchestration of the plant response to abiotic stress and the driving of changes in transcriptomic, metabolic, and proteomic networks that lead to plant acclimation and survival is mediated by the integration of these signalling events, which is mediated by an interplay between ROS and different plant hormones. Depending on their abundance in plants, ROS can play both detrimental and helpful effects. They destroy biomolecules and trigger genetically programmed cell suicide

A. Hussain
Plant Physiology Lab, Department of Botany, Faculty of Life Sciences, Aligarh Muslim University, Aligarh, India

C. Chen
School of Landscape and Horticulture, Yangzhou Polytechnic College, Yangzhou, Jiangsu, China

F. Yu
Collaborative Innovation Center of Sustainable Forestry in Southern China, College of Forest Science, Nanjing Forestry University, Nanjing, China

S. M. Ahmed · M. Faizan (✉)
Botany Section, School of Sciences, Maulana Azad National Urdu University, Hyderabad, India
e-mail: mohammadfaizan@manuu.edu.in

M. Faizan et al. (eds.), *Reactive Oxygen Species*,
https://doi.org/10.1007/978-981-19-9794-5_4

events at high concentrations, but operate as a second messenger in intracellular signalling cascades that mediate a variety of responses in plant cells at low concentrations. ROS level changes in space and time are viewed as signals for a variety of biological activities, including growth, development, tolerance to abiotic stress factors, and correct response to stress. One of the particular early stress responses in the plant's acclamatory performance is the molecular language linked with ROS-mediated signal transduction, which leads to gene expression modulation. By altering the cell's redox balance, ROS may serve as a "second messenger", influencing the activities of specific proteins or gene expression. At every stage of plant growth, the network of redox signals orchestrates metabolism to regulate energy production and use, interfering with major signalling agents (hormones) to respond to changing environmental stimuli. The outcome or fine-tuning of biological responses to changed ROS levels is determined by interactions with other signalling molecules. Despite the recent identification of numerous components of the ROS signalling network, understanding how ROS-derived signals are integrated to eventually regulate biological processes as plant growth, development, stress adaption, and programmed cell death remains a challenge.

Keywords

Abiotic stress · Plant growth · Proteomic networks · Second messenger

4.1 Introduction

In the regulation of several biological processes, including plant growth, development, and responses to biotic and/or abiotic stimuli, reactive oxygen species (ROS) play a crucial role as signalling molecules. They serve as the primary regulator of cell physiology and cellular responses to the environment and are positioned at the centre of a complex network of signalling pathways in plants (Liu and He 2016). In the chloroplasts, photosynthesis is the primary source of singlet oxygen ($^1O_2$), superoxide radical ($O_2^{\bullet}$) and hydrogen peroxide (H_2O_2). $O_2^{\bullet}$ and H_2O_2 are also produced by mitochondrial electron transport. The antioxidant systems, including antioxidants and antioxidative enzymes, eliminate these oxidants in these organelles (Noctor and Foyer 2016). Plants have developed a complex network of ROS-producing and -detoxifying enzymes to regulate ROS levels in accordance with the physiological requirements in various cell types and organs at a given moment and at various developmental stages. ROS were able to be co-opted as signalling molecules that regulate cell growth and cell death thanks to this evolutionary advancement (Yi et al., 2016). Indeed, there is a possibility that the temporal and spatial coordination of ROS and other signals can control a variety of biological processes that take place in distinct tissue types and under a variety of environmental conditions. Early signalling events in plants in response to stress stimuli include an increase in Ca^{2+} flow into the cytosol, activation of mitogen-activated protein kinases (MAPKs), and protein phosphorylation. Each of these control mechanisms can be activated in a matter of seconds or minutes. Long-term responses regulate phenotypic changes such as cell growth, development, and survival after these early signalling events (Mullineaux and Baker 2010; Dubiella et al., 2013). By altering the redox balance of the cell, ROS may function as a "second messenger" to mediate the stressful situations by influencing the activities of particular proteins or the expression of genes (Bhattacharjee 2012). Arabidopsis flu mutant plants increase their antioxidative defence in response to ROS signalling by upregulating the expression of antioxidative genes and turning on the genes for inducible stress proteins (Op Den Camp et al., 2003; Laloi et al., 2007). In order to use ROS as biological signals that regulate multiple genetic stress processes, plant cells have developed techniques. This view is predicated on the implicit premise that a particular ROS can interact specifically with a target molecule that detects an increase in ROS concentration and converts this information into a change in gene expression. A redox-sensitive transcription factor (TF) can be directly modified to cause such a shift in transcriptional activity, or signalling pathway components that activate TFs can be oxidised (Laloi et al., 2004). The impact of ROS extends far beyond toxicity and the relatively direct regulation of transcription or translation, as demonstrated by the influence of H_2O_2 on Protein Synthesis Elongation Factor G (EF-G) in Chloroplasts, which slows translation as a defence mechanism during ROS stress (Nishiyama et al., 2011; Murata et al., 2012).

4.2 Production of ROS

Free radicals, reactive molecules, and ions collectively known as ROS are produced when oxygen is consumed. In numerous subcellular locations, including the cytosol, nucleus, endoplasmic reticulum, plasma membranes, peroxisomes, glyoxysomes, apoplast, and cell walls, about 1% of the O_2 that plants take in is diverted to produce ROS. Indeed, the main redox-active compartments in plant cells are chloroplasts, mitochondria, and peroxisomes. An array of enzymes in chloroplasts and mitochondria enable cellular processes like carbon absorption, respiration, photorespiration, and gene expression by processing ROS to maintain their steady-state concentrations low (Noctor and Foyer 2016). When under abiotic stress, electron leakage from complex I to III normally causes the production of O_2, which MnSOD can then convert to H_2O_2 (Huang et al., 2016). Changes in the quantities of ROS generated by mitochondria during abiotic stress can cause retrograde signalling between mitochondria and nuclei and regulate plant acclimatisation (Woodson and Chory 2008). Due to increased photorespiration, glycolate oxidases enzymatic activity in peroxisomes causes ROS generation (Sarvajeet and Narendra 2010; Baishnab and Ralf 2012). Other significant ROS generators in plants include the endoplasmic reticulum and cytochrome P450-catalyzed detoxification processes in the cytoplasm (ER). Plants can produce ROS at the plasma membrane level or extracellularly in the apoplast. Much emphasis has been given to plasma membrane NADPH-dependent oxidase (NADPH oxidase) as a source of ROS for oxidative burst (Dybing et al., 1976). Four distinct processes each mediate a portion of the apoplast's ROS generation during abiotic stress. The plasma membrane (PM) NADPH oxidase-RBOH (respiratory burst oxidase homolog) proteins that link calcium and ROS signalling during the stress and generate superoxide in the apoplast are the most researched of the four (Gilory et al., 2014, 2016). In phospholipids of plant cell membranes, peroxidation of lipids is another possible cause of ROS generation. Lipoxygenase, an enzyme, can also cause lipid peroxidation in plant cells (Winston 1990). This enzyme has the ability to start the peroxidation process and the production of fatty acid hydroperoxides.

4.3 ROS Signalling

Recently, it was shown that reactive oxygen species, in conjunction with calcium signalling and potentially even electric waves, promote rapid systemic signalling in plants in response to abiotic stress (Mittler et al., 2011; Gilroy et al., 2014, 2016). The reaction of all of a plant's organs and leaves to abiotic stress must be coordinated for a plant to achieve maximum fitness in the field. An auto-propagating wave of ROS production (the ROS wave) that begins in a group of cells that first detects the stress and spreads to the entire plant at a rate of up to 8.4 cm min^{-1} is assumed to be responsible for this coordination (Miller et al., 2009). According to the fundamental molecular explanation explaining the ROS wave, abiotic stress on nearby cells causes a flow of calcium into the cytosol. This flux either directly activates RBOHs or sets off a chain of events that phosphorylates and activates RBOHs through activating calcium-dependent protein kinases (Gilroy et al., 2014). As a result of the surrounding cells sensing the ROS produced by the activated RBOHs at the apoplast, these cells experience a calcium flux, which in turn activates their own RBOHs. This state of calcium flux from ROS combined with calcium-derived activation of RBOHs subsequently spreads throughout the entire plant from cell to cell, inducing systemic responses to abiotic stress (Miller et al., 2009). Recent research has demonstrated that the ROS wave in Arabidopsis is controlled by RBOHD and that a systemic calcium wave coordinates it (Gilroy et al., 2016). Additionally, it was shown that the creation of some electric signals under abiotic stress depends on the ROS wave (Suzuki et al., 2013). Additionally, it was demonstrated that the ROS wave is necessary to trigger a systemic acclimation response to light or heat stress (Suzuki et al., 2013). The ROS wave was also demonstrated to be coordinated with ABA function in systemic leaves, at least in terms of heat stress (Suzuki et al., 2013). Following the discovery of the ROS wave (Miller et al., 2009) and its interplay with the calcium wave (Choi et al., 2014; Gilroy et al., 2014), it was suggested that abiotic stress reactions in each leaf are managed independently and connected to stomata function, and that each leaf's response is transmitted to all other leaves via the combined function of the ROS, calcium, hydraulic (León et al., 2001), and electric waves (Mittler and Blumwald, 2010). This model suggests a major role for ROS, ABA, and stomatal responses in the systemic acclimation response of plants to abiotic stress, and it explains many of the prior observations involving systemic plant responses to stress (Mittler and Blumwald, 2010). Although it is believed that RBOHs play a major role in controlling the ROS wave, current research has identified other ROS types and ROS producers as potential contributors to fast systemic signalling. For instance, it has been demonstrated that singlet oxygen generated in the chloroplast is necessary to start the RBOH-derived ROS wave in response to severe light stress (Carmody et al., 2016). Additionally, a potential function for NO and glutamate receptor-like channels in integrating the ROS, calcium, and electric waves during systemic acquired acclimation was suggested (SAA; Gilroy et al., 2016). According to the studies mentioned above, the ROS wave is crucial in setting up the entire plant for the induction of the SAA state. The ROS wave is extremely necessary for the systemic response even though it does not provide abiotic stress specificity (Suzuki et al., 2013).

4.4 Role of ROS in Plant Growth and Development

The preservation of an optimal equilibrium between cell division and differentiation is crucial for growth in multicellular organisms. In plants, disruption of this balance between cell division and differentiation can result in an early end to organogenesis or as a result of aberrant growth, whereas disruption in mammals can result in tumour growth and disease (Zhang et al., 2008). The change from cellular proliferation to elongation, which is controlled by ROS homeostasis, denotes the beginning of differentiation (Tsukagoshi et al., 2010). O_2 and H_2O_2 are the two primary ROS that are variably distributed in the root tissues of the model plant Arabidopsis, according to Dunand et al. (2007). According to Wells et al. (2010), H_2O_2 accumulates in the elongation zone while $O_2^{\bullet-}$ mostly accumulates in growing meristem cells, with both kinds of ROS overlapping in the "transition zone". The delicate balance between $O_2^{\bullet-}$ and H_2O_2, which is governed by a TF:UPBEAT1, is primarily responsible for controlling the transition between root cell proliferation and differentiation (UPB1; Tsukagoshi et al., 2010). In the root transition zone, TheUPB1, a member of the basic/helix loop-helix (bHLH) TF family, exhibits enhanced expression (Tsukagoshi et al., 2010). According to Tsukagoshi et al. (2010), whereas plants missing UPB1 (upb1 mutant) had longer roots with increased meristem size and longer root cells, plants over-expressing UPB1 had shorter roots due to a decrease in both meristem size and mature cells. The mechanisms of seed dormancy and seed germination are crucial steps in plant growth and are thought to be connected by a complex regulatory network. As observed in *A. thaliana* (Leymarie et al., 2012), barley (Bahin et al., 2011), wheat (Ishibashi et al., 2008), and sunflower (Ishibashi et al., 2008), ROS are regarded as major signalling agents in this. Water intake, the beginning of cell division, and radical protrusion mark the end of seed germination (Holdsworth et al., 2008; Bewley et al., 2013). Due to greatly diminished enzyme activity in dry seeds, ROS are likely produced by non-enzymatic processes like lipid peroxidation that take place even at very low moisture concentrations (McDonald 1999). Although all metabolically active compartments, including chloroplasts (by electron transfer in photosystems), glyoxysomes (by lipid catabolism), mitochondria (via respiratory activity), peroxisomes (by purine catabolism), and plasma membranes, can produce ROS in hydrated seeds (which occurs after imbibition) (by NADPH oxidase; Bailly 2004). Because the electrons in the mitochondrial electron transfer chain (mETC) have enough free energy to directly decrease the O_2, continuing respiration in hydrated seeds can result in electron leakage. This can be viewed as an inevitable source of enhanced ROS production in mitochondria (Rhoads et al., 2006; El-Maarouf-Bouteau and Bailly, 2008). For this reason, during germination of hydrated seeds, mitochondrial activity is thought to be a primary source of ROS (such as H_2O_2) generation (Noctor et al., 2007). The intricate interplay between cell expansion and proliferation determines the ultimate size of a single organ or organism (Lu et al., 2014). Meristematic cell proliferation initiates the development of plant leaves, followed by a second phase of unbroken cell growth (Beemster et al., 2005). Numerous genetic pathways control both phases in an equilibrium between negative and positive regulators. For instance, TFs are essential (Townsley and Sinha 2012). Of course, modifications to cell wall composition and design have an impact on cell expansion (Rubio-Diaz et al., 2012). Such modifications could result from Prxs that change the ROS levels in leaves (Passardi et al., 2004). The Prxs, in particular the apoplastic ones, directly regulate the rigidity of the cell wall by either encouraging or inhibiting cellular expansion (Lee et al., 2013). According to Müller et al. (2009), the first scenario involves $O_2^{\bullet-}$ produced by cell wall peroxidases acting as cell wall loosening agents and promoting growth by cleaving cell wall polysaccharides. Contrarily, H_2O_2 generation in the cell wall encourages rigid cross-linking of cell wall constituents, which limits development or results in the cell wall being stiffer (Passardi et al., 2004; Lu et al., 2014). KUODA1 (KUA1), an MYB-like TF that was discovered to act as a favourable regulator of cell growth during leaf development by modifying apoplastic ROS homeostasis in *A. thaliana*, inhibits the activity and expression of Prxs (Lu et al., 2014). Without changing the number of cells, overexpression of KUA1 causes bigger leaves and greater cell sizes (Lu et al., 2014). The kua1–1 mutant, on the other hand, has smaller leaves than the wild type as a result of a reduction in cell size, although cell number was once more unaltered. Additionally, the kua1 mutant exhibits higher H2O2 levels and class III Prx activity. KUA1 disruption resulted in smaller leaf cells and an increase in Prxs activity. Therefore, ROS homeostasis, which is regulated by KUA1, controls both cell growth and the eventual size of the organ (Lu et al., 2014). It is notable that variations in apoplastic H_2O_2 levels may be related to this positive regulation or promotion and may also have an impact on the $O_2^{\bullet-}$ pool (Liszkay et al., 2003).

4.5 ROS Roles During Exposure to Multiple Stresses

In the natural world, plants are frequently subjected to multiple stresses at once. For instance, a stress combination—drought, intense light, and heat—activates numerous signalling pathways. ROS were discovered to be crucial for plants to adapt to such conditions4. Indeed, mutants without ASCORBATE PEROXIDASE 1 (APX1) are more vulnerable to conditions of multifactorial stress combination, while mutants lacking RBOHD are more vulnerable to conditions of drought and heat stress (Koussevitzky et al., 2008; Zandalinas et al., 2021). The ROS-regulated MPK3/ MPK6 and MPK4 cascades play opposing roles in the triggering of defence and acclimatisation networks during the integration of cold stress and pathogen responses (Jagodzik et al., 2018). Thus, ROS play a significant part in the assimilation of various signals produced during the combination of stresses. Multiple stresses may affect a plant sequentially or simultaneously, producing different ROS signatures, and the combination of them may attenuate or control the plant's overall response to challenging environmental conditions. Integration of two distinct ROS signatures could potentially take place during interactions with the plant microbiome or when a specific stimulus, like heat, occurs against the backdrop of a certain developmental stage, like plant reproduction (Muhlemann et al., 2018). Under these circumstances, the total ROS levels are combined to cause a particular or general state of plant tolerance or sensitivity to stress.

4.6 Crosstalk Between ROS and Phytohormones Under Various Stresses

4.6.1 Salinity Stress

The inactivation of various proteins and severe ion imbalances brought on by excessive Na + ion buildup in plant tissues restrict survival and growth. Osmotic stress, nutritional imbalances, oxidative damage, decreased ROS-scavenging ability, decreased stomatal aperture, and reduced photosynthetic activity are other effects of salinity stress (Mittler and Blumwald, 2015; Shabala et al., 2015). Several signalling pathways, such as hormone/ROS signalling, the salt-overly-sensitive pathway, or MAPK signalling cascades that culminate in the activation of stress-related genes, are triggered in plants as a response to salinity stress (Zhu, 2016; Yang and Guo, 2018). By controlling Na^+ and K^+ transport, accumulating various osmoprotectants, and lowering ROS toxicity, salt tolerance is attained. A wide range of adaptive responses, including ABA-induced RBOHD-dependent ROS generation, can be facilitated by phytohormone-mediated ROS synthesis in response to salt stress (Kurusu et al., 2015; Julkowska and Testerink, 2015). Salt stress causes ABA buildup, which alters the expression of genes. Previously, it was demonstrated that during salt stress, ABA-induced ROS-regulated Na^+/K^+ homeostasis (Ma et al., 2012). According to a different study, the administration of NaCl caused RBOHs to produce H2O2, which led to proline buildup and antioxidant defences in Arabidopsis (Rejeb et al., 2015a, 2015b). It has also been demonstrated that salt stress adversely impacts lateral root growth, which is dependent on ABA signalling (Ma et al., 2012; Duan et al., 2013). SLs were found to be crucial in the control of redox homeostasis during salt stress, working in conjunction with ABA. Studies utilising MAX2 (an F-box protein necessary for SL signalling) deletion and overexpression mutants revealed increased resistance to salt stress linked to controlling redox homeostasis and ABA (Ha et al., 2014;). Salt stress tolerance has been demonstrated to be regulated by CKs and ROS. To create a tolerance for salt stress, plants control the production of CK and ROS equilibrium. For instance, increased ROS generation and decreased expression of ROS-scavenging enzymes come from overexpression of the CK biosynthetic gene IPT8 in Arabidopsis, which reduces the plant's ability to adjust to salt stress (Wang et al., 2015). JA was also demonstrated to play a role in the reactions to salt stress. Wheat's ability to tolerate salt was improved by exogenous application of JA or methyl-jasmonate, which reduced H2O2 concentrations and raised antioxidant enzyme transcript levels (Qiu et al., 2014).

4.6.2 Heat Stress

Most elements of plant growth and development are negatively impacted by high temperatures (Lippmann et al., 2019). Histones sensors, endoplasmic reticulum (ER) unfolded protein response sensors, plasma membrane channels (which momentarily open to cause inward flux of Ca^{2+} into the cytosol), and phytochrome B are examples of potential heat sensors in plants (Jung et al., 2016; Dai Vu et al., 2019). Additionally, a class of HEAT SHOCK FACTOR (HSF) family transcription factors (TFs) (such as HSFA1s) control how plants react to extreme temperatures (Ohama et al., 2017). Numerous studies indicate that pretreating plants with either phytohormones or H_2O_2 enhances plant thermotolerance by increasing the expression of genes for ROS-scavenging enzymes like catalase and redox regulators like glutaredoxin (Wang et al., 2014). ABA biosynthesis is triggered by heat stress. Its detection and subsequent signalling activate TFs like AREB/ABFs that control the expression of genes that are ABA-dependent and, in turn, cause the creation of ROS or increase antioxidant capacity (Kuromori et al., 2018; Yoshida et al., 2019). In a study employing ABA, it was discovered that the RNA binding protein Flowering Control locus A (FCA), which regulates the expression of genes encoding antioxidants such 1-cysteine peroxiredoxin 1 (PER 1) necessary for promoting tolerance under heat stress, interacts with ABI5 (Lee et al., 2015). It has been suggested that interactions between ABA and RBOHD-dependent ROS generation are engaged in long-distance communication in response to heat stress via the association of SAA to heat stress with activation of the ROS wave and temporary accumulation of ABA in systemic tissues (Suzuki et al., 2013; Suzuki and Katano, 2018). According to this study, systemic stomatal opening responses were induced when localised application of heat stress to a single leaf caused the ROS wave to occur (Devireddy et al., 2020; McLachlan, 2020). ROS-induced TFs are essential for triggering the effects of heat stress. For instance, in Arabidopsis, ABA-induced ERF74 positively controls RBOHD by physically attaching to the RbohD promoter and promoting heat stress tolerance. Studies utilising ERF74 mutants with lower basal thermotolerance revealed that their overexpression increased tolerance to heat and other stressors (Yao et al., 2017). It was discovered that H_2O_2 mediates a crosstalk between ABA and BRs in response to heat stress. The synthesis of H_2O_2 by NADPH oxidase was demonstrated to be temporarily increased by BRs, which improved tomato plant stress tolerance to heat and increased ABA accumulation (Zhou et al., 2014). Ascorbate peroxidase 6 (APX6) was also discovered to have a role in protecting seeds from heat stress in the crosstalk between ABA, auxin, and ROS regulatory networks, and RBOH-dependent H_2O_2 was discovered to be crucial for the preservation of learned thermotolerance following acclimation (Sun et al., 2018).

4.6.3 Drought Stress

One of the main abiotic factors that impair plant growth, performance, and output is drought stress (Gupta et al., 2020). ABA is one of the primary phytohormones involved in modulating drought-induced stress responses. An ABA-PYR/PYL complex is created when ABA is sensed by PYR-like proteins, also known as PYR-resistance proteins (PYR/PYL receptors), which interact to inhibit the activity of protein phosphatase 2Cs (PP2Cs). SNF1-related protein kinase 2 (SnRK2)/OPEN STOMATA 1 (OST1) kinases, which act as positive regulators of the ABA response, are released as a result of this interaction. Released SnRK2s autophosphorylate and activate downstream signalling elements, including various TFs' activity. For instance, SnRK2s activates the ABA-responsive transcription factors (TFs) ABARESPONSIVE PROMOTER ELEMENTS (ABREs) BINDING FACTORS (ABFs) (AREB/ABFs), which control the expression of ABA-dependent genes implicated in the responses to drought stress (Kuromori et al., 2018; Yoshida et al., 2019). ABA was discovered to play a crucial role in the regulation of ROS formation during responses to drought, in addition to directly controlling TF function through phosphorylation-driven cascades. H2O2 is created as part of the ABA signalling pathway when SnRK2/OST1 interacts with and phosphorylates RBOHs, and OST1-dependent H_2O_2 production may indicate the release of additional active OST1 by inactivating PP2C. In order to regulate gene expression and modulate cellular responses to drought stress, ABA and ROS work together in a positive feedback loop that results in greater ROS/ABA levels (Yoshida et al., 2019). Through ROS signalling, TFs produced by ABA can also have a significant impact on the development of drought stress tolerance. The AP2/ERF family member Redox Responsive Transcription Factor 1 (RRTF1), which is a part of the core redox signalling network in Arabidopsis, is activated by ABA and ROS in response to several stressors, including drought. Further ROS accumulation is caused by elevated RRTF1 levels in plants, which is necessary to trigger further acclimation responses (Matsuo et al., 2015; He et al., 2018). Drought stress tolerance is mediated by CKs and ROS. The amount of CKs that build up during a drought varies on how severe and long the stress is. So, CKs can affect how well a person can tolerate drought in both good and bad ways (Zwack and Rashotte, 2015; Li et al., 2016). Plants with drought stress have higher antioxidant levels and antioxidant capacity thanks to CKs, which shields cells from excessive stress-related ROS buildup

(Rivero et al. 2009). In especially during systemic acquired resistance, but also SAA, SA is implicated in controlling plant responses to biotic and abiotic stressors (Zandalinas et al., 2020). SA stimulates the synthesis of reactive oxygen species (ROS) in at least two ways: I by modifying NADPH oxidases by CDPKs, which is facilitated by an intracellular rise in cytosolic Ca2+; and (ii) by producing ROS independently of NADPH oxidase via extracellular peroxidases that regulate stomatal function during drought (Miura et al., 2013; Prodhan et al., 2018). It has been demonstrated that SA and glutathione interact to regulate plant antioxidant responses (Foyer and Noctor, 2011; Han et al., 2013). In response to drought and other challenges, the balance between SA and glutathione activates redox-regulated mechanisms that are involved in the transcription of defence genes. Under ideal circumstances, SA levels regulate the production of ROS, but excessive SA concentrations encourage their synthesis, leading to oxidative stress and lowered resistance to salinity and drought (Miura and Tada, 2014).

4.7 Induction of Plant Resilience Through Transcriptional Regulation by ROS

Receptors and ROS-activated redox sensors that sense stress cause a variety of transcription factor networks to be triggered and modulated, allowing the plant to react to a variety of different situations. In plants, there are two different pathways that control transcriptional responses: (1) Changes in phosphorylation, Ca2+—binding, sumoylation, or other signal transduction pathways brought on by stress or ROS that modify the activity of transcription factors, and (2) direct or indirect ROS-induced redox modulation (Lee et al., 2021; D'alessandro et al., 2018). Because ROS signalling and other signalling events (such those mediated by Ca2+ and phosphorylation) are also coupled, for example through RBOHs and AQPs, these two processes are intertwined. Other techniques can also be used to modify gene expression in a redox-dependent manner in response to stress. The plant Mediator complex's subunits are redox controlled, and ROS can affect the expression and function of various microRNAs as well as mRNA splicing (He et al., 2021; Iyer et al., 2012). Plant stress responses are further tuned by the impact of ROS on these systems, which are linked to cellular ROS levels. For instance, a rise in ROS levels might prevent the production of particular groups of housekeeping genes that depend on complicated splicing, microRNA activity, or interactions with the Mediator complex (for example, during heat stress) (Ohama et al., 2017). The redox-regulated transcriptional regulators NPR1, HSFA8/HSFA1A, MBF1C, and ANAC013/ANAC017/ANAC 089, which are involved in responses to biotic and abiotic stresses, are affected by ROS during stress in addition to regulating transcription through genetic and/or epigenetic mechanisms. ROS also affect the translocation of these transcriptional regulators from the cytosol or the outer membranes of the endoplasmic reticulum (D'alessandro et al., 2018; Albertos et al., 2021). These transcriptional regulators are then translocated into the nucleus, where they activate gene expression networks and improve plant stress resistance. A recent study identified several new ROS-regulated transcription factors and defined some of the regulatory networks and hubs they control using a supervised learning approach to create a ROS-response integrated gene regulatory network using DNA motifs, open chromatin regions, transcription factor-binding sites, and expression-based regulatory interactions (De Clercq et al., 2021).It was also discovered by transcriptomic analyses of mutants lacking regulatory hubs such RBOHs, MAPK cascades, HSFs, and various Ca2+ signalling pathways how these hubs integrate ROS signals with other signal transduction networks that are activated under stress. For instance, the transcription factor MYB30, which was discovered to be crucial in plants' responses to oxidative stress, was revealed to be essential for the production of several early response transcripts in a study analysing the transcriptome response of the rbohD mutant to light exposure. In addition, MYB30 controls thousands of transcripts in response to mild stress by acting upstream of numerous other transcription factors (Fichman et al., 2020).

4.8 Conclusion

Despite incredible progress in our understanding of ROS biology, the precise makeup of the ROS-signalling network is still mostly unknown. The regulating role of ROS in plant growth and development is an issue that is attempted to be addressed in this chapter. These substances were once thought to be hazardous byproducts alone, but it has now been discovered that they play a crucial role in intricate signalling networks. Here, we have made an effort to clarify the useful function of ROS as signalling molecules. Unquestionably, over the past ten years, our knowledge of ROS signalling has greatly increased, but there is still much to learn. A fascinating era of ROS signalling in plants is about to begin. All cellular compartments, cell types, and organs during various embryonic stages must maintain a finely adjusted balance of ROS production, conversions, and metabolism for signalling molecules to function as intended. Changes to this balance, in either direction, have a significant impact on a plant's ability to grow or survive. Today, new ROS-responsive genes, their transcriptional regulators, ROS-directed regulatory mechanisms, and target molecules are continuously being identified. The issue of incorporating these new players into the ROS signalling network still exists, despite the fact that a clearer understanding of the interaction between ROS and signal transduction components is starting to emerge.

References

Albertos P, Tatematsu K, Mateos I, Sánchez-Vicente I, Fernández-Arbaizar A, Nakabayashi K, Nambara E, Godoy M, Franco JM, Solano R, Gerna D (2021) Redox feedback regulation of ANAC089 signaling alters seed germination and stress response. Cell Rep 35(11):109263

Bahin E, Bailly C, Sotta B, Kranner I, Corbineau F, Leymarie J (2011) Crosstalk between reactive oxygen species and hormonal signalling pathways regulates grain dormancy in barley. Plant Cell Environ 34(6):980–993

Bailly C (2004) Active oxygen species and antioxidants in seed biology. Seed Sci Res 14:93–107. https://doi.org/10.1079/SSR2004159
Baishnab CT, Ralf O (2012) Reactive oxygen species generation and signaling in plants. Plant Signal Behav 7:1621–1633
Beemster GT, De Veylder L, Vercruysse S, West G, Rombaut D, Van Hummelen P, Galichet A, Gruissem W, Inzé D, Vuylsteke M (2005) Genome-wide analysis of gene expression profiles associated with cell cycle transitions in growing organs of Arabidopsis. Plant Physiol 138(2): 734–743
Rejeb KB, Lefebvre-De Vos D, Le Disquet I, Leprince AS, Bordenave M, Maldiney R, Jdey A, Abdelly C, Savouré A (2015a) Hydrogen peroxide produced by NADPH oxidases increases proline accumulation during salt or mannitol stress in A rabidopsis thaliana. New Phytol 208(4): 1138–1148
Bewley JD, Bradford KJ, Hilhorst HWM, Nonogaki H (2013) Seeds: physiology of development, germination and dormancy. Springer, Berlin
Bhattacharjee S (2012) The language of reactive oxygen species signaling in plants. J Bot 2012: 985298
Carmody M, Crisp PA, d'Alessandro S, Ganguly D, Gordon M, Havaux M, Albrecht-Borth V, Pogson BJ (2016) Uncoupling high light responses from singlet oxygen retrograde signaling and spatial-temporal systemic acquired acclimation. Plant Physiol 171(3):1734–1749
Choi WG, Toyota M, Kim SH, Hilleary R, Gilroy S (2014) Salt stress-induced Ca^{2+} waves are associated with rapid, long-distance root-to-shoot signaling in plants. Proc Natl Acad Sci 111(17):6497–6502
D'alessandro S, Ksas B, Havaux M (2018) Decoding β-cyclocitral-mediated retrograde signaling reveals the role of a detoxification response in plant tolerance to photooxidative stress. Plant Cell 30(10):2495–2511
Dai Vu L, Gevaert K, De Smet I (2019) Feeling the heat: searching for plant thermosensors. Trends Plant Sci 24(3):210–219
De Clercq I, Van de Velde J, Luo X, Liu L, Storme V, Van Bel M, Pottie R, Vaneechoutte D, Van Breusegem F, Vandepoele K (2021) Integrative inference of transcriptional networks in Arabidopsis yields novel ROS signalling regulators. Nat Plants 7(4):500–513
Devireddy AR, Arbogast J, Mittler R (2020) Coordinated and rapid whole-plant systemic stomatal responses. New Phytol 225(1):21
Duan L, Dietrich D, Ng CH, Chan PMY, Bhalerao R, Bennett MJ, Dinneny JR (2013) Endodermal ABA signaling promotes lateral root quiescence during salt stress in Arabidopsis seedlings. Plant Cell 25(1):324–341
Dubiella U, Seybold H, Durian G, Komander E, Lassig R, Witte CP, Schulze WX, Romeis T (2013) Calcium-dependent protein kinase/NADPH oxidase activation circuit is required for rapid defense signal propagation. Proc Natl Acad Sci 110(21):8744–8749
Dunand C, Penel C (2007) Localization of superoxide in the root apex of Arabidopsis. Plant Signal Behav 2(2):131–132
Dybing E, Nelson JR, Mitchell JR, Sesame HA, Gillette JR (1976) Oxidation of a methyldopa and other catechols by cytochromes R450 generated superoxide anion: possible mechanism of methyldopa hepatitis. Mol Pharamacol 12:911–920
El-Maarouf-Bouteau H, Bailly C (2008) Oxidative signaling in seed germination and dormancy. Plant Signal Behav 3:175–182
Fichman Y, Zandalinas SI, Sengupta S, Burks D, Myers RJ Jr, Azad RK, Mittler R (2020) MYB30 orchestrates systemic reactive oxygen signaling and plant acclimation. Plant Physiol 184(2): 666–675
Foyer CH, Noctor G (2011) Ascorbate and glutathione: the heart of the redox hub. Plant Physiol 155(1):2–18
Gilroy S, Białasek M, Suzuki N, Górecka M, Devireddy AR, Karpiński S, Mittler R (2016) ROS, calcium, and electric signals: key mediators of rapid systemic signaling in plants. Plant Physiol 171(3):1606–1615

Gilroy S, Suzuki N, Miller G, Choi WG, Toyota M, Devireddy AR, Mittler R (2014) A tidal wave of signals: calcium and ROS at the forefront of rapid systemic signaling. Trends Plant Sci 19(10):623–630

Gupta A, Rico-Medina A, Caño-Delgado AI (2020) The physiology of plant responses to drought. Science 368(6488):266–269

Ha CV, Leyva-González MA, Osakabe Y, Tran UT, Nishiyama R, Watanabe Y, Tanaka M, Seki M, Yamaguchi S, Dong NV, Yamaguchi-Shinozaki K (2014) Positive regulatory role of strigolactone in plant responses to drought and salt stress. Proc Natl Acad Sci 111(2):851–856

Han Y, Chaouch S, Mhamdi A, Queval G, Zechmann B, Noctor G (2013) Functional analysis of Arabidopsis mutants points to novel roles for glutathione in coupling H_2O_2 to activation of salicylic acid accumulation and signaling. Antioxid Redox Signal 18(16):2106–2121

He H, Denecker J, Van Der Kelen K, Willems P, Pottie R, Phua SY, Hannah MA, Vertommen D, Van Breusegem F, Mhamdi A (2021) The Arabidopsis mediator complex subunit 8 regulates oxidative stress responses. Plant Cell 33(6):2032–2057

He H, Van Breusegem F, Mhamdi A (2018) Redox-dependent control of nuclear transcription in plants. J Exp Bot 69(14):3359–3372

Holdsworth MJ, Bentsink L, Soppe WJ (2008) Molecular networks regulating Arabidopsis seed maturation, after-ripening, dormancy and germination. New Phytol 179(1):33–54

Huang S, Van Aken O, Schwarzländer M, Belt K, Millar AH (2016) The roles of mitochondrial reactive oxygen species in cellular signaling and stress response in plants. Plant Physiol 171(3): 1551–1559

Ishibashi Y, Yamamoto K, Tawaratsumida T, Yuasa T, Iwaya-Inoue M (2008) Hydrogen peroxide scavenging regulates germination ability during wheat (*Triticum aestivum* L.) seed maturation. Plant Signal Behav 3(3):183–188

Iyer NJ, Jia X, Sunkar R, Tang G, Mahalingam R (2012) microRNAs responsive to ozone-induced oxidative stress in Arabidopsis thaliana. Plant Signal Behav 7(4):484–491

Jagodzik P, Tajdel-Zielinska M, Ciesla A, Marczak M, Ludwikow A (2018) Mitogen-activated protein kinase cascades in plant hormone signaling. Front Plant Sci 9:1387

Julkowska MM, Testerink C (2015) Tuning plant signaling and growth to survive salt. Trends Plant Sci 20(9):586–594

Jung JH, Domijan M, Klose C, Biswas S, Ezer D, Gao M, Khattak AK, Box MS, Charoensawan V, Cortijo S, Kumar M (2016) Phytochromes function as thermosensors in Arabidopsis. Science 354(6314):886–889

Koussevitzky S, Suzuki N, Huntington S, Armijo L, Sha W, Cortes D, Shulaev V, Mittler R (2008) Ascorbate peroxidase 1 plays a key role in the response of Arabidopsis thaliana to stress combination. J Biol Chem 283(49):34197–34203

Kuromori T, Seo M, Shinozaki K (2018) ABA transport and plant water stress responses. Trends Plant Sci 23(6):513–522

Kurusu T, Kuchitsu K, Tada Y (2015) Plant signaling networks involving Ca2+ and Rboh/Nox-mediated ROS production under salinity stress. Front Plant Sci 6:427

Laloi C, Apel K, Danon A (2004) Reactive signalling: the latest news. Curr Opin Plant Biol 7(3): 323–328

Laloi C, Stachowiak M, Pers-Kamczyc E, Warzych E, Murgia I, Apel K (2007) Cross-talk between singlet oxygen-and hydrogen peroxide-dependent signaling of stress responses in *Arabidopsis thaliana*. Proc Natl Acad Sci 104(2):672–677

Lee ES, Park JH, Wi SD, Kang CH, Chi YH, Chae HB, Paeng SK, Ji MG, Kim WY, Kim MG, Yun DJ (2021) Redox-dependent structural switch and CBF activation confer freezing tolerance in plants. Nat Plants 7(7):914–922

Lee S, Lee HJ, Jung JH, Park CM (2015) The A rabidopsis thaliana RNA-binding protein FCA regulates thermotolerance by modulating the detoxification of reactive oxygen species. New Phytol 205(2):555–569

Lee Y, Rubio MC, Alassimone J, Geldner N (2013) A mechanism for localized lignin deposition in the endodermis. Cell 153:402–412

León J, Rojo E, Sánchez-Serrano JJ (2001) Wound signalling in plants. J Exp Bot 52(354):1–9
Leymarie J, Vitkauskaité G, Hoang HH, Gendreau E, Chazoule V, Meimoun P, Corbineau F, El-Maarouf-Bouteau H, Bailly C (2012) Role of reactive oxygen species in the regulation of Arabidopsis seed dormancy. Plant Cell Physiol 53(1):96–106
Li W, Herrera-Estrella L, Tran LSP (2016) The Yin–Yang of cytokinin homeostasis and drought acclimation/adaptation. Trends Plant Sci 21(7):548–550
Lippmann R, Babben S, Menger A, Delker C, Quint M (2019) Development of wild and cultivated plants under global warming conditions. Curr Biol 29(24):R1326–R1338
Liszkay A, Kenk B, Schopfer P (2003) Evidence for the involvement of cell wall peroxidase in the generation of hydroxyl radicals mediating extension rowth. Planta 217(4):658–667
Liu Y, He C (2016) Regulation of plant reactive oxygen species (ROS) in stress responses: learning from AtRBOHD. Plant Cell Rep 35(5):995–1007
Lu D, Wang T, Persson S, Mueller-Roeber B, Schippers JH (2014) Transcriptional control of ROS homeostasis by KUODA1 regulates cell expansion during leaf development. Nat Commun 5(1): 1–9
Ma L, Zhang H, Sun L, Jiao Y, Zhang G, Miao C, Hao F (2012) NADPH oxidase AtrbohD and AtrbohF function in ROS-dependent regulation of Na+/K+ homeostasis in Arabidopsis under salt stress. J Exp Bot 63(1):305–317
Matsuo M, Johnson JM, Hieno A, Tokizawa M, Nomoto M, Tada Y, Godfrey R, Obokata J, Sherameti I, Yamamoto YY, Böhmer FD (2015) High REDOX RESPONSIVE TRANSCRIPTION FACTOR1 levels result in accumulation of reactive oxygen species in Arabidopsis thaliana shoots and roots. Mol Plant 8(8):1253–1273
McDonald MB (1999) Seed deterioration: physiology, repair and assessment. Seed Sci Technol 27(1):177–237
McLachlan DH (2020) Systemic signalling, and the synchronization of stomatal response. New Phytol 225(1):5–6
Miller G, Schlauch K, Tam R, Cortes D, Shulaev V, Dangl JL, Mittler R (2009) The plant NADPH oxidase RBOHD mediates rapid systemic signaling in response to diverse stimuli. Sci Signal 2(84):ra45
Mittler R, Blumwald E (2010) Genetic engineering for modern agriculture: challenges and perspectives. Annu Rev Plant Biol 61(1):443–462
Mittler R, Blumwald E (2015) The roles of ROS and ABA in systemic acquired acclimation. Plant Cell 27(1):64–70
Mittler R, Vanderauwera S, Suzuki N, Miller GAD, Tognetti VB, Vandepoele K, Gollery M, Shulaev V, Van Breusegem F (2011) ROS signaling: the new wave? Trends Plant Sci 16(6): 300–309
Miura K, Tada Y (2014) Regulation of water, salinity, and cold stress responses by salicylic acid. Front Plant Sci 5:4
Miura K, Okamoto H, Okuma E, Shiba H, Kamada H, Hasegawa PM, Murata Y (2013) SIZ1 deficiency causes reduced stomatal aperture and enhanced drought tolerance via controlling salicylic acid-induced accumulation of reactive oxygen species in *Arabidopsis*. Plant J 73(1): 91–104
Muhlemann JK, Younts TL, Muday GK (2018) Flavonols control pollen tube growth and integrity by regulating ROS homeostasis during high-temperature stress. Proc Natl Acad Sci 115(47): E11188–E11197
Müller K, Linkies A, Vreeburg RAM, Fry SC, Krieger-Liszkay A, Luebner-Metzger G (2009) In vivo cell wall loosening by hydroxyl radicals during cress seed germination and elongation growth. Plant Physiol 150:1855–1865
Mullineaux PM, Baker NR (2010) Oxidative stress: antagonistic signaling for acclimation or cell death? Plant Physiol 154(2):521–525
Murata N, Allakhverdiev SI, Nishiyama Y (2012) The mechanism of photoinhibition in vivo: re-evaluation of the roles of catalase, α-tocopherol, non-photochemical quenching, and electron transport. Biochim Biophy Acta Bioenerg 1817(8):1127–1133

Nishiyama Y, Allakhverdiev SI, Murata N (2011) Protein synthesis is the primary target of reactive oxygen species in the photoinhibition of photosystem II. Physiol Plant 142(1):35–46

Noctor G, Foyer CH (2016) Intracellular redox compartmentation and ROS-related communication in regulation and signaling. Plant Physiol 171(3):1581–1592

Noctor G, De Paepe R, Foyer CH (2007) Mitochondrial redox biology and homeostasis in plants. Trends Plant Sci 12(3):125–134

Ohama N, Sato H, Shinozaki K, Yamaguchi-Shinozaki K (2017) Transcriptional regulatory network of plant heat stress response. Trends Plant Sci 22(1):53–65

op den Camp RG, Przybyla D, Ochsenbein C, Laloi C, Kim C, Danon A, Wagner D, Hideg É, Göbel C, Feussner I, Nater M (2003) Rapid induction of distinct stress responses after the release of singlet oxygen in Arabidopsis. Plant Cell 15(10):2320–2332

Passardi F, Penel C, Dunand C (2004) Performing the paradoxical: how plant peroxidases modify the cell wall. Trends Plant Sci 9(11):534–540

Prodhan MY, Munemasa S, Nahar MNEN, Nakamura Y, Murata Y (2018) Guard cell salicylic acid signaling is integrated into abscisic acid signaling via the Ca2+/CPK-dependent pathway. Plant Physiol 178(1):441–450

Qiu Z, Guo J, Zhu A, Zhang L, Zhang M (2014) Exogenous jasmonic acid can enhance tolerance of wheat seedlings to salt stress. Ecotoxicol Environ Saf 104:202–208

Rejeb KB, Benzarti M, Debez A, Bailly C, Savouré A, Abdelly C (2015b) NADPH oxidase-dependent H2O2 production is required for salt-induced antioxidant defense in Arabidopsis thaliana. J Plant Physiol 174:5–15

Rhoads DM, Umbach AL, Subbaiah CC, Siedow JN (2006) Mitochondrial reactive oxygen species. Contribution to oxidative stress and interorganellar signaling. Plant Physiol 141(2):357–366

Rivero R, Peleg Z, Szczerba M, Tumimbang E, Jauregui RN, Liu L, Blumwald E (2009) Delayed leaf senescence induces extreme drought tolerance in crop plants. Comp Biochem Physiol 153(2):S188

Rubio-Diaz S, Perez-Perez JM, Gonzalez-Bayon R, Munoz-Viana R, Borrega N, Mouille G, Hernandez-Romero D, Robles P, Hoefte H, Ponce MR, Micol JL (2012) Cell expansion-mediated organ growth is affected by mutations in three EXIGUA genes. PLoS One 7(5):e36500

Sarvajeet SG, Narendra T (2010) Reactive oxygen species and antioxidant machinery in abiotic stress tolerance in crop plants. Plant Physiol Biochem 48:909–930

Shabala S, Wu H, Bose J (2015) Salt stress sensing and early signalling events in plant roots: Current knowledge and hypothesis. Plant Sci 241:109–119

Sun M, Jiang F, Cen B, Wen J, Zhou Y, Wu Z (2018) Respiratory burst oxidase homologue-dependent H2O2 and chloroplast H_2O_2 are essential for the maintenance of acquired thermotolerance during recovery after acclimation. Plant Cell Environ 41(10):2373–2389

Suzuki N, Katano K (2018) Coordination between ROS regulatory systems and other pathways under heat stress and pathogen attack. Front Plant Sci 9:490

Suzuki N, Miller G, Salazar C, Mondal HA, Shulaev E, Cortes DF, Shuman JL, Luo X, Shah J, Schlauch K, Shulaev V (2013) Temporal-spatial interaction between reactive oxygen species and abscisic acid regulates rapid systemic acclimation in plants. Plant Cell 25(9):3553–3569

Townsley BT, Sinha NR (2012) A new development: evolving concepts in leaf ontogeny. Annu Rev Plant Biol 63(1):535–562

Tsukagoshi H, Busch W, Benfey PN (2010) Transcriptional regulation of ROS controls transition from proliferation to differentiation in the root. Cell 143(4):606–616

Wang L, Guo Y, Jia L, Chu H, Zhou S, Chen K, Wu D, Zhao L (2014) Hydrogen peroxide acts upstream of nitric oxide in the heat shock pathway in Arabidopsis seedlings. Plant Physiol 164(4):2184–2196

Wang Y, Shen W, Chan Z, Wu Y (2015) Endogenous cytokinin overproduction modulates ROS homeostasis and decreases salt stress resistance in Arabidopsis thaliana. Front Plant Sci 6:1004

Wells DM, Wilson MH, Bennett MJ (2010) Feeling UPBEAT about growth: linking ROS gradients and cell proliferation. Dev Cell 19(5):644–646

Winston GW (1990) Free radicals in cells. In: Stress responses in plants: adaptations and acclimation mechanisms. Willy-Liss, New York, pp 57–86

Woodson JD, Chory J (2008) Coordination of gene expression between organellar and nuclear genomes. Nat Rev Genet 9(5):383–395

Yang Y, Guo Y (2018) Elucidating the molecular mechanisms mediating plant salt-stress responses. New Phytol 217(2):523–539

Yao Y, He RJ, Xie QL, Zhao XH, Deng XM, He JB, Song L, He J, Marchant A, Chen XY, Wu AM (2017) ETHYLENE RESPONSE FACTOR 74 (ERF74) plays an essential role in controlling a respiratory burst oxidase homolog D (RbohD)-dependent mechanism in response to different stresses in Arabidopsis. New Phytol 213(4):1667–1681

Yi J, Moon S, Lee YS, Zhu L, Liang W, Zhang D, Jung KH, An G (2016) Defective tapetum cell death 1 (DTC1) regulates ROS levels by binding to metallothionein during tapetum degeneration. Plant Physiol 170(3):1611–1623

Yoshida T, Obata T, Feil R, Lunn JE, Fujita Y, Yamaguchi-Shinozaki K, Fernie AR (2019) The role of abscisic acid signaling in maintaining the metabolic balance required for Arabidopsis growth under nonstress conditions. Plant Cell 31(1):84–105

Zandalinas SI, Fichman Y, Devireddy AR, Sengupta S, Azad RK, Mittler R (2020) Systemic signaling during abiotic stress combination in plants. Proc Natl Acad Sci 117(24):13810–13820

Zandalinas SI, Sengupta S, Fritschi FB, Azad RK, Nechushtai R, Mittler R (2021) The impact of multifactorial stress combination on plant growth and survival. New Phytol 230(3):1034–1048

Zhang L, Ren F, Zhang Q, Chen Y, Wang B, Jiang J (2008) The TEAD/TEF family of transcription factor Scalloped mediates Hippo signaling in organsize control. Dev Cell 14:377–387

Zhou J, Wang J, Li X, Xia XJ, Zhou YH, Shi K, Chen Z, Yu JQ (2014) H_2O_2 mediates the crosstalk of brassinosteroid and abscisic acid in tomato responses to eat and oxidative stresses. J Exp Bot 65(15):4371–4383

Zhu JK (2016) Abiotic stress signaling and responses in plants. Cell 167(2):313–324

Zwack PJ, Rashotte AM (2015) Interactions between cytokinin signalling and abiotic stress responses. J Exp Bot 66(16):4863–4871

The Ecology of Reactive Oxygen Species Signalling

5

Muhammad Mohiuddin, Sidra tul Muntha, Abid Ali, Mohammad Faizan, and Samrana Samrana

Abstract

In the environment, plant growth is affected and controlled by various biotic and abiotic factors. The global climatic factors and abiotic stressors such as extreme temperatures, salinity, droughts, and heavy metal contamination have all had a significant impact on plant growth and development, influencing crop output and quality, as well as agriculture's long-term viability. Abiotic stress causes plant cells to create oxygen radicals and their derivatives, known as reactive oxygen species (ROS). Furthermore, in higher plants, the creation of ROS is a critical mechanism that transfers cellular signalling information in response to changing environmental circumstances. Abiotic stress disrupts the equilibrium between ROS production and antioxidant defence mechanisms, causing excessive ROS to build up and oxidative stress in plants. Plants under stress can even maintain

M. Mohiuddin (✉)
Department of Environmental Sciences, Kohsar University, Murree, Pakistan

Faculty of Biosciences, Kohsar University, Murree, Pakistan

S. t. Muntha
Faculty of Biosciences, Kohsar University, Murree, Pakistan

Laboratory of Germplasm Innovation and Molecular Breeding, College of Agriculture and Biotechnology, Zhejiang University, Hangzhou, China

A. Ali
Laboratory of Germplasm Innovation and Molecular Breeding, College of Agriculture and Biotechnology, Zhejiang University, Hangzhou, China

M. Faizan
Botany Section, School of Sciences, Maulana Azad National Urdu University, Hyderabad, India

S. Samrana
Department of Botany, University of Swabi, Swabi, Pakistan

M. Faizan et al. (eds.), *Reactive Oxygen Species*,
https://doi.org/10.1007/978-981-19-9794-5_5

the balance between detoxification and ROS generation that is controlled by the enzymatic and non-enzymatic defence systems. Despite the high level of interest in this field, it is relatively unexplored, and our understanding of ROS signalling is limited. The harmful effects of ROS, antioxidant defence systems involved in ROS detoxification under various abiotic stresses, and molecular crosstalk with other important signalling molecules such as reactive nitrogen, sulphur, and carbonyl species will all be discussed in this chapter. Furthermore, cutting-edge molecular strategies for ROS-mediated enhancement of plant antioxidant defence during abiotic stress adaption will be explored.

Keywords

Antioxidant defence system · Climatic factors · Signalling molecules

5.1 Introduction

Threats to global food security are posed by pathogens, insects, and various abiotic stresses, including flooding, protracted droughts, and heat waves. The alarming increase in these stresses' frequency and intensity, which is a result of climate change and global warming, emphasises how crucial it is to comprehend the processes that make plants more resilient to such stresses. In stress sensing, the integration of several stress-response signalling networks, the activation of plant defence systems, and acclimatisation, reactive oxygen species (ROS) play critical roles. We can boost plant stress tolerance and our capacity to reduce crop damage when crops are exposed to harsh environmental conditions by analysing and comprehending how ROS organise plant responses to stress. While most biological components are largely unaffected by O2, ROS can induce the oxidation of lipids, proteins, RNA, DNA, and many other tiny molecules in cells. The changed chemistry of ROS in comparison to O_2 that enables them to donate an electron or transfer an excited energy state to an acceptor molecule is the cause of their elevated reactivity towards these biological components. Hydrogen peroxide (H_2O_2), superoxide (O_2), singlet oxygen (O_2), the hydroxyl radical (HO), and numerous types of organic and inorganic peroxides are among the most prevalent ROS in cells, with their characteristics and chemical reactivity varying widely. The production of ROS, which is highly reactive and created independently in all or most cell compartments, must be kept under check in order to prevent unintentional cellular oxidation. This is accomplished by balancing the synthesis, scavenging, and transport of ROS, which collectively maintain ROS at low concentrations and regulate the course of ROS signalling processes.

5.2 ROS Production and Scavenging during Stress

The plant's developmental stage, environmental factors, and interactions with its microbiomes, all have an impact on the baseline level of ROS, which is a sign of cellular equilibrium. This balance can be upset by various biotic and abiotic stimuli, which can also decouple metabolic activity and cause the production of ROS in cell organelles. O_2 and $^1O_2$, for instance, are principally generated in the chlorophyll-containing organelle especially by photosystems I and II during periods of excessive light stress. If photorespiration is triggered, H_2O_2 will also be produced in the peroxisomes (Wang et al. 2020a, 2020b; Galvez-Valdivieso et al. 2009; Kerchev et al. 2016). When stomata close, it limits CO_2 availability, thus extra energy absorbed by the photosystem cannot be converted into CO_2 fixation, and the production of ROS may be further increased (Hipsch et al. 2021). O_2 and H_2O_2 are formed in mitochondria and chloroplasts under heat stress when membrane complexes involved in various electron transport chains are damaged (Schwarzländer et al. 2009). Because of the imbalance and disruption of metabolic pathways, O_2 and H_2O_2 are mostly produced in chloroplasts and the apoplast, where specific oxidases such as RBOH are activated (Kámán-Tóth et al. 2019; Liu et al. 2021b, 2021a). In contrast, it has recently been demonstrated that viral infection results in the reduction of peroxisomal ROS generation as a result of interactions between viral proteins and glycolate oxidase). Recent advancements in the development of genetically encoded ROS sensors revealed that distinct cell compartments accumulate different types of ROS depending on the stressor (Bratt et al. 2016; Ugalde et al. 2021; Nietzel et al. 2019). As a result, distinct ROS accumulation patterns or fingerprints are induced in cells in a stress-specific way. Additionally, recent research has shown that distinct cell compartments and the nucleus can activate various retrograde and anterograde signalling pathways when ROS is carried into or out of those compartments (Exposito-Rodriguez et al. 2017; Shapiguzov et al. 2019; Estavillo et al. 2011). Different ROS and other signals generated in various cell compartments in response to various stimuli may activate signal transduction pathways specific to stress that activate defence and acclimation mechanisms specialised to stress (Inupakutika et al. 2016; Jabłońska and Tawfik 2021; Mittler 2017).

5.3 ROS in Disease Resistance

Co-evolutionary strategies and plant resistance to disease attacks have always been gradual and essential for agricultural production systems over time. Plant susceptibility and resistance to disease invasion from the outside are controlled by recognition and signalling processes between the host and the pathogen. A signalling cascade that produces ROS, phytoalexin, and anti-microbial genes are activated by microbe/pathogen-associated molecular patterns, or MAMPs/PAMPs, and Avr (avirulent) gene products that correlate to their host receptors (plants). This also turns on a variety of plant defence genes that are effective against a variety of diseases. Peroxisomal photorespiration and the electron transport chain in chloroplasts and mitochondria in plants both produce ROS (Goraya and Asthir 2016). The four ROS that are most persistent and prevalent in plants are hydrogen peroxide, superoxide, hydroxyl, and singlet oxygen. Between the four, there is a quick interconversion that offers more functional versatility. H2O2 is the most stable of the four classes and is carried by ROS through aquaporin membranes (Foyer 2018). Differing physiological effects are caused by different ROS generation concentrations. ROS can act as a signalling agent in low doses. Due to the damaging oxidative effects that ROS have on lipids, proteins, and nucleic acids, an excessive buildup of ROS may, however, cause cell necrosis. When under oxidative stress, calcium channels, NADPH oxidases, and calcium fluxes interact to produce a ROS wave that can transmit messages across great distances (Smirnoff and Arnaud 2019). A ROS surge is a respiratory burst-homologue-D-mediated (RBOHD), cell-to-cell self-replicating process of ROS generation. Once activated, it causes a single cell to produce more ROS than usual, acting as a sensory signal to neighbouring cells to also produce more ROS. There is a coordination between various generated stresses and the systemic stomatal response across the entire plant, according to a recently discovered unique function of the ROS surge (Fichman and Mittler 2020; Devireddy et al. 2020a, 2020b). The RBOHD protein mediates the ROS wave, a cell-to-cell auto-propagating process of ROS generation (Jan et al. 2020; Yamasaki et al. 2019; McLachlan 2020). It causes the cell to produce more ROS when it is activated in a single cell. As a result, ROS buildup at the apoplast occurs.

5.4 ROS Perception and Redox Regulation

ROS awareness and redox control changes in ROS levels in cells can affect many different signal transduction pathways by changing the structure and function of several proteins, unlike most "traditional" signal transduction agents like hormones or peptides that have a predetermined set of receptors. This "many pathway" signalling ability of ROS is mainly mediated by oxidative post-translational modifications (oxi-PTMs) (Huang et al. 2019; Nietzel et al. 2020; De Smet et al. 2019; Leferink et al. 2009; Chan et al. 2016) and enables ROS to be a versatile and dynamic regulator of different responses to stress. Various proteins' oxidative PTMs when under stress. Many proteins' methionine and cysteine residues contain thiols that are vulnerable to nucleophilic assault by ROS (Zaffagnini et al. 2016). Sulfenic acid (–SOH), a highly reactive and reversible intermediate of the Cys thiol's first ROS-initiated oxidation, is the product. Sulfenic acid can undergo further oxidation to generate sulfinic ($-SO_2H$) and sulfonic (SO_3H) acids, both of which are thought to be largely irreversible modifications that cause protein breakdown ($-SO_2H$ formation can occasionally be reversed by the action of sulfiredoxin) (Akter et al. 2018; Puerto-Galán et al. 2015; Iglesias-Baena et al. 2011). The interactions of sulfenic acids with nearby proteinaceous thiols that are either inter- or intramolecular or with small molecules like glutathione are the most frequent in the context of ROS signalling events (Bender et al. 2015; Niazi et al. 2019). Cys thiols can be modified by reactive electrophilic species besides ROS. Nitric oxide (NO), for instance, can cause the synthesis of S-nitrosothiols (–SNO), while hydrogen sulphide (H_2S) can combine with -SOH to produce persulfides (–SSH). Proteins' methionine residues can also be oxidised to generate methionine sulfoxides, which methionine sulfoxide reductases (Jacques et al. 2015) can then convert back to methionine. Methionine sulfoxide can be further transformed into methionine sulfone if they are not reduced back. Most of the oxi-PTMs mentioned above have been shown in recent studies to alter protein structure. In order to activate or repress stress-response signal transduction mechanisms, these proteins' activity, specificity, and localisation can all be altered by ROS. A key component of ROS signalling is the reversibility in oxi-PTMs that it causes. The ability to controllably reverse an oxi-PTM provides plasticity to ROS signalling under stress when it comes to absorbing different stress or developmental stimuli and recovering from stress. While glutaredoxins (GRXs) frequently reverse glutathionylation activities to produce the original thiol, protein disulfides are largely reduced back by thioredoxins (TRXs) (Trnka et al. 2020; Tarrago et al. 2009; Daloso et al. 2015). GRXs act as oxidoreductases that control the redox state of thiol groups or exchange a glutathionylated moiety with a protein, whereas TRXs contain at least one conserved redox-active dithiol and form a mixed disulfide bond with their target proteins, regulating their structure and function. Because of their potential for high selectivity, these reactions can complicate redox signalling under stress. Reversing the oxi-PTM may reactivate or inhibit protein function, which in turn may activate, suppress, or modify stress-response pathways, depending on the initial context of the oxi-PTM.

5.5 Defence System Against ROS Production and Accumulation

The end of the twentieth century saw the discovery of the ROS defence system in plants, along with its many mechanisms and elements (Baba et al. 2019; Bowler et al. 1994; Foyer and Halliwell 1976; Hossain and Asada 1987; Karpinska et al. 2001). Recently, it was determined that the defence mechanism against ROS includes both enzymatic and non-enzymatic defence mechanisms, and that these defence mechanisms are triggered by various environmental stimuli from the outside world (Noctor et al. 2018). Various enzymatic and non-enzymatic defence mechanisms control the formation of ROS, which is a byproduct of routine cellular metabolism. In contrast to non-enzymatic defence mechanisms, which include glutathione (GSH), ascorbic acid (AA), phenolic compounds, and tocopherols (TOCs), enzymatic defence mechanisms include APX, CAT SOD, and GPX (Karpinska et al. 2001; Apel and Hirt 2004; Munné-Bosch and Alegre 2004). Plants have numerous ROS defence mechanisms that include both enzymatic and non-enzymatic components. Peroxisomes, chloroplasts, and mitochondria are examples of cell components where ROS production and scavenging can occur. In the case of such pathways, there is substantial coordination between these organelles (Pang and Wang 2008). Plants produce and scavenge ROS in equilibrium under normal circumstances, but under stress, this equilibrium is upset, causing an increase in ROS levels (Pang and Wang 2008), which causes oxidative stress to cell components. More advanced plants, however, have a natural defence system to counteract this increase in ROS levels (Sharma et al. 2019). To lower ROS levels in plants during abiotic stressors, the ROS defence mechanism is crucial. Plants have developed advanced defence mechanisms over time to combat the buildup and generation of ROS (Berni et al. 2019).

5.6 Regulation of Plant Defence, Enzyme Defence, and Acclimatisation by ROS

Defence and acclimation responses are triggered by changes in ROS levels in various cell compartments and the integration of such signals during stress. Different sensors and receptors are present in plants that may detect changes in osmotic pressure, temperature, and light. These include receptor-like kinases and ROP proteins that detect osmotic changes, Ca^{2+} – permeable channels that detect osmotic changes, and photoreceptors that detect changes in light quality and intensity. Plants are assumed to sense light stress by retrograde signalling, the release of ROS and Ca2+ from chloroplasts, and PHYB also detects changes in temperature (Jung et al. 2016; Legris et al. 2016; Finka et al. 2012). Some of these receptors may be physically close to RBOHs during the early phases of stress sensing and responses, such as when they share nanodomains at the plasma membrane or when chloroplasts are close to the plasma membrane or nucleus. Therefore, ROS generation may be directly related to the first perception of biotic stressors by plants through various receptors and sensors that results in quick alterations in Ca^{2+} signalling and phosphorylation responses. This process is comparable to the detection of pathogens in plants, in which Ca^{2+} dependent and/or phosphorylation-dependent activation of RBOHs quickly initiates the production of ROS highlighting the evolutionary significance of ROS signalling for plants and the crucial function of RBOHs in these processes. One of the most exciting discoveries in recent years is that light stress does not cause fast ROS formation in plants (Yuan et al. 2014; Basu and Haswell 2020; Griffin et al. 2020) in the absence of specific RBOHs. The likelihood that chloroplasts can regulate their internal ROS levels under light stress and that ROS build up in cells is predominantly the result of ROS generation for RBOH signalling is suggested by this finding. It was once thought that excess ROS created during light stress in chloroplasts diffused to the cytosol through AQPs. Furthermore, it is likely that two different populations of chloroplasts, including those associated with the nucleus, which promote chloroplast-to-nucleus signalling, and those associated with the plasma membrane, which initiate RBOH-driven ROS signals, are involved in ROS signalling under light stress. Numerous retrograde signals that stimulate ROS generation by RBOHs are sent if O_2, H_2O_2, or $^1O_2$ build up in chloroplasts under light stress (even at modest levels). According to this new theory of how plants react to stress, various types of stress are quickly detected by stress-specific receptors, which then induce RBOHs to produce ROS or lead to stress-specific imbalances that change the amounts of ROS and other stress-related metabolites. With changes in redox, Ca^{2+} levels, phosphorylation, and other signalling events that activate stress-specific signal transduction pathways, this process takes place seconds to minutes after stress is initiated. Rapid increases in the levels of hormones, such as freshly produced jasmonic acid or the release of abscisic acid and salicylic acid from conjugated forms (Glauser et al. 2009; Yuan et al. 2017; Suzuki et al. 2013a, 2013b; Yang et al. 2017), are also associated with the activation of these pathways. These early signalling events also change ROS signatures, boosting plant stress resistance, by activating defence and acclimatisation networks (Suzuki et al. 2013a, 2013b; Karpinski et al. 1999). Through ROS-related epigenetic mechanisms 172, some characteristics of this heightened state of resistance may be long-lasting or passed on to the following generation. As a result, practically all phases of early and late responses to stress are influenced by ROS, and many of the pathways, networks, and hormones necessary for plant survival under stress are closely related to ROS. ROS function when subjected to various stressors. In the wild, plants are frequently subjected to a variety of stresses at the same time. For instance, a stress combination of drought, intense light, and heat—activates a number of signalling pathways. ROS were discovered to be crucial for a plant's ability to adapt to these conditions. The ROS-regulated MPK3/MPK6 and MPK4 cascades have opposing roles in the triggering of defence and acclimatisation networks during the integration of cold stress and pathogen responses (Jagodzik et al. 2018). Thus, ROS play a significant part in the assimilation of various signals produced during the combination of stresses. Diverse stresses may affect a plant sequentially or simultaneously, resulting in different ROS signatures and the combination of them may attenuate or control the plant's overall response to challenging environmental conditions. When a specific stress, like heat, occurs against the backdrop of a certain developmental stage, like plant reproduction, for example, integration of two distinct ROS signatures may also take place or during interactions with the plant microbiome. Under these circumstances, the total ROS levels are combined to cause a particular or general state of plant tolerance or sensitivity to stress. The activation and control of multiple transcription factor networks by the sensing of stress via receptors and ROS-activated redox sensors enables the plant to respond to a number of varied conditions. The mechanisms that regulate transcriptional responses in plants include changes in phosphorylation, Ca2+ binding, SUMOylation, and other signal transduction mechanisms that modify transcription factor function, as well as direct and indirect ROS-induced redox regulation (Schmidt et al. 2013; Pérez-Salamó et al. 2014). These two processes are interconnected as a result of the connections between ROS signalling and other signalling events (e.g. those mediated by Ca2+ and phosphorylation), such as through RBOHs and Aquaporins (AQP). In response to stress, gene expression can be altered via other methods in a redox-dependent manner. The components of the plant Mediator complex are redox regulated, and ROS can influence mRNA splicing as well as the production and function of a variety of microRNAs (Tran et al. 2013; Iyer et al. 2012). For example, an increase in ROS levels may inhibit the expression of specific housekeeping gene families that require intense splicing, microRNA activity, or interactions with the Mediator complex to function. The redox-regulated transcriptional regulators involved in responses to biotic and abiotic stresses are affected by ROS when they move from the cytosol or the outer membranes of the endoplasmic reticulum to the nucleus, in addition to controlling transcription through genetic or epigenetic mechanisms during stress. In a recent study, a ROS-response integrated gene regulatory network was developed using DNA patterns, transcription factor-binding sites, and expression-based regulatory interactions. Once inside the nucleus, these transcriptional regulators activate gene expression networks and increase plant stress tolerance. This matrix contained some of the regulatory hubs and pathways that some of the new ROS-regulated transcription factors regulate (De Clercq et al. 2021).

5.7 Enzymatic Defence Systems

The enzymatic defence mechanisms against ROS involve a number of enzymes, such as glutathione reductases (GRs), dehydroascorbate reductases (DHARs), superoxide dismutases (SODs), monodehydroascorbate reductases (MDHARs), glutathione peroxidases (GPXs), catalases (CATs), and ascorbate peroxidases (APXs). The defence system's several enzymes are discussed, including SOD from the metalloenzyme family. The elimination of O_2, which is converted into O_2 and H_2O_2 via a dismutation process that lessens the creation of OH, is one of the mechanisms by which they lower ROS (Lu et al. 2017). SODs are found in plants' peroxisomes, chloroplasts, and mitochondria. Abiotic and biotic stressors both cause plants to produce more SOD (Lu et al. 2017; Ahmad et al. 2010). Two molecules of H_2O_2 are broken down by heme-containing CAT enzymes into H_2O and O_2 in a dismutation process. The location of their presence affects their tissue selectivity (Mhamdi et al. 2010). Drought, cold, heat, and UV radiation frequently increase the expression of CAT (Caverzan et al. 2016; Sofo et al. 2015). H_2O_2 can be eliminated by a variety of processes that ascorbate-glutathione (AsA-GSH) undergoes. The cytosol, mitochondria, and chloroplasts are where this enzyme is primarily active (Nishimura et al. 2011; Jimenez et al. 1997). Plant glutathione-S-transferases (GSTs) are engaged in biotic and abiotic stress response as well as programmed cell death (Dixon et al. 2010; Gong et al. 2005). Depending on the type of stress and the location of oxidative damage, the enzymatic defence system against ROS is highly evolved.

Without enzymes plants are equipped with non-enzymatic and low-molecular antioxidants such as proline, GSH, AsA, and carotenoids. These have a role in retrograde signalling as well (König et al. 2018; Agati et al. 2012). The primary components that lower H_2O_2 are GSH and AsA because they renew fast. The elimination of ROS is carried out by the GSH-AsA cycle (Zhang 2012). In addition to being the catalyst that creates tocopherol, AsA also functions as a cofactor. Additionally, GSH is crucial for scavenging ROS and serves as a starting point for different peroxidases. GSH can be found in vacuoles, mitochondria, nuclei, and the endoplasmic reticulum (Koffler et al. 2013; Cheng et al. 2015; Hasanuzzaman et al. 2014). Since tocopherols are fat-hating substances that break down ROS, they also provide membrane protection (Kamal-Eldin and Appelqvist 1996). Carotenoids are mostly found in plastids and are lipid soluble. A higher concentration of these promotes the development of resistance to diverse abiotic stressors (Nisar et al. 2015). Flavonoids found in seeds, flowers, and fruits are responsible for the pigmentation of blue, red, and purple. Additionally, they are said to defend plants from a variety of abiotic challenges such as high temperatures and UV rays (Petrussa et al. 2013; Winkel-Shirley 2002; Das and Roychoudhury 2014).

5.8 Transcriptional Control of Biotic and Abiotic Stress Responses

Reactive oxygen species (ROS) is the name for a group of molecules that are produced from molecular oxygen. While most cellular components are generally unaffected by O_2, ROS may cause the oxidation of proteins, lipids, nucleic acids, and other small molecules in cells. Due to their different chemistry from O_2, ROS have a high reactivity towards these cellular components because they can transfer an electron to an acceptor molecule (Halliwell and Gutteridge 2015) 5. The primary ROS in cells are superoxide (O_2), singlet oxygen ($^1O_2$), hydrogen peroxide (H_2O_2), the hydroxyl radical (HO), and various types of organic and inorganic peroxides. ROS have a wide range of properties and chemical reactivity (Halliwell and Gutteridge 2015; Mittler 2017; Waszczak et al. 2018; Smirnoff and Arnaud 2019; Sies and Jones 2020). The production of ROS, which are highly reactive and produced independently in all most all cell compartments, must be kept in check in order to avoid unintended cellular oxidation. This is achieved by balancing ROS generation, scavenging, and transport, which together maintain ROS at low concentrations and regulate the course of ROS signalling reactions. The redox condition of numerous proteins, including receptors, enzymes, and small molecules, is impacted by accumulation of ROS in cells during stress, which activates, modifies, or integrates a range of stress-response signal transduction pathways. They change gene expression and improve a plant's ability to withstand stress (Zou et al. 2015; Nie et al. 2013; Liu et al. 2021b, 2021a; Hu et al. 2017; Yin et al. 2018; Shi et al. 2020; Martins et al. 2020). The discovery of particular ROS sensors and regulatory hubs that link ROS signalling with other stress-response signal transduction pathways and hormones are recent developments in our understanding of these crucial processes.

5.9 Induction of Plant Adaptability Through Transcriptional Regulation by ROS

The ability of the plant to respond to various conditions is enabled by the activation and modulation of different transcription factor networks by the detection of stress via receptors: ROS-activated redox sensors interactions. Both direct and indirect ROS-induced redox regulation and stress- or ROS-derived changes in phosphorylation, sumoylation, Ca2+–binding, and/or other signal transduction mechanisms that affect Transfer Function (TF) control transcription responses in plants (Lee et al. 2021; Fu and Dong 2013; Giesguth et al. 2015; Meng et al. 2019; D'Alessandro et al. 2018; Pérez-Salamó et al. 2014; De Clercq et al. 2021; Ohama et al. 2017). Because of the connections between ROS signalling and other signalling events (like those mediated by Ca2+ and phosphorylation), such as through RBOHs and AQPs, these two processes are interconnected. Other techniques can also be used to modify gene expression in a redox-dependent manner in response to stress. The plant Mediator complex's subunits are redox controlled, and ROS can affect the expression and activity of several microRNAs as well as mRNA splicing (Tran et al. 2013; Iyer et al. 2012; He et al. 2021). Plant stress responses are further tuned by the impact of ROS on these systems, which are linked to cellular ROS levels. For instance, a rise in ROS levels may prevent the production of particular housekeeping gene families that need intensive splicing, microRNA activity, or interactions with the Mediator complex to operate (for instance, in conditions of heat stress) (Ohama et al. 2017). After being activated, different redox-regulated transcriptional regulators move from the cytosol or the endoplasmic reticulum's outer membranes to the nucleus, which are involved in responses to biotic and abiotic stimuli (Albertos et al. 2021; Fu and Dong 2013;Giesguth et al. 2015; Meng et al. 2019; D'alessandro et al. 2018; Suzuki et al. 2013a, 2013b). These regulators include HSFA8/HSFA1A, MBF1C, NPR1, and ANAC013/ANAC/ANAC017 (Charbonnel et al. 2017). These transcriptional regulators are subsequently translocated into the nucleus, where they activate gene expression networks and improve plant stress resistance. A recent study that used supervised learning to construct a ROS-response integrated gene regulatory network discovered several new transcription factors that are regulated by ROS and defined some of the regulatory networks and hubs they control. This was made possible by the use of DNA motifs, transcription factor-binding sites, open chromatin regions, and expression-based regulatory interactions (De Clercq et al. 2021).

The integration of ROS signals with other signal transduction networks triggered during stress was also shown by transcriptomic analyses of mutants lacking regulatory hubs such as RBOHs, HSFs, MAPK cascades, and other Ca2+ signalling pathways. As an illustration, a study explaining the transcriptome analysis of the rbohD mutant to light stress revealed that RBOHD is necessary for the expression of a number of early response genes (Zandalinas et al. 2019), such as the transcription factor MYB30, which was discovered to be significant in plants' responses to oxidative stress (Mabuchi et al. 2018). MYB30 also controls hundreds of transcripts in response to mild stress by working upstream of several other transcription factors (Fichman et al. 2020).

As will be addressed under, redox-regulated transcription factors play roles in the response to heat stress, infections, and light.

5.9.1 Responses to High Temperature

Plant cytosol and nuclei accumulate ROS in response to rising temperatures (Babbar et al. 2021). Redox-dependent activation and translocation of MBF1C and HSFA1 from the cytoplasm to the nucleus are caused by the increased ROS levels (Giesguth et al. 2015; Suzuki et al. 2013a, 2013b). The endoplasmic reticulum's BZIP28 is also activated and moved to the nucleus (Srivastava et al. 2013). The collaboration of HSFA and MBF1C in the transcriptional activation of various heat shock proteins and other transcription factors, such as dehydration-responsive element-binding factors, is integral to the development of thermotolerance (Ohama et al. 2017). In response to thermo-stress, BZIP28 and BZIP60 work together to transcriptionally activate the unfolded protein response (Srivastava et al. 2013). Furthermore, the induction of thermo tolerance depends on the chloroplast's redox status, indicating that chloroplasts are involved in these responses (Dickinson et al. 2018).

5.9.2 Biotic Responses

Following pathogen identification (for instance, by plasma membrane-localised pattern recognition receptors), responses to pathogen infection are frequently started by a brief oxidative burst, which is mediated by RBOHs or peroxidases at the apoplast (Torres et al. 2002; Daudi et al. 2012). Following this burst, the cytoplasm's reduced condition increases, salicylic acid, a plant hormone, accumulates, and callose is deposited at the cell wall and plasmodesmata, which inhibits the spread of pathogens (Daudi et al. 2012). Following pathogen detection, there is an increased buildup of ROS and salicylic acid, which results in a redox-regulated transcriptional response mediated by NPR1. NPR1 is confined to the cytoplasm under regulated growth circumstances as an oligomer held by intermolecular Cys bonds involving Cys82 and Cys216 (Després et al. 2003; Lindermayr et al. 2010).

Thioredoxins (TRX3 and TRX5) are involved in the decrease of these bonds that salicylic acid causes, which leads to the monomerization of NPR1 (Tada et al. 2008). Monomeric NPR1 is transported to the nucleus, where it interacts with TGA1 in a redox-dependent way and promotes the transcription of several genes that code for proteins involved in pathogenesis, including WRKY (Fu and Dong 2013). Jasmonic acid, a plant hormone, interestingly inhibits this process by encouraging the S nitrosylation of NPR1 on Cys (Yuan et al. 2014), which results in its oligomerization (Tada et al. 2008). NPR1 may play a key integrative role between daily variations in redox levels and plant responses to biotic and abiotic stresses 141. It is also implicated in the response to other abiotic stimuli (such as salinity) (Jayakannan et al. 2015).

5.9.3 Excess Light Stress

Overexposure to light stresses the chloroplast, apoplast, and cytosol, which in turn oxidises them. Multiple redox-response transcription factors, including as MYB30, ZAT10, ZAT12, RELATED TO APETALA 2 (RAP2), and other HSFs (Davletova et al. 2005; Fichman et al. 2020; Shaikhali et al. 2008) control nuclear transcription as a result. The chloroplastic 3′-phosphoadenosine 5′-phosphate (PAP) phosphatase SAL1 goes through redox-dependent oxidative inactivation during light exposure. PAP builds up as a result, and PAP acts as a retrograde signal to control gene expression in the nuclei (Estavillo et al. 2011; Chan et al. 2016). It is interesting to note that PAP is connected to another mitochondria-based retrograde signalling system. The redox-activated ANAC transcription factors that go from the endoplasmic reticulum to the cytoplasm and the antagonist RCD146 interact in this pathway to influence PAP levels.

As a result, ROS and retrograde signals are coupled and mediate a variety of signal transduction reactions to stress, and this integration may be crucial in the event of pathogen infection or heat stress (for example, under conditions of excess light). The increased accumulation of various antioxidants, osmoprotectants, molecular chaperones, pathogen-response proteins, and many other enzymes and proteins as a result of the transcriptional changes triggered in response to elevated ROS levels during stress allows the plant to withstand the stress and survive (Zou et al. 2015; Nie et al. 2013; Liu et al. 2021b, 2021a; Hu et al. 2017; Yin et al. 2018; Shi et al. 2020; Martins et al. 2020; Suzuki et al. 2013a, 2013b; Zandalinas et al. 2021b, 2021a; Fichman et al. 2020; Zandalinas et al. 2020). The ability of ROS signals to spread from their localised source of production to other plant cells and tissues and coordinate the systemic, whole-plant responses to stress is covered in the section that follows.

5.10 Roles of ROS at Integrated Points of Biotic and Abiotic Stress-Response Pathways

Multiple cellular functions in plants entail the strict control of the steady-state levels of ROS (Zandalinas et al. 2021b, 2021a). While ROS also serve as signalling molecules, certain ROS species are harmful byproducts of aerobic metabolism (Zandalinas et al. 2021b, 2021a). Both ABA signalling and disease resistance responses depend heavily on rapid ROS generation (Shapiguzov et al. 2019; Nomura et al. 2012). The NADPH-dependent respiratory burst oxidase homologue genes (AtrbohD and AtrbohF) may be necessary for ROS formation, which results in ABA-induced stomatal closure and hypersensitive cell death in response to avirulent pathogen assault, according to a number of lines of evidence (Halliwell and Gutteridge 2015; Bienert and Chaumont 2014). Under diverse stress circumstances, ROS scavengers are hypothesised to detoxify the harmful effects of ROS (Zandalinas et al. 2021b, 2021a; Inupakutika et al. 2016). A significant number of genes that encode ROS-scavenging enzymes are induced by various abiotic and biotic stress treatments, according to large-scale transcriptome investigations of plants that had undergone these treatments (Mittler 2017;Smirnoff and Arnaud 2019; Inupakutika et al. 2016). Ascorbate peroxidase, glutathione peroxidase, and superoxide dismutase are a few examples of scavenging enzymes that have been used to create plants that are resistant to abiotic stressors (Jabłońska and Tawfik 2021; Huang et al. 2019). Numerous ABA-inducible genes are triggered by oxidative stress, according to microarray investigation of Arabidopsis cultivated cells (Nietzel et al. 2020). A C2H2-type zincfinger transcription factor called Zat12 has recently been proposed as a potential regulator of the ROS-scavenging process involved in biotic and abiotic stress responses.

ASCORBATE PEROXIDASE 1 (APX1) gene expression is suppressed by Zat12 deficiency, which is highly susceptible to a variety of stimuli including wounding, pathogen infection, and abiotic stresses (Mittler 2017; Wang et al. 2020a, 2020b; De Smet et al. 2019;Leferink et al. 2009). This raises the degree of H2O2-induced protein oxidation (Chan et al. 2016). Genes that are susceptible to oxidative and light stress are upregulated as a result of Zat12 overexpression, which also improves tolerance to high light, freezing, and oxidative stressors (De Smet et al. 2019; Leferink et al. 2009; Chan et al. 2016; Zaffagnini et al. 2016). It is interesting to note that HEAT SHOCK FACTOR (HSF) 21, a redox-sensitive transcription factor, controls the expression of Zat12. HSF 21 is likely to be an early sensor for H2O2 that builds up in response to diverse stressors (Rodrigues et al. 2017). These results imply that the ROS may facilitate interaction between networks of genes that are sensitive to biotic and abiotic stress.

5.11 Impact of ROS on Biomolecules

5.11.1 Impact on Proteins

Proteins may be altered as a result of ROS attack in a variety of ways, some of which are direct and others indirect. Nitrosylation, carbonylation, disulphide bond formation, and glutathionylation are examples of direct modifications that alter a protein's activity. Through conjugation with the byproducts of fatty acid peroxidation, proteins can undergo indirect modification (Yamauchi et al. 2008). Site-specific amino acid modification, peptide chain fragmentation, aggregation of cross-linked reaction products, altered electric charge, and increased susceptibility of proteins to proteolysis all result from excessive ROS production. Increased levels of carbonylated proteins, a common indicator of protein oxidation, are typically found in tissues damaged by oxidative stress (Møller and Kristensen 2004). Under various stresses, plants have been found to have increased protein modification (Sharma and Dubey 2005; Maheshwari and Dubey 2009; Tanou et al. 2009; Romero-Puertas et al. 2002).

The susceptibility of the amino acids in a peptide to ROS attack varies. Thiol groups and amino acids that contain sulphur are particularly vulnerable to attack by ROS. An H atom from cysteine residues can be removed by activated oxygen to create a thiyl radical, which will cross-link with another thiyl radical to create a disulphide bridge. It has been demonstrated that a number of metals, including Cd, Pb, and Hg, deplete protein-bound thiol groups (Stohs and Bagchi 1995). A methionine can also be combined with oxygen to create a derivative known as methionine sulphoxide (Brot and Weissbach 1982). In the presence of ROS, tyrosine readily cross-link's to produce bityrosine products (Davies 1987).

The irreversible oxidation of iron-sulfur centres by O2• results in the inactivation of enzymes (Gardner and Fridovich 1991). In these circumstances, the metal (Fe) binds to a protein's divalent cation-binding site. Then, the metal (Fe) reacts in a Fenton reaction to produce a •OH, which quickly oxidises an amino acid residue at or close to the protein's cation-binding site (Stadtman 1986). Proteins that have been oxidised make for better proteolytic substrates. Protein oxidation has been hypothesised to make it more susceptible to ubiquitination, which would make it a target for proteasomal degradation (Cabiscol et al. 2000). The amount of carbonyl increased when pea leaf crude extracts with increasing H2O2 concentrations, plants treated with Cd, and purified pea leaf peroxisomes were incubated. The metal treatment increased the proteolytic activity by 20% and improved the efficiency with which oxidised proteins were broken down (Romero-Puertas et al. 2002). According to several studies, after a certain point, additional damage results in highly cross-linked and aggregated products, which not only make poor substrates for protein degradation but also can prevent proteases from breaking down other oxidised proteins (Grune et al. 1997).

5.11.2 Protein Carbonylation

Protein carbonylation is a sign of protein oxidation and is an irreversible protein oxidation reaction brought on by ROS attacks on proteins. Under the catalysis of the transitional metal ion system, ROS can directly oxidise arginine, lysine, threonine, proline, and other residues of protein side chains to form carbonyl groups (Møller et al. 2011; Requena et al. 2001). The -carbon of a protein chain can also be used by ROS to extract hydrogen atoms, which results in the oxidative hydrolysis of the carbon atoms and the formation of a carbonyl group (Bizzozero 2009; Zheng and Bizzozero 2010). Additionally, ROS can catalyse lipid peroxidation and non-enzymatic glycosylation to create some active carbonyl compounds. These compounds are then cross-linked to protein side chains through the carbonyl ammonia reaction to form carbonyl groups, which then undergo additional reactions to create carbonyl proteins (Madian and Regnier 2010). The carbonyl proteins can only be destroyed in a 20S proteasome-dependent manner; they cannot be repaired. However, if the carbonyl protein is not removed from the body quickly enough, it builds up inside the cell and causes cell apoptosis (Yamaguchi et al. 2012; Griesser et al. 2017; Chen et al. 2018). The germination capacity is decreased by carbonylation, which also affects the synthesis and conformational processes of proteins involved in responses to stresses brought on by ageing (Nguyen et al. 2015; Yin et al. 2017).

5.11.3 Impact on Lipids

Increased lipid peroxidation occurs in cellular and organellar membranes when ROS levels rise above the threshold, which in turn affects normal cellular function. Lipid peroxidation worsens oxidative stress by generating lipid-derived radicals, which have the potential to interact with proteins and DNA and cause damage. Indicators of ROS-mediated damage to cell membranes under stressful circumstances have frequently been based on the degree of lipid peroxidation. Plants growing in stressful environments have been shown to have increased lipid peroxidation (degradation) (Sharma and Dubey 2005; Han et al. 2009; Tanou et al. 2009; Mishra et al. 2011). Increased production of ROS coincides with increased lipid peroxidation under these stresses. Damage to cell membranes is caused by malondialdehyde (MDA), one of the byproducts of peroxidation of unsaturated fatty acids in phospholipids (Halliwell and Gutteridge 1989).

The unsaturated (double) bond between two carbon atoms and the ester linkage between glycerol and the fatty acid are two common places where ROS attack phospholipid molecules. In particular, membrane phospholipids' polyunsaturated fatty acids (PUFAs) are vulnerable to attack from ROS. As a result of the reactions involved in this process being a part of a cyclic chain reaction, a single •OH can cause the peroxidation of numerous polyunsaturated fatty acids. Initiation, progression, and termination steps are the three distinct stages that make up the overall process of lipid peroxidation. The rate-limiting step in the first stage of lipid peroxidation is O2 activation. PUFA methylene groups can be reacted with by O2• and •OH to produce conjugated dienes, lipid peroxy radicals, and hydroperoxides (Hu et al. 2008; Han et al. 2009; Smirnoff 1995).

$$\mathrm{PUFA-H+X\bullet \longrightarrow PUFA+X-H}$$

$$\mathrm{PUFA+O2 \longrightarrow PUFA-OO\bullet}$$

The resulting peroxy radical is extremely reactive and capable of starting a chain reaction (Maheshwari and Dubey 2009):

$$\mathrm{PUFA-OO\bullet + PUFA-OOH \longrightarrow PUFA-OOH+PUFA\bullet}$$

Free radicals attack the hydrogens of methylene groups that separate double bonds, causing the bonds to rearrange, which leads to the formation of conjugated dienes (Sharma et al. 2012). According to the following reaction, reduced metals like Fe2+ can reductively cleave the lipid hydroperoxides created (PUFA-OOH) (Tanou et al. 2009):

$$\mathrm{Fe2+complex+PUFA-OOH \longrightarrow Fe3+complex+PUFA-O\bullet}$$

Lipid hydroperoxide can easily break down to form a number of reactive species, such as lipid alkoxyl radicals, alcohols, aldehydes (such as malondialdehyde, acrolein, and crotonaldehyde), alkanes, and lipid epoxides. The lipid alkoxy radical that is created, (PUFA-O•), has the potential to start more chain reactions (Buettner 1993; Mishra et al. 2011):

$$\text{PUFA} - \text{O}\bullet + \text{PUFA} - \text{H} \rightarrow \text{PUFA} - \text{OH} + \text{PUFA}\bullet$$

Chain breakage brought on by polyunsaturated fatty acid peroxidation caused by ROS attack can increase membrane fluidity and permeability.

5.11.4 Impact on Nucleic Acid (DNA)

DNA damage is primarily caused by ROS (Imlay and Linn 1988). DNA in the mitochondria, the nucleus, and chloroplasts can all suffer oxidative damage from ROS. The genetic makeup of a cell is represented by DNA, and any damage to this material can alter the encoded proteins, causing malfunctions or even their total inactivation. Deoxyribose oxidation, strand breakage, nucleotide removal, a variety of modifications to the organic bases of the nucleotides, and DNA-protein cross-links are all effects of oxidative attack on DNA. Additionally, modifications to one strand's nucleotides may cause mismatches with those of the other strand, resulting in mutations. Salinity (Liu et al. 2000) and metal toxicity (Meriga et al. 2004) are two environmental stresses that have been linked to increased DNA degradation in plants.

DNA's sugar and base moieties are both prone to ROS-induced oxidation. While sugar damage primarily results from hydrogen abstraction from deoxyribose, oxidative attack on DNA bases typically involves •OH addition to double bonds (Dizdaroglu 1993). All purine and pyrimidine bases, as well as the deoxyribose backbone, have been shown to react with the hydroxyl radical (Halliwell and Gutteridge 2015). •OH produces a number of products from DNA bases, the most notable of which are 8-oxo-7,8 dehydro-2′-deoxyguanosine, hydroxymethyl urea, urea, thymine glycol, and saturated products for thymine and adenine (Tsuboi et al. 1998). The product that is most frequently seen is 8-hydroxyguanine. H2O2 and O2• do not react with bases at all, while 1O2 only reacts with guanine (Dizdaroglu 1993; Halliwell and Aruoma 1991). Numerous mutagenic changes are also included in the list of DNA harms caused by ROS. An oxidative attack on DNA by ROS, for instance, is indicated by mutations that arise from the selective modification of G:C sites. Through reactive byproducts of ROS attack on other macromolecules, like lipid, ROS indirectly attack DNA bases (Fink et al. 1997).

Single-strand breaks occur as a result of ROS attack on DNA sugars. Deoxyribose is formed when ROS remove a hydrogen atom from the C4 position, which causes the radical to react and cause DNA strand breaks (Evans et al. 2004). Neither H2O2 alone nor O2• can break in vitro strands under physiological circumstances. As a result, it was determined that the Fenton reaction is most likely to be the cause of the toxicity linked to these ROS in vivo. DNA-protein cross-link's are created when •OH attacks DNA or the proteins that are linked to it (Oleinick et al. 1987). If replication or transcription takes place before repair, DNA-protein cross-link's cannot be easily repaired and may even be fatal. Because they lack the protective protein histones and are situated close to the ROS-producing systems, mitochondrial and chloroplast DNA are more vulnerable to oxidative damage than nuclear DNA (Richter 1992). Although there is a repair system for damaged DNA, excessive changes brought on by ROS result in permanent DNA damage, which could have a negative impact on the cell.

5.12 Future Perspectives and Conclusion

ROS biology in plants began with an emphasis on ROS-scavenging and generation processes in chloroplasts. This emphasis has shifted to researching active ROS generation, such as RBOHs, and how it is regulated by various post-translational modifications. Because ROS levels are determined by the interaction of generation, scavenging, and transport, it is critical to identify the mechanisms that govern ROS transport, such as AQPs or other transporters. Furthermore, our understanding of how ROS are generated in cells during stress should be reconsidered. It should be examined if the fact that most ROS formation in plants during excess light exposure is dependent on RBOHs 152,160 rather than emanating from chloroplasts applies to other stressors and plant species. Furthermore, this discovery suggests that ROS may not be as hazardous to cells as previously thought6. More study is needed to identify how organelle ROS signalling is connected to the cytosol, nucleus, and apoplast, as well as how information in the form of ROS signatures is conveyed across these compartments. The discovery of novel ROS and redox sensors, redox relays and hubs, and the investigation of ROS responsive transcriptional networks will help us better understand how ROS signals are integrated in response to stress. Understanding the activities of ROS in plants in response to stress requires acquiring a comprehensive picture of the stress-induced ROS signalling landscape of the cell and connecting it to plant transcriptional, metabolic, and proteomic networks.

References

Agati G, Azzarello E, Pollastri S, Tattini M (2012) Flavonoids as antioxidants in plants: location and functional significance. Plant Sci 196:67–76

Ahmad P, Jaleel CA, Salem MA, Nabi G, Sharma S (2010) Roles of enzymatic and nonenzymatic antioxidants in plants during abiotic stress. Crit Rev Biotechnol 30(3):161–175

Akter S, Fu L, Jung Y, Conte ML, Lawson JR, Lowther WT et al (2018) Chemical proteomics reveals new targets of cysteine sulfinic acid reductase. Nat Chem Biol 14(11):995–1004

Albertos P et al (2021) Redox feedback regulation of ANAC089 signaling alters seed germination and stress response. Cell Rep 35:109263

Apel K, Hirt H (2004) Reactive oxygen species: metabolism, oxidative stress, and signaling transduction. Annu Rev Plant Biol 55:373

Baba AI, Andrási N, Valkai I, Gorcsa T, Koczka L, Darula Z et al (2019) AtCRK5 protein kinase exhibits a regulatory role in hypocotyl hook development during skotomorphogenesis. Int J Mol Sci 20(14):3432

Babbar R, Karpinska B, Grover A, Foyer CH (2021) Heat-induced oxidation of the nuclei and cytosol. Front Plant Sci 11:617779

Basu D, Haswell ES (2020) The mechanosensitive ion channel MSL10 potentiates responses to cell swelling in Arabidopsis seedlings. Curr Biol 30(14):2716–2728

Bender KW, Wang X, Cheng GB, Kim HS, Zielinski RE, Huber SC (2015) Glutaredoxin AtGRXC2 catalyses inhibitory glutathionylation of Arabidopsis BRI1-associated receptor-like kinase 1 (BAK1) in vitro. Biochem J 467:399–413

Berni R, Luyckx M, Xu X, Legay S, Sergeant K, Hausman JF et al (2019) Reactive oxygen species and heavy metal stress in plants: impact on the cell wall and secondary metabolism. Environ Exp Bot 161:98–106

Bienert GP, Chaumont F (2014) Aquaporin-facilitated transmembrane diffusion of hydrogen peroxide. Biochim Biophys Acta 1840:1596–1604

Bizzozero OA (2009) Protein carbonylation in neurodegenerative and demyelinating CNS diseases. In: Handbook of neurochemistry and molecular neurobiology. Springer, New York

Bowler C, Van Camp W, Van Montagu M, Inzé D, Asada K (1994) Superoxide dismutase in plants. Crit Rev Plant Sci 13(3):199–218

Bratt A, Rosenwasser S, Meyer A, Fluhr R (2016) Organelle redox autonomy during environmental stress. Plant Cell Environ 39(9):1909–1919

Brot N, Weissbach H (1982) The biochemistry of methionine sulfoxide residues in proteins. Trends Biochem Sci 7(4):137–139

Buettner GR (1993) The pecking order of free radicals and antioxidants: lipid peroxidation, α-tocopherol, and ascorbate. Arch Biochem Biophys 300(2):535–543

Cabiscol E, Piulats E, Echave P, Herrero E, Ros J (2000) Oxidative stress promotes specific protein damage in Saccharomyces cerevisiae. J Biol Chem 275(35):27393–27398

Caverzan A, Casassola A, Brammer SP (2016) Reactive oxygen species and antioxidant enzymes involved in plant tolerance to stress. Abiotic and biotic stress in plants-recent advances and future perspectives 17:463–480

Chan KX, Mabbitt PD, Phua SY, Mueller JW, Nisar N, Gigolashvili T et al (2016) Sensing and signaling of oxidative stress in chloroplasts by inactivation of the SAL1 phosphoadenosine phosphatase. Proc Natl Acad Sci 113(31):E4567–E4576

Charbonnel C, Niazi AK, Elvira-Matelot E, Nowak E, Zytnicki M, de Bures A et al (2017) The siRNA suppressor RTL1 is redox-regulated through glutathionylation of a conserved cysteine in the double-stranded-RNA-binding domain. Nucleic Acids Res 45(20):11891–11907

Chen X, Yin G, Börner A, Xin X, He J, Nagel M et al (2018) Comparative physiology and proteomics of two wheat genotypes differing in seed storage tolerance. Plant Physiol Biochem 130:455–463

Cheng MC, Ko K, Chang WL, Kuo WC, Chen GH, Lin TP (2015) Increased glutathione contributes to stress tolerance and global translational changes in Arabidopsis. Plant J 83(5): 926–939

D'alessandro S, Ksas B, Havaux M (2018) Decoding β-cyclocitral-mediated retrograde signaling reveals the role of a detoxification response in plant tolerance to photooxidative stress. Plant Cell 30(10):2495–2511

Daloso DM, Müller K, Obata T, Florian A, Tohge T, Bottcher A et al (2015) Thioredoxin, a master regulator of the tricarboxylic acid cycle in plant mitochondria. Proc Natl Acad Sci 112(11): E1392–E1400

Das K, Roychoudhury A (2014) Reactive oxygen species (ROS) and response of antioxidants as ROS-scavengers during environmental stress in plants. Front Environ Sci 2:53

Daudi A, Cheng Z, O'Brien JA, Mammarella N, Khan S, Ausubel FM, Bolwell GP (2012) The apoplastic oxidative burst peroxidase in Arabidopsis is a major component of pattern-triggered immunity. Plant Cell 24(1):275–287

Davies KJ (1987) Protein damage and degradation by oxygen radicals. I. General aspects. J Biol Chem 262(20):9895–9901

Davletova S, Schlauch K, Coutu J, Mittler R (2005) The zinc-finger protein Zat12 plays a central role in reactive oxygen and abiotic stress signaling in Arabidopsis. Plant Physiol 139(2): 847–856

De Clercq I, Van de Velde J, Luo X, Liu L, Storme V, Van Bel M et al (2021) Integrative inference of transcriptional networks in Arabidopsis yields novel ROS signalling regulators. Nat Plants 7(4):500–513

De Smet B, Willems P, Fernandez-Fernandez AD, Alseekh S, Fernie AR, Messens J, Van Breusegem F (2019) In vivo detection of protein cysteine sulfenylation in plastids. Plant J 97(4):765–778

Després C, Chubak C, Rochon A, Clark R, Bethune T, Desveaux D, Fobert PR (2003) The Arabidopsis NPR1 disease resistance protein is a novel cofactor that confers redox regulation of DNA binding activity to the basic domain/leucine zipper transcription factor TGA1. Plant Cell 15(9):2181–2191

Devireddy AR, Arbogast J, Mittler R (2020a) Coordinated and rapid whole-plant systemic stomatal responses. New Phytol 225(1):21

Devireddy AR, Liscum E, Mittler R (2020b) Phytochrome B is required for systemic stomatal responses and reactive oxygen species signaling during light stress. Plant Physiol 184(3): 1563–1572

Dickinson PJ, Kumar M, Martinho C, Yoo SJ, Lan H, Artavanis G et al (2018) Chloroplast signaling gates thermotolerance in Arabidopsis. Cell Rep 22(7):1657–1665

Dixon DP, Skipsey M, Edwards R (2010) Roles for glutathione transferases in plant secondary metabolism. Phytochemistry 71(4):338–350

Dizdaroglu M (1993) Chemistry of free radical damage to DNA and nucleoproteins. In: DNA and free radicals. Ellis Horwood, New York, pp 19–39

Estavillo GM, Crisp PA, Pornsiriwong W, Wirtz M, Collinge D, Carrie C et al (2011) Evidence for a SAL1-PAP chloroplast retrograde pathway that functions in drought and high light signaling in Arabidopsis. Plant Cell 23(11):3992–4012

Evans MD, Dizdaroglu M, Cooke MS (2004) Oxidative DNA damage and disease: induction, repair and significance. Mutat Res/Rev Mutat Res 567(1):1–61

Exposito-Rodriguez M, Laissue PP, Yvon-Durocher G, Smirnoff N, Mullineaux PM (2017) Photosynthesis-dependent H2O2 transfer from chloroplasts to nuclei provides a high-light signalling mechanism. Nat Commun 8(1):1–11

Fichman Y, Zandalinas SI, Sengupta S, Burks D, Myers RJ Jr, Azad RK, Mittler R (2020) MYB30 orchestrates systemic reactive oxygen signaling and plant acclimation. Plant Physiol 184(2): 666–675

Fichman Y, Mittler R (2020) Rapid systemic signaling during abiotic and biotic stresses: is the ROS wave master of all trades? Plant J 102(5):887–896

Fink SP, Reddy GR, Marnett LJ (1997) Mutagenicity in Escherichia coli of the major DNA adduct derived from the endogenous mutagen malondialdehyde. Proc Natl Acad Sci 94(16):8652–8657

Finka A, Cuendet AFH, Maathuis FJ, Saidi Y, Goloubinoff P (2012) Plasma membrane cyclic nucleotide gated calcium channels control land plant thermal sensing and acquired thermotolerance. Plant Cell 24(8):3333–3348

Foyer CH (2018) Reactive oxygen species, oxidative signaling and the regulation of photosynthesis. Environ Exp Bot 154:134–142

Foyer CH, Halliwell B (1976) The presence of glutathione and glutathione reductase in chloroplasts: a proposed role in ascorbic acid metabolism. Planta 133(1):21–25

Fu ZQ, Dong X (2013) Systemic acquired resistance: turning local infection into global defense. Annu Rev Plant Biol 64:839–863

Galvez-Valdivieso G, Fryer MJ, Lawson T, Slattery K, Truman W, Smirnoff N et al (2009) The high light response in Arabidopsis involves ABA signaling between vascular and bundle sheath cells. Plant Cell 21(7):2143–2162

Gardner PR, Fridovich I (1991) Superoxide sensitivity of the Escherichia coli 6-phosphogluconate dehydratase. J Biol Chem 266(3):1478–1483

Giesguth M, Sahm A, Simon S, Dietz KJ (2015) Redox-dependent translocation of the heat shock transcription factor AtHSFA8 from the cytosol to the nucleus in Arabidopsis thaliana. FEBS Lett 589(6):718–725
Glauser G, Dubugnon L, Mousavi SA, Rudaz S, Wolfender JL, Farmer EE (2009) Velocity estimates for signal propagation leading to systemic jasmonic acid accumulation in wounded Arabidopsis. J Biol Chem 284(50):34506–34513
Gong H, Jiao Y, Hu WW, Pua EC (2005) Expression of glutathione-S-transferase and its role in plant growth and development in vivo and shoot morphogenesis in vitro. Plant Mol Biol 57(1): 53–66
Goraya GK, Asthir B (2016) Magnificant role of intracellular reactive oxygen species production and its scavenging encompasses downstream processes. J Plant Biol 59(3):215–222
Griesser E, Vemula V, Raulien N, Wagner U, Reeg S, Grune T, Fedorova M (2017) Cross-talk between lipid and protein carbonylation in a dynamic cardiomyocyte model of mild nitroxidative stress. Redox Biol 11:438–455
Griffin JHC, Prado K, Sutton P, Toledo-Ortiz G (2020) Coordinating light responses between the nucleus and the chloroplast, a role for plant cryptochromes and phytochromes. Physiol Plant 169(4):515–528
Grune T, Reinheckel T, Davies KJ (1997) Degradation of oxidized proteins in mammalian cells. FASEB J 11(7):526–534
Halliwell B, Aruoma OI (1991) DNA damage by oxygen-derived species its mechanism and measurement in mammalian systems. FEBS Lett 281(1–2):9–19
Halliwell B, Gutteridge JM (2015) Free radicals in biology and medicine. Oxford University Press, Oxford
Halliwell B, Gutteridge JMC (1989) Free radicals in biology and medicine. Oxford University Press, Oxford. Google Scholar, 1–81
Han C, Liu Q, Yang Y (2009) Short-term effects of experimental warming and enhanced ultraviolet-B radiation on photosynthesis and antioxidant defense of Picea asperata seedlings. Plant Growth Regul 58(2):153–162
Hasanuzzaman M, Alam M, Rahman A, Hasanuzzaman M, Nahar K, Fujita M (2014) Exogenous proline and glycine betaine mediated upregulation of antioxidant defense and glyoxalase systems provides better protection against salt-induced oxidative stress in two rice (Oryza sativa L.) varieties. BioMed Res Int 2014:757219
He H, Denecker J, Van Der Kelen K, Willems P, Pottie R, Phua SY et al (2021) The Arabidopsis mediator complex subunit 8 regulates oxidative stress responses. Plant Cell 33(6):2032–2057
Hipsch M, Lampl N, Zelinger E, Barda O, Waiger D, Rosenwasser S (2021) Sensing stress responses in potato with whole-plant redox imaging. Plant Physiol 187(2):618–631
Hossain M, Asada K (1987) Ascorbate-regenerating enzymes in chloroplasts. Indian J Biochem Biophys 24(5):52–56
Hu WH, Song XS, Shi K, Xia XJ, Zhou YH, Yu JQ (2008) Changes in electron transport, superoxide dismutase and ascorbate peroxidase isoenzymes in chloroplasts and mitochondria of cucumber leaves as influenced by chilling. Photosynthetica 46(4):581–588
Hu Y, Wu Q, Peng Z, Sprague SA, Wang W, Park J et al (2017) Silencing of OsGRXS17 in rice improves drought stress tolerance by modulating ROS accumulation and stomatal closure. Sci Rep 7(1):1–14
Huang J, Willems P, Wei B, Tian C, Ferreira RB, Bodra N et al (2019) Mining for protein S-sulfenylation in Arabidopsis uncovers redox-sensitive sites. Proc Natl Acad Sci 116(42): 21256–21261
Iglesias-Baena I, Barranco-Medina S, Sevilla F, Lázaro JJ (2011) The dual-targeted plant sulfiredoxin retroreduces the sulfinic form of atypical mitochondrial peroxiredoxin. Plant Physiol 155(2):944–955
Imlay JA, Linn S (1988) DNA damage and oxygen radical toxicity. Science 240(4857):1302–1309
Inupakutika MA, Sengupta S, Devireddy AR, Azad RK, Mittler R (2016) The evolution of reactive oxygen species metabolism. J Exp Bot 67(21):5933

Iyer NJ, Jia X, Sunkar R, Tang G, Mahalingam R (2012) microRNAs responsive to ozone-induced oxidative stress in Arabidopsis thaliana. Plant Signal Behav 7(4):484–491

Jabłońska J, Tawfik DS (2021) The evolution of oxygen-utilizing enzymes suggests early biosphere oxygenation. Nat Ecol Evol 5(4):442–448

Jacques S, Ghesquière B, De Bock PJ, Demol H, Wahni K, Willems P et al (2015) Protein methionine sulfoxide dynamics in Arabidopsis thaliana under oxidative stress [S]. Mol Cell Proteomics 14(5):1217–1229

Jagodzik P, Tajdel-Zielinska M, Ciesla A, Marczak M, Ludwikow A (2018) Mitogen-activated protein kinase cascades in plant hormone signaling. Front Plant Sci 9:1387

Jan B, Bhat TA, Sheikh TA, Wani OA, Bhat MA, Nazir A et al (2020) Agronomic bio-fortification of rice and maize with iron and zinc: a review. Int Res J Pure Appl Chem 21:28–37

Jayakannan M, Bose J, Babourina O, Shabala S, Massart A, Poschenrieder C, Rengel Z (2015) The NPR1-dependent salicylic acid signalling pathway is pivotal for enhanced salt and oxidative stress tolerance in Arabidopsis. J Exp Bot 66(7):1865–1875

Jimenez A, Hernandez JA, del Río LA, Sevilla F (1997) Evidence for the presence of the ascorbate-glutathione cycle in mitochondria and peroxisomes of pea leaves. Plant Physiol 114(1):275–284

Jung JH, Domijan M, Klose C, Biswas S, Ezer D, Gao M et al (2016) Phytochromes function as thermosensors in Arabidopsis. Science 354(6314):886–889

Kamal-Eldin A, Appelqvist LÅ (1996) The chemistry and antioxidant properties of tocopherols and tocotrienols. Lipids 31(7):671–701

Kámán-Tóth E, Dankó T, Gullner G, Bozsó Z, Palkovics L, Pogány M (2019) Contribution of cell wall peroxidase-and NADPH oxidase-derived reactive oxygen species to Alternaria brassicicola-induced oxidative burst in Arabidopsis. Mol Plant Pathol 20(4):485–499

Karpinska B, Karlsson M, Schinkel H, Streller S, Suss KH, Melzer M, Wingsle G (2001) A novel superoxide dismutase with a high isoelectric point in higher plants. Expression, regulation, and protein localization. Plant Physiol 126(4):1668–1677

Karpinski S, Reynolds H, Karpinska B, Wingsle G, Creissen G, Mullineaux P (1999) Systemic signaling and acclimation in response to excess excitation energy in Arabidopsis. Science 284(5414):654–657

Kerchev P, Waszczak C, Lewandowska A, Willems P, Shapiguzov A, Li Z et al (2016) Lack of GLYCOLATE OXIDASE1, but not GLYCOLATE OXIDASE2, attenuates the photorespiratory phenotype of CATALASE2-deficient Arabidopsis. Plant Physiol 171(3): 1704–1719

Koffler BE, Bloem E, Zellnig G, Zechmann B (2013) High resolution imaging of subcellular glutathione concentrations by quantitative immunoelectron microscopy in different leaf areas of Arabidopsis. Micron 45:119–128

König K, Vaseghi MJ, Dreyer A, Dietz KJ (2018) The significance of glutathione and ascorbate in modulating the retrograde high light response in Arabidopsis thaliana leaves. Physiol Plant 162(3):262–273

Lee ES, Park JH, Wi SD, Kang CH, Chi YH, Chae HB et al (2021) Redox-dependent structural switch and CBF activation confer freezing tolerance in plants. Nat Plants 7(7):914–922

Leferink NG, van Duijn E, Barendregt A, Heck AJ, van Berkel WJ (2009) Galactonolactone dehydrogenase requires a redox-sensitive thiol for optimal production of vitamin C. Plant Physiol 150(2):596–605

Legris M, Klose C, Burgie ES, Rojas CCR, Neme M, Hiltbrunner A et al (2016) Phytochrome B integrates light and temperature signals in Arabidopsis. Science 354(6314):897–900

Lindermayr C, Sell S, Müller B, Leister D, Durner J (2010) Redox regulation of the NPR1-TGA1 system of Arabidopsis thaliana by nitric oxide. Plant Cell 22:2894–2907

Liu D, Li YY, Zhou ZC, Xiang X, Liu X, Wang J et al (2021b) Tobacco transcription factor bHLH123 improves salt tolerance by activating NADPH oxidase NtRbohE expression. Plant Physiol 186(3):1706–1720

Liu R, Chen T, Yin X, Xiang G, Peng J, Fu Q et al (2021a) A Plasmopara viticola RXLR effector targets a chloroplast protein PsbP to inhibit ROS production in grapevine. Plant J 106(6): 1557–1570

Liu T, Van Staden J, Cress WA (2000) Salinity induced nuclear and DNA degradation in meristematic cells of soybean (Glycine max (L.)) roots. Plant Growth Regul 30(1):49–54

Lu T, Meng Z, Zhang G, Qi M, Sun Z, Liu Y, Li T (2017) Sub-high temperature and high light intensity induced irreversible inhibition on photosynthesis system of tomato plant (Solanum lycopersicum L.). Front Plant Sci 8:365

Mabuchi K et al (2018) MYB30 links ROS signaling, root cell elongation, and plant immune responses. Proc Natl Acad Sci U S A 115:4710–4719

Madian AG, Regnier FE (2010) Proteomic identification of carbonylated proteins and their oxidation sites. J Proteome Res 9(8):3766–3780

Maheshwari R, Dubey RS (2009) Nickel-induced oxidative stress and the role of antioxidant defence in rice seedlings. Plant Growth Regul 59(1):37–49

Martins L, Knuesting J, Bariat L, Dard A, Freibert SA, Marchand CH et al (2020) Redox modification of the iron-sulfur glutaredoxin GRXS17 activates holdase activity and protects plants from heat stress. Plant Physiol 184(2):676–692

McLachlan DH (2020) Systemic signalling, and the synchronization of stomatal response. New Phytol 225(1):5–6

Meng X, Li L, De Clercq I, Narsai R, Xu Y, Hartmann A et al (2019) ANAC017 coordinates organellar functions and stress responses by reprogramming retrograde signaling. Plant Physiol 180(1):634–653

Meriga B, Reddy BK, Rao KR, Reddy LA, Kishor PK (2004) Aluminium-induced production of oxygen radicals, lipid peroxidation and DNA damage in seedlings of rice (Oryza sativa). J Plant Physiol 161(1):63–68

Mhamdi A, Queval G, Chaouch S, Vanderauwera S, Van Breusegem F, Noctor G (2010) Catalase function in plants: a focus on Arabidopsis mutants as stress-mimic models. J Exp Bot 61(15): 4197–4220

Mishra S, Jha AB, Dubey RS (2011) Arsenite treatment induces oxidative stress, upregulates antioxidant system, and causes phytochelatin synthesis in rice seedlings. Protoplasma 248(3): 565–577

Mittler R (2017) ROS are good. Trends Plant Sci 22(1):11–19

Møller IM, Kristensen BK (2004) Protein oxidation in plant mitochondria as a stress indicator. Photochem Photobiol Sci 3(8):730–735

Møller IM, Rogowska-Wrzesinska A, Rao RSP (2011) Protein carbonylation and metal-catalyzed protein oxidation in a cellular perspective. J Proteome 74(11):2228–2242

Munné-Bosch S, Alegre L (2004) Die and let live: leaf senescence contributes to plant survival under drought stress. Funct Plant Biol 31(3):203–216

Nguyen TP, Cueff G, Hegedus DD, Rajjou L, Bentsink L (2015) A role for seed storage proteins in Arabidopsis seed longevity. J Exp Bot 66(20):6399–6413

Niazi AK, Bariat L, Riondet C, Carapito C, Mhamdi A, Noctor G, Reichheld JP (2019) Cytosolic isocitrate dehydrogenase from Arabidopsis thaliana is regulated by glutathionylation. Antioxidants 8(1):16

Nie WF, Wang MM, Xia XJ, Zhou YH, Shi K, Chen Z, Yu JQ (2013) Silencing of tomato RBOH1 and MPK2 abolishes brassinosteroid-induced H2O2 generation and stress tolerance. Plant Cell Environ 36(4):789–803

Nietzel T, Elsässer M, Ruberti C, Steinbeck J, Ugalde JM, Fuchs P et al (2019) The fluorescent protein sensor roGFP 2-Orp1 monitors in vivo H2O2 and thiol redox integration and elucidates intracellular H2O2 dynamics during elicitor-induced oxidative burst in Arabidopsis. New Phytol 221(3):1649–1664

Nietzel T, Mostertz J, Ruberti C, Née G, Fuchs P, Wagner S et al (2020) Redox-mediated kick-start of mitochondrial energy metabolism drives resource-efficient seed germination. Proc Natl Acad Sci 117(1):741–751

Nisar N, Li L, Lu S, Khin NC, Pogson BJ (2015) Carotenoid metabolism in plants. Mol Plant 8(1): 68–82
Nishimura K, Sano M, Ohtaka M, Furuta B, Umemura Y, Nakajima Y et al (2011) Development of defective and persistent Sendai virus vector: a unique gene delivery/expression system ideal for cell reprogramming. J Biol Chem 286(6):4760–4771
Noctor G, Reichheld JP, Foyer CH (2018, August) ROS-related redox regulation and signaling in plants. In: Seminars in cell & developmental biology, vol 80. Academic Press, London, pp 3–12
Nomura H, Komori T, Uemura S, Kanda Y, Shimotani K, Nakai K et al (2012) Chloroplast-mediated activation of plant immune signalling in Arabidopsis. Nat Commun 3(1):1–11
Ohama N, Sato H, Shinozaki K, Yamaguchi-Shinozaki K (2017) Transcriptional regulatory network of plant heat stress response. Trends Plant Sci 22(1):53–65
Oleinick NL, Chiu SM, Ramakrishnan N, Xue LY (1987) The formation, identification, and significance of DNA-protein cross-links in mammalian cells. Br J Cancer Suppl 8:135
Pang CH, Wang BS (2008) Oxidative stress and salt tolerance in plants. In: Progress in botany. Springer, Berlin, pp 231–245
Pérez-Salamó I, Papdi C, Rigó G, Zsigmond L, Vilela B, Lumbreras V et al (2014) The heat shock factor A4A confers salt tolerance and is regulated by oxidative stress and the mitogen-activated protein kinases MPK3 and MPK6. Plant Physiol 165(1):319–334
Petrussa E, Braidot E, Zancani M, Peresson C, Bertolini A, Patui S, Vianello A (2013) Plant flavonoids—biosynthesis, transport and involvement in stress responses. Int J Mol Sci 14(7): 14950–14973
Puerto-Galán L, Pérez-Ruiz JM, Guinea M, Cejudo FJ (2015) The contribution of NADPH thioredoxin reductase C (NTRC) and sulfiredoxin to 2-Cys peroxiredoxin overoxidation in Arabidopsis thaliana chloroplasts. J Exp Bot 66(10):2957–2966
Requena JR, Chao CC, Levine RL, Stadtman ER (2001) Glutamic and aminoadipic semialdehydes are the main carbonyl products of metal-catalyzed oxidation of proteins. Proc Natl Acad Sci 98(1):69–74
Richter C (1992) Reactive oxygen and DNA damage in mitochondria. Mutat Rese/DNAging 275(3–6):249–255
Rodrigues O, Reshetnyak G, Grondin A, Saijo Y, Leonhardt N, Maurel C, Verdoucq L (2017) Aquaporins facilitate hydrogen peroxide entry into guard cells to mediate ABA-and pathogen-triggered stomatal closure. Proc Natl Acad Sci 114(34):9200–9205
Romero-Puertas MC, Palma JM, Gómez M, Del Rio LA, Sandalio LM (2002) Cadmium causes the oxidative modification of proteins in pea plants. Plant Cell Environ 25(5):677–686
Schmidt R, Mieulet D, Hubberten HM, Obata T, Hoefgen R, Fernie AR et al (2013) SALT-RESPONSIVE ERF1 regulates reactive oxygen species–dependent signaling during the initial response to salt stress in rice. Plant Cell 25(6):2115–2131
Schwarzländer M, Fricker MD, Sweetlove LJ (2009) Monitoring the in vivo redox state of plant mitochondria: effect of respiratory inhibitors, abiotic stress and assessment of recovery from oxidative challenge. Biochim Biophys Acta (BBA) Bioenerg 1787(5):468–475
Shaikhali J, Heiber I, Seidel T, Ströher E, Hiltscher H, Birkmann S et al (2008) The redox-sensitive transcription factor Rap2. 4a controls nuclear expression of 2-Cys peroxiredoxin A and other chloroplast antioxidant enzymes. BMC Plant Biol 8(1):1–14
Shapiguzov A, Vainonen JP, Hunter K, Tossavainen H, Tiwari A, Järvi S et al (2019) Arabidopsis RCD1 coordinates chloroplast and mitochondrial functions through interaction with ANAC transcription factors. elife 8:e43284
Sharma P, Dubey RS (2005) Drought induces oxidative stress and enhances the activities of antioxidant enzymes in growing rice seedlings. Plant Growth Regul 46(3):209–221
Sharma P, Jha AB, Dubey RS (2019) Oxidative stress and antioxidative defense system in plants growing under abiotic stresses. In: Handbook of plant and crop stress, 4th edn. CRC Press, Boca Raton, pp 93–136
Sharma P, Jha AB, Dubey RS, Pessarakli M (2012) Reactive oxygen species, oxidative damage, and antioxidative defense mechanism in plants under stressful conditions. J Bot 2012:217037

Shi Y, Chang YL, Wu HT, Shalmani A, Liu WT, Li WQ et al (2020) OsRbohB-mediated ROS production plays a crucial role in drought stress tolerance of rice. Plant Cell Rep 39(12): 1767–1784

Sies H, Jones DP (2020) Reactive oxygen species (ROS) as pleiotropic physiological signalling agents. Nat Rev Mol Cell Biol 21(7):363–383

Smirnoff H (1995) Antioxidant systems and plant response to the environment. In: Environment and plant metabolism. Bios Scientific Publishers, Oxford

Smirnoff N, Arnaud D (2019) Hydrogen peroxide metabolism and functions in plants. New Phytol 221(3):1197–1214

Sofo A, Scopa A, Nuzzaci M, Vitti A (2015) Ascorbate peroxidase and catalase activities and their genetic regulation in plants subjected to drought and salinity stresses. Int J Mol Sci 16(6): 13561–13578

Srivastava R, Deng Y, Shah S, Rao AG, Howell SH (2013) BINDING PROTEIN is a master regulator of the endoplasmic reticulum stress sensor/transducer bZIP28 in Arabidopsis. Plant Cell 25(4):1416–1429

Stadtman ER (1986) Oxidation of proteins by mixed-function oxidation systems: implication in protein turnover, ageing and neutrophil function. Trends Biochem Sci 11(1):11–12

Stohs SJ, Bagchi D (1995) Oxidative mechanisms in the toxicity of metal ions. Free Radic Biol Med 18(2):321–336

Suzuki N, Miller G, Salazar C, Mondal HA, Shulaev E, Cortes DF et al (2013a) Temporal-spatial interaction between reactive oxygen species and abscisic acid regulates rapid systemic acclimation in plants. Plant Cell 25(9):3553–3569

Suzuki N, Miller G, Sejima H, Harper J, Mittler R (2013b) Enhanced seed production under prolonged heat stress conditions in Arabidopsis thaliana plants deficient in cytosolic ascorbate peroxidase 2. J Exp Bot 64(1):253–263

Tada Y, Spoel SH, Pajerowska-Mukhtar K, Mou Z, Song J, Wang C et al (2008) Plant immunity requires conformational charges of NPR1 via S-nitrosylation and thioredoxins. Science 321(5891):952–956

Tanou G, Molassiotis A, Diamantidis G (2009) Induction of reactive oxygen species and necrotic death-like destruction in strawberry leaves by salinity. Environ Exp Bot 65(2–3):270–281

Tarrago L, Laugier E, Zaffagnini M, Marchand C, Le Maréchal P, Rouhier N et al (2009) Regeneration mechanisms of Arabidopsis thaliana methionine sulfoxide reductases B by glutaredoxins and thioredoxins. J Biol Chem 284(28):18963–18971

Torres MA, Dangl JL, Jones JD (2002) Arabidopsis gp91phox homologues AtrbohD and AtrbohF are required for accumulation of reactive oxygen intermediates in the plant defense response. Proc Natl Acad Sci 99(1):517–522

Tran D, El-Maarouf-Bouteau H, Rossi M, Biligui B, Briand J, Kawano T et al (2013) Post-transcriptional regulation of GORK channels by superoxide anion contributes to increases in outward-rectifying K+ currents. New Phytol 198(4):1039–1048

Trnka D, Engelke AD, Gellert M, Moseler A, Hossain MF, Lindenberg TT et al (2020) Molecular basis for the distinct functions of redox-active and FeS-transfering glutaredoxins. Nat Commun 11(1):1–12

Tsuboi H, Kouda K, Takeuchi H, Takigawa M, Masamoto Y, Takeuchi M, Ochi H (1998) 8-Hydroxydeoxyguanosine in urine as an index of oxidative damage to DNA in the evaluation of atopic dermatitis. Br J Dermatol 138(6):1033–1035

Ugalde JM, Fuchs P, Nietzel T, Cutolo EA, Homagk M, Vothknecht UC et al (2021) Chloroplast-derived photo-oxidative stress causes changes in H2O2 and E GSH in other subcellular compartments. Plant Physiol 186(1):125–141

Wang H, Schoebel S, Schmitz F, Dong H, Hedfalk K (2020a) Characterization of aquaporin-driven hydrogen peroxide transport. Biochim Biophys Acta (BBA) Biomembr 1862(2):183065

Wang L, Leister D, Guan L, Zheng Y, Schneider K, Lehmann M et al (2020b) The Arabidopsis SAFEGUARD1 suppresses singlet oxygen-induced stress responses by protecting grana margins. Proc Natl Acad Sci 117(12):6918–6927

Waszczak C, Carmody M, Kangasjärvi J (2018) Reactive oxygen species in plant signaling. Annu Rev Plant Biol 69:209–236

Winkel-Shirley B (2002) Biosynthesis of flavonoids and effects of stress. Curr Opin Plant Biol 5(3): 218–223

Yamaguchi O, Taneike M, Otsu K (2012) Cooperation between proteolytic systems in cardiomyocyte recycling. Cardiovasc Res 96(1):46–52

Yamasaki H, Ogura MP, Kingjoe KA, Cohen MF (2019) d-cysteine-induced rapid root abscission in the water fern Azolla Pinnata: implications for the linkage between d-amino acid and reactive sulfur species (RSS) in plant environmental responses. Antioxidants 8(9):411

Yamauchi Y, Furutera A, Seki K, Toyoda Y, Tanaka K, Sugimoto Y (2008) Malondialdehyde generated from peroxidized linolenic acid causes protein modification in heat-stressed plants. Plant Physiol Biochem 46(8–9):786–793

Yang C, Li W, Cao J, Meng F, Yu Y, Huang J et al (2017) Activation of ethylene signaling pathways enhances disease resistance by regulating ROS and phytoalexin production in rice. Plant J 89(2):338–353

Yin G, Xin X, Fu S, An M, Wu S, Chen X et al (2017) Proteomic and carbonylation profile analysis at the critical node of seed ageing in Oryza sativa. Sci Rep 7(1):1–12

Yin Y, Qin K, Song X, Zhang Q, Zhou Y, Xia X, Yu J (2018) BZR1 transcription factor regulates heat stress tolerance through FERONIA receptor-like kinase-mediated reactive oxygen species signaling in tomato. Plant Cell Physiol 59(11):2239–2254

Yuan F, Yang H, Xue Y, Kong D, Ye R, Li C et al (2014) OSCA1 mediates osmotic-stress-evoked Ca2+ increases vital for osmosensing in Arabidopsis. Nature 514(7522):367–371

Yuan LB, Dai YS, Xie LJ, Yu LJ, Zhou Y, Lai YX et al (2017) Jasmonate regulates plant responses to postsubmergence reoxygenation through transcriptional activation of antioxidant synthesis. Plant Physiol 173(3):1864–1880

Zaffagnini M, Fermani S, Calvaresi M, Orrù R, Iommarini L, Sparla F et al (2016) Tuning cysteine reactivity and sulfenic acid stability by protein microenvironment in glyceraldehyde-3-phosphate dehydrogenases of Arabidopsis thaliana. Antioxid Redox Signal 24(9):502–517

Zandalinas SI, Fichman Y, Devireddy AR, Sengupta S, Azad RK, Mittler R (2020) Systemic signaling during abiotic stress combination in plants. Proc Natl Acad Sci 117(24):13810–13820

Zandalinas SI, Sengupta S, Fritschi FB, Azad RK, Nechushtai R, Mittler R (2021b) The impact of multifactorial stress combination on plant growth and survival. New Phytol 230(3):1034–1048

Zandalinas SI, Fritschi FB, Mittler R (2021a) Global warming, climate change, and environmental pollution: recipe for a multifactorial stress combination disaster. Trends Plant Sci 26:588–599

Zandalinas SI, Sengupta S, Burks D, Azad RK, Mittler R (2019) Identification and characterization of a core set of ROS wave-associated transcripts involved in the systemic acquired acclimation response of Arabidopsis to excess light. Plant J 98:126–141

Zhang Y (2012) Ascorbic acid in plants: biosynthesis, regulation and enhancement. Springer Science & Business Media, New York

Zheng J, Bizzozero OA (2010) Traditional reactive carbonyl scavengers do not prevent the carbonylation of brain proteins induced by acute glutathione depletion. Free Radic Res 44(3): 258–266

Zou JJ, Li XD, Ratnasekera D, Wang C, Liu WX, Song LF et al (2015) Arabidopsis CALCIUM-DEPENDENT PROTEIN KINASE8 and CATALASE3 function in abscisic acid-mediated signaling and H2O2 homeostasis in stomatal guard cells under drought stress. Plant Cell 27(5):1445–1460

Physiological Impact of Reactive Oxygen Species on Leaf

6

Shareen, Ahmad Faraz, and Mohammad Faizan

Abstract

In this chapter, we summarized our current understanding of the various ROS types, their production mechanisms, and their physiological effects on plants exposed to abiotic stress. Since ROS are highly reactive, they have toxic qualities. The most frequent ROS generated by aerobic metabolism is typically hydrogen peroxide (H_2O_2), singlet oxygen ($^1O_2$), superoxide anion ($O2^{\bullet-}$), and hydroxyl radicals ($OH^{\bullet}$). Additionally, ROS are crucial signalling molecules in plant science that control stress responses, plant growth, and development in response to alterations in environmental conditions like high light intensity, high temperature, drought, salinity, pathogen attack, and so forth. Abiotic stress causes high levels of ROS in cells, which demarcates crop productivity and yield. Similarly, it has been reported that ROS molecules are scavenged by several antioxidative defense mechanisms, thereby protecting plants from abiotic stresses. Antioxidants involved in ROS detoxification include superoxide dismutase (SOD), peroxidases (POX), ascorbic acid (AA), catalase (CAT), and others. Despite significant research to better understand ROS's role and impact on plants, ROS signalling is still not fully understood.

Shareen
Environmental Biotechnology Lab, College of Biology and Environment, Nanjing Forestry University, Nanjing, China

A. Faraz
School of Life Sciences, Glocal University, Saharanpur, UP, India

M. Faizan (✉)
Botany Section, School of Sciences, Maulana Azad National Urdu University, Hyderabad, India
e-mail: mohammadfaizan@manuu.edu.in

M. Faizan et al. (eds.), *Reactive Oxygen Species*,
https://doi.org/10.1007/978-981-19-9794-5_6

Keywords

Hydroxyl radicals · Physiological effects · Pathogen attack · ROS detoxification

6.1 Introduction

Reactive Oxygen Species (ROS) are constantly generated as a derivative of several constructive and destructive metabolisms such as respiration, photorespiration, and photosynthesis that takes place in different plant cells (Foyer et al. 1994a, b). In plants, ROS production occurs on the sites of mitochondria, peroxisomes, chloroplasts, endoplasmic reticulum, cell wall, and plasma membrane where vigorous electron flow occurs (Das and Roychoudhury 2014). Various forms of ROS generally include free radicals such as superoxide anion ($O_2^{\bullet -}$) and hydroxyl radical ($OH^{\bullet}$) as well as non-radical molecules such as hydrogen peroxide (H_2O_2), and singlet oxygen ($^1O_2$). In plant physiology, ROS are defined as a "double-edged sword or a dagger" because of its property of reacting with a wide range of biomolecules causing an irreversible damage which lead to necrosis and death of plant cells (Rebeiz et al. 1988; Girotti 2001). On the other hand, ROS also acts as signalling molecules for maintaining several biological processes such as abiotic stress response, programmed cell death, signal transduction, and pathogen resistance (Mittler 2002). A certain amount of ROS is essential for plants to regulate their natural biological processes. The higher concentrations of ROS can lead to the disruptions of the intracellular redox balance, thus causing oxidative damage to plant cells (Jithesh et al. 2006; Zhang et al. 2001). Environmental stresses such as low or high temperature, salinity, drought, chilling, UV-B radiation, metal toxicity, nutrient deficiency, and pathogen infection trigger the ROS production in plants as a result of disturbance in cellular homeostasis (Mittler 2002; Hu et al. 2008; Han et al. 2009; Tanou et al. 2009a, b). Increased ROS production under stress conditions can cause a risk to plant cells by inducing oxidation of proteins, peroxidation of lipids, enzyme inhibition, damage to nucleic acids, and activation of programmed cell death pathways, as a result causing death of plant cells, this condition is referred to as oxidative stress (Shah et al. 2001; Meriga et al. 2004; Maheshwari and Dubey 2009; Srivastava and Dubey 2011). Consequently, excess amount of ROS accumulating in plants under environmental stresses is believed to be an essential factor affecting normal plant growth and crop yield (Apel and Hirt 2004).

6.2 Types of ROS

Reactive oxygen species (ROS) have become an essential substance of our daily life with the development of molecular oxygen (O_2) in the environment by O_2-yielding photosynthetic microorganisms. The reactive ROS is produced in response to the reduction or stimulation of oxygen which includes singlet oxygen ($^1O_2$), hydrogen peroxide (H_2O_2), superoxide anion ($O_2^{\bullet -}$), and hydroxyl radical ($OH^{\bullet}$). In the oxidizing environment, plants continuously generate ROS in membrane-bound cell organelles (mitochondria, peroxisomes, chloroplasts) and other main sites of the cell in response to their metabolic pathways such as photosynthesis, photorespiration, and respiration (Tripathy and Oelmüller 2012).

6.2.1 Singlet Oxygen ($^1O_2$)

Singlet oxygen ($^1O_2$) or commonly known as a triplet oxygen, is considered to be the first excited electronic state of molecular oxygen which is produced in plant leaves in sunlight by the reaction chlorophylls triplet state with O_2 in the chloroplasts antenna system.

$$\mathrm{Chl} \rightarrow {}^3\mathrm{Chl}$$

$$^3\mathrm{Chl} + {}^3\mathrm{O}_2 \rightarrow \mathrm{Chl} + {}^1\mathrm{O}_2$$

$^1O_2$ has a short lifespan, which suggests a small diffusion route in plant cells. Depending on environmental conditions, singlet oxygen can persist for one hour at room temperature, due to its exceptional properties. Singlet and triplet oxygen has different chemical properties due to the differences in their electron shells (Edreva 2005). The major source for the generation of singlet oxygen is the chlorophyll pigments which are associated with the electron transport process. During photosynthesis, insufficient energy disintegration can cause the chlorophyll triplet state formation which can transfer its excitation energy onto the ground state O_2 to make $^1O_2$ (Holt et al. 2005). On the contrary, $^1O_2$ can also be produced as a derivative of lipoxygenase activity. Like other ROS, $^1O_2$ is extremely catastrophic in nature as it can react with most of the biomolecules and, hence, can be fatal to plant cells and can activate two types of responses involving signalling of acclimation processes or programmed cell death (Cadenas 1989). In electron transport chain of photosynthesis, the transfer of energy or electrons takes place by the excited singlet state of chlorophyll. Moreover, $^1O_2$ has the ability to diffuse from the chloroplasts into the apoplast and the cytoplasm. Exposures to environmental stresses increase levels of $^1O_2$ in root cells of the plants (Dmitrieva et al. 2020). Recent studies of photo-oxidative damage to plant tissues, in association with the lipid peroxidation processes reported that the main ROS involved in the leaves destruction is $^1O_2$ (Triantaphylides et al. 2008).

6.2.2 Superoxide Radical ($O_2^{\bullet-}$)

Superoxide radical is an oxygen compound produced and the primary form of ROS produced by the mitochondria, exist commonly in nature. The formation of superoxide radical takes place in the photosystem I (PSI) of the thylakoids membrane, produced by the non-cyclic electron transport chain. The factors that determine the redox state of the ferredoxin pool include the controlled activation of Calvin cycle and regulated electron flow rate. It is important for the electron carriers and the ferredoxin present on the reduced side of PSI, possessing negative electrochemical potentials necessary for donating electrons to oxygen atoms, thus, leading to the superoxide radical formation.

$$O_2^{\bullet-} + Fe^{3+} \rightarrow {}^1O_2 + Fe^{2+}$$

$$O_2^{\bullet-} + 2H^+ \rightarrow O_2 + H_2O_2\ Fe^{3+}$$

$$Fe^{2+} + H_2O_2 + Fe^{3+} \rightarrow OH^- + OH^\bullet$$

$O_2^{\bullet-}$ have a short half-life of 2–4 μs and is highly reactive in nature. It changes to hydroxy radicals which results in membrane dissolution and immense lipid peroxidation (Halliwell 2006).

6.2.3 Hydrogen Peroxide (H_2O_2)

Hydrogen Peroxide (H_2O_2) is a compound with an oxygen-oxygen single bond. H_2O_2 is a slightly reactive oxygen species as well as the simplest form of peroxide. H_2O_2 is a clear liquid that is somewhat more viscous than water and is colourless in dilute solution (Asada 1999). It is formed through monovalent reduction and protonation of the $O_2^{\bullet-}$. Superoxide radical generation is caused by the reaction catalysed by the superoxide dismutase (SOD) activity:

$$2O_2^{\bullet-} + 2H^+ \rightarrow H_2O_2 + O_2$$

In plant cells, multiple sources caused the production of H_2O_2 such as the electron transport chain (ETC) in mitochondria or chloroplasts, cell membrane, endoplasmic reticulum (ER), photorespiration, and lipid peroxidation (Miller et al. 2010). It is also produced in response to the oxidative stress caused by the several environmental factors like UV radiation, high light intensity, salinity, drought, and pathogen attack (Sharma et al. 2012). H_2O_2 like every other ROS acts as a double-edged sword; it is favourable at lower concentrations, but at the same time can be toxic at higher concentrations in the plant cells. It also acts as a key regulator for several important physiological processes like photosynthesis and photorespiration (Noctor et al. 2002), senescence (Chang-lian et al. 2005), stomatal movement (Bright et al. 2006), cell cycle, and plant growth and development (Tanou et al. 2009a, b). Compared to other ROS members, it has a half-life of 1 ms, due to which it can travel longer distances and can traverse plant cell membranes.

6.2.4 Hydroxyl Radical ($OH^•$)

Among all the ROS members, hydroxyl radical ($OH^•$) is the neutral form of the hydroxide ion (Saed-Moucheshi et al. 2014), highly reactive due to its high reduction potential (Mohammed et al. 2004), and the most carcinogenic ROS known till now (Abd El 2012). $OH^•$ has a very short half-life of 10^{-9} s (Sies 1993). $OH^•$ is generated by the Fenton reaction between H_2O_2 and $O_2^{•-}$ catalysed by the Fe^{2+} or Fe^{3+} at neutral pH.

$$O_2^{•-} + H^+ + H_2O_2 \rightarrow O_2 + OH^• + H_2O$$

Moreover, the initial formation of O_2 occurs when $O_2^{•-}$ undergoes stepwise monovalent reduction. This $O_2^{•-}$ formed then acts as an electron donor in the hydroxyl radical production by the Haber–Weiss reaction. Since Haber–Weiss reaction has a finite role in the generation of superoxide and $OH^•$, formulated as the H_2O_2 source which causes hydroxyl radicals production via the Fenton reactions (Koppenol 2001). $OH^•$ has the great potential of damaging different cellular components by cell membrane destruction, protein damage, and β-oxidation of fatty acids. In addition, the increase in the concentration of $OH^•$ in the plant cells causes cellular death (Pinto et al. 2003).

6.3 Location of ROS Production and their Effects

Under both natural and stressful conditions, ROS is being produced in different compartments of the plants including chloroplasts, mitochondria, plasma membrane, peroxisomes, endoplasmic reticulum, and the cell wall, all of which possess high metabolic activity or have an intense electron flow rate . Moreover, NADPH oxidase present in the plasma membrane and amine oxidases and peroxidases situated in cell walls generates ROS in plants in response to environmental stresses (Tripathy and Oelmüller 2012). The major sources of the ROS generation during photosynthesis are chloroplasts and peroxisomes which occurs in the presence of the sunlight, whereas mitochondria is the current ROS producer in the absence of light (Choudhury et al. 2013). Photosynthesized leaves trap sunlight energy to fix and reduce CO_2 which is followed by the electron transfer via electron transport chain and by the translocation between chlorophyll molecules, which occurs through the production of chlorophyll excited states. Both processes result in the ROS generation either by electron transfer or by energy to oxygen molecules.

6.3.1 Chloroplast

According to Foyer et al. (1994a, b), Chloroplast is one of the main cellular sites which is responsible for ROS generation. In the chloroplast, supremely ordered system of thylakoid membranes dwells the light trapping photosynthetic machinery as well as structural and functional requirements for efficient light harvesting (Pfannschmidt 2003). During Photosynthesis, PSI and PSII (Photosystem I and photosystem II), the two light-harvesting complexes in the chloroplast thylakoidal membranes are the major sources of ROS generation. Drought, salinity, high temperature, etc., are some of the abiotic stress factors which are responsible for the water stress and minimize CO_2 concentrations, associated with excess light, ultimately leads to the $O_2^{\bullet-}$ formation at the PS through the Mehler reaction. Eventually, $O^{\bullet-}$ is converted to H_2O_2 by a membrane-bound Cu/Zn SOD at the PSI (Miller et al. 2010).

$$2O + 2Fd_{red} \rightarrow 2O^{\bullet -} + 2Fd_{ox}$$

Electron discharge occurs through the Q^A and Q^B electron acceptors leading to the production of $O^{\bullet-}$ at the PSII. Then, $O_2^{\bullet-}$ converts itself into more fatal ROS via H_2O_2 intermediate by the Fenton reaction at the Fe-S centres. The $^1O_2$ is also produced at the PSII which occurs in two ways: primarily, when the balance is disrupted between the light harvesting and energy utilization under abiotic stress cause the formation of triplet Chl ($^3Chl^*$), reacts with dioxygen ($^3O_2$) releases singlet oxygen ($^1O_2$) (Karuppanapandian et al. 2011). Secondarily, the light-harvesting complex (LHC) produced $^1O_2$ at the PSII when the electron transport chain is reduced (Asada 2006). The peroxidation of membrane lipids, especially Polyunsaturated Fatty Acids (PUFA) and disruption of membrane proteins caused by the stockpiling of $^1O_2$ in the chloroplast impose risk on the P680 reaction centre of PSII. This could ultimately lead to death of the plant cells (Møller et al. 2007; Triantaphylides et al. 2008).

6.3.2 Mitochondria

Mitochondrial electron transport chain also produces toxic ROS like $O^{\bullet-}$ and H_2O_2 under favourable conditions, but to a lesser extent as compared to chloroplasts and peroxisomes in the presence of sunlight (Navrot et al. 2007; Foyer and Noctor 2003). Plant mitochondria differs greatly from that of animal mitochondria in having the carbohydrate and O_2 rich environment as well as electron transport chain components and functions involved in photorespiration process (Rhoads et al. 2006). In other words, the mitochondrial contribution for ROS in plant tissues is not very effective as compared with mammalian cells (Purvis 1997). The mitochondrial electron transport chain (mtETC) generates ROS which can transfer electron to reduce O_2 and produce $O^{\bullet-}$. The ROS production especially superoxide formation is likely to occur in the two major components of the mtETC, Complex I and Complex III (Møller 2001; Møller et al. 2007; Noctor et al. 2007). The mitochondrial NADH-dependent dehydrogenases or Complex I which consists of a multimeric flavocytochrome, precisely reduces O_2 to $O_2^{\bullet-}$. The reverse flow of electrons arise from Complex III to Complex I because of the absence of NAD^+-linked substrates, consequently accelerate the generation of ROS at Complex I, which is controlled by

ATP hydrolysis (Turrens 2003). In Complex III, electron is being transferred to Cytochrome C_1 by the fully reduced form of ubiquinone, which promotes electron leakage to O_2, inevitably producing $O_2^{\bullet -}$ (Murphy 2009).

Other mitochondrial sources of ROS formation include several enzymes present in the mitochondrial matrix such as aconitase and 1-Galactono-γ-lactone dehydrogenase (GAL), the former directly produces ROS whereas the latter indirectly produce ROS by transferring electrons to electron transport chain (Rasmusson et al. 2008).

6.3.3 Peroxisomes

Peroxisomes, the major sites for the production of intracellular H_2O_2 through their integral oxidative metabolism are single-membrane bound organelles (del Río et al. 2006; Palma et al. 2009). They also produce $O_2^{\bullet -}$ at two different sites. First is the Xanthine oxidase situated in the peroxisomal matrix, which breaks down both xanthine and hypoxanthine into uric acid generating $O_2^{\bullet -}$ and H_2O_2 in small amounts (Navrot et al. 2007). Second is the NADPH-dependent small electron transport chain that is composed of NADH and Cyt b placed in peroxisomal membrane uses O_2 as an electron acceptor and liberates $O_2^{\bullet -}$ into the cytosol (Das and Roychoudhury 2014). The photoreduction of O_2 to $O_2^{\bullet -}$ happens due to the reduced electron transport components connected with PSI in the peroxisome. Several environmental stresses limit the availability of CO_2 inside the leaves due to which ribulose-1,5-bisphosphate carboxylase/oxygenase metabolizes a dynamic reaction, favours oxygen as a substrate rather than CO_2. This reaction causes glycolate formation which is transported to peroxisomes and after subsequent oxidation by glycolate oxidases produces H_2O_2 (Tripathy and Oelmüller 2012).

6.3.4 Plasma Membrane

Plasma membrane is just like other cellular membranes, fully surrounds the cell surface. It plays a main role in reciprocity with the abiotic stress conditions and feed essential information for uninterrupted survival of the plant cell. The NADPH-dependent oxidases (NOX), also known as respiratory burst oxidase homologue (RBOH) located in the plasma membrane are known for their efficient role due to the gene expression and the presence of different homologues under environmental stresses (Apel and Hirt 2004). NOX can translocate electrons from cytosolic NADPH across the plasma membrane to molecular O2 in the apoplast region, generate $O_2^{\bullet -}$ which is either catalysed by SOD or dismutate to H_2O_2 through different mechanisms. Moreover, NOX is crucial for plant defence against pathogen attack and abiotic stresses (Kwak et al. 2003). On the other hand, RBOH genes belong to a polygenic family, 10 members were identified from *Arabidopsis thaliana* (RBOHA-RBOH) and 9 were identified from Rice (*Oryza sativa*) (Torres et al. 2002; Sagi and Fluhr 2006; Skelly and Loake 2013).

The membrane-bound respiratory burst oxidase homologue (RBOH) with a molecular weight of about 105–112 kDa and its cytosolic regulator ROP (Rho-like protein) is a Rac homologue of plants, are the two main components of the plant RBOH protein (Yoshie et al. 2005). RBOH plays an important role in several plant processes like cell growth (Foreman et al. 2003), pollen tube growth (Kaya et al. 2014), plant development and stomatal closure (Shi et al. 2012), abiotic stress and pathogen attack (Wojtaszek 1997; Daudi et al. 2012), and symbiotic relationship (Marino et al. 2012).

6.4 Targets of ROS

The detrimental effects of ROS on the plant cell generally include:

6.4.1 Oxidation of Amino Acids in Proteins

Under abiotic stresses, enhanced level of ROS generation causes the oxidation of proteins. The protein alteration is caused either by direct or indirect method. Direct alteration of protein is carried out by the mechanisms involving nitrosylation, glutathionylation, disulphide bond formation, and carboxylation (Møller et al. 2007). Indirect alteration of proteins is caused by fusion with the decomposed products of lipid peroxidation (LPO) (Yamauchi et al. 2008). The higher concentration of ROS production in plant cell leads to the site-specific amino acid modification like Proline, Lysine, Arginine, Threonine, and Tryptophan, increased susceptibility of proteins to proteolysis, modified electric charge, disintegration of the peptide chain (Sharma et al. 2012; Møller et al. 2007). During various abiotic stresses, escalated protein alteration has been studied in plants (Sharma and Dubey 2005; Han et al. 2009; Maheshwari and Dubey 2009; Romero-Puertas et al. 2002). The sites which are highly susceptible to attack by ROS belong to the thiol groups and sulphur containing amino acids. Metals like Cd, Hg, and Pb are responsible for the depletion of protein bound thiol groups (Stohs and Bagchi 1995). In the presence of ROS, cross-linking of tyrosine leads to the formation of bityrosine products (Davies 1987). Inactivation of enzymes causes oxidation of iron-sulphur (Fe-S) clusters by $O_2^{\bullet -}$, which cannot be reversed later (Gardner and Fridovich 1991).

6.4.2 Damage of DNA

ROS are the major source of DNA damage, causes oxidative damage to mitochondrial as well as chloroplastic DNA as a result of close proximity to ROS generation machinery and lack of protective histones. Any damage caused to DNA by ROS results in the modification of the nucleotide base, oxidation of deoxyribose sugar residue, removal of nucleotide, cross-linking of DNA, and protein and strand breakage, which may ultimately lead to deactivation or malfunctioning of the proteins. The hydroxy radical damages the deoxyribose sugar backbone by extracting H-atom and also known to react with all purine and pyrimidine bases double bonds (Halliwell 2006). Oxidative damage to DNA bases results from addition of $OH^{\bullet}$ to double bonds and damage to sugar generally includes removal of hydrogen from deoxyribose forming a deoxyribose radical which further reacts to cause single strand breaks in the DNA (Dizdaroglu 1993; Evans et al. 2004). The products generated from the DNA bases by $OH^{\bullet}$ commonly include C-8 hydroxylation of guanine to form 8-oxo-7,8 dehydro-2′-deoxyguanosine, thymine glycol, thymine, hydroxymethyl urea, and adenine ring-opened and saturated products (Tsuboi et al. 1998). When $OH^{\bullet}$ reacts with either DNA or associated proteins are well-known for creating DNA-protein cross links.

6.4.3 Oxidations of Polyunsaturated Fatty Acids in Lipids

Lipids account for large portion of the plasma membrane surrounding the cell which helps plant cell to adapt to the changing environmental conditions. Despite, during stressful conditions, when the level of ROS increases, accelerated lipid peroxidation occurs in both organellar and cellular membranes degrading normal cellular functions. LPO damages DNA and proteins by initiating a chain reaction which infuriate the oxidative stress through the generation of lipid-derived radicals. Elevated LPO with increased production of ROS has been studied in plants during environmental stresses (Sharma and Dubey 2005; Tanou et al. 2009a, b; Mishra et al. 2011). In phospholipids, the final products of unsaturated fatty acids are malondialdehyde which is responsible for cell membrane damage (Halliwell and Gutteridge 2015). The ester linkage between glycerol and the fatty acids and the double bond between carbon atoms are two main sites of ROS attack in membrane phospholipid molecules. The polyunsaturated fatty acids (PUFAs) are essential components of the plasma membrane which are considered to be the hotspots for ROS attack. ROS damage by $OH^{\bullet}$ and $^1O_2$ can result in the peroxidation of several PUFAs like linoleic and linolenic acid. Initiation, progression, and termination are the three main steps involved in the lipid peroxidation process. Initiation phase includes triggering of O_2, which is a rate limiting step, form radicals like $OH^{\bullet}$ and $O_2^{\bullet-}$, which in turn reacts with methylene groups of PUFA yielding lipid peroxyl radicals, conjugated dienes, and hydroperoxides (Smirnoff 2000).

$$\text{PUFA} - \text{H} + \text{OH}^{\bullet} \rightarrow \text{PUFA}^{\bullet}\ (\text{PUFA alkyl radical}) + \text{H}_2\text{O}$$

$$\text{PUFA}^{\bullet} + \text{O}_2 \rightarrow \text{PUFA} - \text{OO}^{\bullet}\ (\text{Peroxyl radical})$$

The formation of PUFA peroxyl radical further proliferates the chain reaction by abstracting one H-atoms from adjoining PUFA side chains.

$$\text{PUFA}-\text{OO}^{\bullet}+\text{PUFA}-\text{H}\ \rightarrow \text{PUFA}-\text{OOH}+\text{PUFA}^{\bullet}$$

The lipid peroxides (PUFA-OOH) produced undergo reductive cleavage by carrying out a reaction with reduced metals like Fe^{2+}.

$$\text{PUFA}-\text{OOH}+\text{Fe}^{2+}\ \rightarrow\ \text{PUFA}-\text{OO}^{\bullet}+\text{Fe}^{2+}$$

The breakdown of lipid hydroperoxides gives rise to several reactive species like alkanes, aldehydes, alcohols, lipid alkoxyl radicals, and lipid epoxides (Buettner 1993). Moreover, LPO allows the entry of substances by increases the membrane fluidity other than through special channels which damages membrane proteins, deactivate the ion channels, membrane receptors and membrane-localized enzymes.

6.5 Interplay Between ROS and Leaf Modifications

Leaf growth is a regulatory process that determines shape and size of the organ which is customized to the changing environment conditions. Golgi-derived vesicle transits building blocks for cell growth that are translocated to the cell surface by the action of the cytoskeleton. The cytoskeleton regulates the direction of cell growth by assisting the transport of Golgi-derived vesicles and the cellulose synthase machinery (Geisler et al. 2008). In isotropic expansion, the cell growth is equal in all the directions, whereas anisotropic expansion is the enlargement of cells in a preferred direction (Crowell et al. 2010). Isotropic expansion occurs in mesophyll cells during leaf growth whereas anisotropic expansion occurs in epidermal cells of leaves (Fu et al. 2002). Still, both isotropic and anisotropic expansion can occur in single cells simultaneously (Crowell et al. 2010). The leaf grows at the flank of the shoot apical meristem through active cell division (Beemster et al. 2005; Polyn et al. 2015; Schippers et al. 2016). Cell expansions in leaves which accounts for about 95% of terminal leaf area determine the final size of the leaf. Turgor pressure and cell wall dynamics are the two major regulators involved in the extension of leaf cells (Gonzalez et al. 2012). Moreover, the synthesis and distribution of new biomaterials are necessary to sustain growth in order to expand leaf cells.

Abiotic stresses such as drought, salinity, high or low temperature, are responsible for an initial growth reduction of leaves and the initiation of programmed cell death during prolonged stress conditions (Hernández et al. 2001; Loggini et al. 1999). Both during controlled conditions and abiotic stress, modifications in the growth rate and cell expansions in leaves are associated with ROS homeostasis which have advanced the fundamental role of ROS in the regulation of plant development. As the ROS level rises under abiotic stress, it does not relate to the growth zone of the leaf. Specifically, it has been reported in maize under saline conditions that a decrease in ROS causes retarded leaf growth rather than increase in ROS levels (Rodríguez et al. 2004). Besides, abiotic stresses like salinity or drought accelerate the antioxidant capacity of the maize leaf, and hence limit cell expansion (Bernstein et al. 2010; Kravchik and Bernstein 2013; Avramova et al. 2015).

Furthermore, also an increase in the levels of ROS can cause restriction of cell growth during unfavourable conditions which indicates that ROS have dual roles in the cell expansion regulation (MacAdam and Grabber 2002; Šimonovičová et al. 2004). However, ROS homeostasis together with abiotic stresses like osmotic stress, drought, or salinity intrudes with the water balance and leads to reduction in cell turgor affecting the mechanical power of the cell to expand (Schopfer 2006).

Metabolic ROS modifications induced alterations in proteins controlling key events of transcription and translation (Foyer and Noctor 2016). Latest findings showed that alterations in ROS levels result in epigenetic modifications such as acylation, modulate the ROS related proteins activity in rice leaves (Zhou et al. 2018). This clearly indicates that the cross-talk between ROS and acylation is necessary for the post-translational modifications of leaf proteins possessing key metabolic functions. However, due to the temporal and spatial variability of ROS alteration and interplay between them, the effects of ROS on plant growth and development are more complex in plants. Still, the role of ROS in the regulation of leaf modification remains uncertain.

6.6 Removal of ROS from the Plant

Plants possess a battery of enzymatic and nonenzymatic antioxidants to scavenge the deleterious effects of ROS. As mentioned above, the specific sites for ROS generation as well as antioxidants machinery are found in various different organelles such as mitochondria, chloroplasts, and peroxisomes (Pang and Wang 2008). The antioxidant defence system includes superoxide dismutase (SOD), enzymes and metabolites from the ascorbate-glutathione cycle, catalase (CAT), peroxidases (POX), ascorbic acid (AA), and α-tocopherol (Bowler et al. 1992; Willekens et al. 1997; Noctor and Foyer 1998). Antioxidants play an important role in the removal of ROS and they are activated in response to unfavourable environmental condition and plant development (Foyer and Noctor 2005). The effective cellular and physiological processes are required to have an efficient antioxidant activity. As ROS are produced constantly in plant cells, any disparity between ROS and antioxidants inculpate to oxidative stress (Scandalios 2002). Also, ROS production is genetically designed, for instance, ROS like O2•- and H2O2 acts as a second messenger (Shao et al. 2008). However, its accumulation at high levels causes oxidative stress which leads to cell death. Catalase (CAT) and ascorbate peroxidase (APX) are the major enzymatic cellular scavengers of H2O2 (Willekens et al. 1997; Noctor and Foyer 1998). Both of these antioxidants have different compatibilities for H2O2 such as CAT does not rely on a reductant to detoxify H2O2 which makes it reducing power-free whereas APX relies on ascorbate. Moreover, CAT has lower affinity for H2O2 (mM range) than APX (μM range) (Mittler 2002). Ascorbic acid is found in all parts of the plants and is synthesized in the mitochondria, which is later translocated to different parts of the plants (Aro and Ohad 2003; Borland et al. 2006; Shao et al. 2008). In the ascorbate-glutathione cycle, ascorbate peroxidase (APX) uses ascorbic acid as a substrate to reduce H2O2 to H2O, forming monodehydroascorbate which is further transformed into ascorbic acid and dehydroascorbate (Gapper and Dolan 2006). In plants, majority of ascorbic acid pool is led by D-mannose/L-galactose also known as Smirnoff-Wheeler pathway that prosecutes through GDP-D-mannose, GDP-L-galactose, L-galactose, and L-galactono-1,4-lactone in the plant mitochondria (Wheeler et al. 1998). Most of the ascorbic acid has been detected in the large number of plant cell types, organelles, and apoplast and is localized in the photosynthetic tissues (Smirnoff et al. 2004; Shao et al. 2008). Glutathione is the low molecular weight nonprotein thiol that plays a key role in the removal of ROS like H2O2 (Foyer and Noctor 2005; Shao and Chu 2005; Shao et al. 2008). It is localized in majority of plant cell compartments such as chloroplasts, mitochondria, cytosol, vacuoles, and endoplasmic reticulum (Asada 1994). GSH in its reduced form is required in several biological processes like cell growth, conjugation of metabolites, synthesis of phytochelatins for metal chelation, synthesis of proteins and nucleic acids, xenobiotics removal, signal transduction, expression of stress-responsive genes, sulfate transport, and enzymatic regulation (Foyer et al. 1997). GSH can react chemically with OH•, O2•-, and H2O2, act as a free radical scavenger. GSH can also support macromolecules by directly forming the adducts with reactive electrophiles or function as an electron donor in the ROS presence or organic free radicals, producing glutathione disulfide (GSSG) (Asada 1994). Nowadays, AA and GSH are studied as important components of redox signalling in plants (Baier et al. 2000; Noctor et al. 2000; Horling et al. 2003). It has been studied in ascorbate/glutathione scavenging pathway that APX and GR activities were promoted under drought stress in cotton and spurred anoda (Ratnayaka et al. 2003), beans (Türkan et al. 2005; Torres-Franklin et al. 2008), alfalfa (Rubio et al. 2002), rice (Sharma and Dubey 2005), wheat seedlings (Keleş and Öncel 2002), cowpea (Contour-Ansel et al. 2006), and in the moss *Tortula ruralis* (Dhindsa 1991).

Superoxide dismutase (SOD) is another major defence system against ROS along with other enzymes like ascorbate peroxidase (APX), catalase (CAT), and glutathione peroxidase (GPX). SOD belongs to the group of metalloenzymes and converts O2•- to H2O2 which is later removed by APX, CAT, and GPX (Apel and Hirt 2004). APX then converts H2O2 to H2O which is called the Mehler-peroxidase reaction or the water-water cycle (Asada 1999). Three isozymes of SOD have been reported in plants, namely, copper/zinc SOD (Cu/Zn-SOD), iron SOD (Fe-SOD), and manganese SOD (Mn-SOD) (Fridovich 1989; Racchi et al. 2001). Three isoforms of Copper/Zinc SOD are found in the cytosol, chloroplast, mitochondria, and peroxisome (). Eukaryotic Copper/Zinc SOD is prone to cyanide and present as dimer, while the other two SOD (Fe-SOD and Mn-SOD) are not prone to cyanide and may be presents as dimer or tetramers (Scandalios 1993; Corpas et al. 2008). Increased activity of SOD has been reported in plants during various environmental stresses, including drought and metal toxicity which is often correlated with increased tolerance of the plant against environmental stresses (Sharma and Dubey 2005; Mishra et al. 2011). Tocopherols are another potential antioxidants against ROS and lipid radicals which are present in all parts of plants (Holländer-Czytko et al. 2005). Tocopherols play an essential role, both as an antioxidant and non-antioxidant components in the biological membrane. α-, β-, γ-, δ- isomers are the four isomers of tocopherols studied in plants of which, α-tocopherol has the highest antioxidant capability because of three methyl groups in its molecular structure (Ashraf 2009). Due to considerable amount of α-tocopherol present in the chloroplast membrane of higher plants protect it from photo-oxidative damage (Munné-Bosch 2005).

6.7 Conclusion

From the last few years, the perspectives of ROS have changed from toxic metabolic by-products which require antioxidant defence system, aiming to protect plant cells from their harmful effects, to the hub of plant science as key regulators of plant growth and development and coordinating responses to biotic and abiotic stress. The production of ROS in various cell compartments is minimal under normal growth conditions. However, environmental stresses like drought, salinity, UV-B, metal toxicity, chilling, and pathogen infection, if extend to certain degrees, interfere with cellular homeostasis and hence, improve the generation of ROS. The extensive damage caused by the ROS attack targets the biomolecules like DNA, proteins, and lipids, resulting in irreversible DNA damage and ultimately leading to its death. However, plants have evolved with a variety of defence mechanism that includes genetic, metabolic, and morphological level to adapt to the unfavourable environmental conditions. Despite the fact that rapid advancements have been made in recent years, there are numerous concerns and gaps in our knowledge on how ROS is formed and their deleterious effects on plants chiefly due to short life span and reactivity of ROS. Research on ROS formation using sophisticated analytical approach will boost to develop broader context of the role of ROS in plants. Further analysis needs to be done to acquire profound knowledge and to decode new measures of regulation in the view of controlling the generation of ROS and their effects on the plant leaves.

References

Abd El, H. A. H. M. (2012). Lipid peroxidation end-products as a key of oxidative stress: effect of antioxidant on their production and transfer of free radicals. In *Lipid peroxidation*. IntechOpen

Apel K, Hirt H (2004) Reactive oxygen species: metabolism, oxidative stress, and signal transduction. Annu Rev Plant Biol 55:373–399

Aro EM, Ohad I (2003) Redox regulation of thylakoid protein phosphorylation. Antioxid Redox Signal 5(1):55–67

Asada K (1994) Production and action of active oxygen species in photosynthetic tissues. In: Foyer CH, Mullineaux PM (eds) Causes of photooxidative stress and amelioration of Defense Systems in Plants. CRC Press, Boca Raton, Fla, USA, pp 77–104

Asada K (1999) The water-water cycle in chloroplasts: scavenging of active oxygens and dissipation of excess photons. Annu Rev Plant Biol 50(1):601–639

Asada K (2006) Production and scavenging of reactive oxygen species in chloroplasts and their functions. Plant Physiol 141:391–396. https://doi.org/10.1104/pp.106.082040

Ashraf M (2009) Biotechnological approach of improving plant salt tolerance using antioxidants as markers. Biotechnol Adv 27(1):84–93

Avramova V, AbdElgawad H, Zhang Z, Fotschki B, Casadevall R, Vergauwen L et al (2015) Drought induces distinct growth response, protection, and recovery mechanisms in the maize leaf growth zone. Plant Physiol 169(2):1382–1396

Baier M, Noctor G, Foyer CH, Dietz KJ (2000) Antisense suppression of 2-cysteine peroxiredoxin in Arabidopsis specifically enhances the activities and expression of enzymes associated with ascorbate metabolism but not glutathione metabolism. Plant Physiol 124(2):823–832

Beemster GT, De Veylder L, Vercruysse S, West G, Rombaut D, Van Hummelen P et al (2005) Genome-wide analysis of gene expression profiles associated with cell cycle transitions in growing organs of Arabidopsis. Plant Physiol 138(2):734–743

Bernstein N, Shoresh M, Xu Y, Huang B (2010) Involvement of the plant antioxidative response in the differential growth sensitivity to salinity of leaves vs roots during cell development. Free Radic Biol Med 49(7):1161–1171

Borland A, Elliott S, Patterson S, Taybi T, Cushman J, Pater B, Barnes J (2006) Are the metabolic components of crassulacean acid metabolism up-regulated in response to an increase in oxidative burden? J Exp Bot 57(2):319–328

Bowler C, Montagu MV, Inze D (1992) Superoxide dismutase and stress tolerance. Annu Rev Plant Biol 43(1):83–116

Bright J, Desikan R, Hancock JT, Weir IS, Neill SJ (2006) ABA-induced NO generation and stomatal closure in Arabidopsis are dependent on H2O2 synthesis. Plant J 45(1):113–122

Buettner GR (1993) The pecking order of free radicals and antioxidants: lipid peroxidation, α-tocopherol, and ascorbate. Arch Biochem Biophys 300(2):535–543

Cadenas E (1989) Biochemistry of oxygen toxicity. Annu Rev Biochem 58(1):79–110

Chang-lian PENG, Zhi-Ying OU, Nan LIU, Gui-zhu LIN (2005) Response to high temperature in flag leaves of super high-yielding rice Pei'ai 64S/E32 and Liangyoupeijiu. Rice Sci 12(3):179

Choudhury S, Panda P, Sahoo L, Panda SK (2013) Reactive oxygen species signaling in plants under abiotic stress. Plant Signal Behav 8(4):e23681

Contour-Ansel D, Torres-Franklin ML, Cruz De Carvalho MH, D'Arcy-Lameta A, Zuily-Fodil Y (2006) Glutathione reductase in leaves of cowpea: cloning of two cDNAs, expression and enzymatic activity under progressive drought stress, desiccation and abscisic acid treatment. Ann Bot 98(6):1279–1287

Corpas FJ, Palma JM, Sandalio LM, Valderrama R, Barroso JB, Luis A (2008) Peroxisomal xanthine oxidoreductase: characterization of the enzyme from pea (Pisum sativum L.) leaves. J Plant Physiol 165(13):1319–1330

Crowell EF, Gonneau M, Vernhettes S, Höfte H (2010) Regulation of anisotropic cell expansion in higher plants. C R Biol 333(4):320–324

Das K, Roychoudhury A (2014) Reactive oxygen species (ROS) and response of antioxidants as ROS-scavengers during environmental stress in plants. Front Environ Sci 2:53

Daudi A, Cheng Z, O'Brien JA, Mammarella N, Khan S, Ausubel FM, Bolwell GP (2012) The apoplastic oxidative burst peroxidase in Arabidopsis is a major component of pattern-triggered immunity. Plant Cell 24(1):275–287

Davies KJ (1987) Protein damage and degradation by oxygen radicals. I. General aspects. J Biol Chem 262(20):9895–9901

Dhindsa RS (1991) Drought stress, enzymes of glutathione metabolism, oxidation injury, and protein synthesis in Tortula ruralis. Plant Physiol 95(2):648–651

Dizdaroglu M (1993) Chemistry of free radical damage to DNA and nucleoproteins. *DNA and free radicals*

Dmitrieva VA, Tyutereva EV, Voitsekhovskaja OV (2020) Singlet oxygen in plants: generation, detection, and signaling roles. Int J Mol Sci 21(9):3237

Edreva A (2005) Generation and scavenging of reactive oxygen species in chloroplasts: a submolecular approach. Agric Ecosyst Environ 106(2–3):119–133

Evans MD, Dizdaroglu M, Cooke MS (2004) Oxidative DNA damage and disease: induction, repair and significance. Mutation Research/Reviews in Mutation Research 567(1):1–61
Foreman J, Demidchik V, Bothwell JH, Mylona P, Miedema H, Torres MA et al (2003) Reactive oxygen species produced by NADPH oxidase regulate plant cell growth. Nature 422(6930): 442–446
Foyer CH, Descourvieres P, Kunert KJ (1994b) Protection against oxygen radicals: an important defence mechanism studied in transgenic plants. Plant Cell Environ 17(5):507–523
Foyer CH, Harbinson J, Mullineaux P (1994a) Oxygen metabolism and the regulation of photosynthetic electron transport. *Causes of photooxidative stress and amelioration of defense systems in plants*, 1–42
Foyer CH, Lopez-Delgado H, Dat JF, Scott IM (1997) Hydrogen peroxide-and glutathione-associated mechanisms of acclimatory stress tolerance and signalling. Physiol Plant 100(2): 241–254
Foyer CH, Noctor G (2003) Redox sensing and signalling associated with reactive oxygen in chloroplasts, peroxisomes and mitochondria. Physiol Plant 119(3):355–364
Foyer CH, Noctor G (2005) Oxidant and antioxidant signalling in plants: a re-evaluation of the concept of oxidative stress in a physiological context. Plant Cell Environ 28(8):1056–1071
Foyer CH, Noctor G (2016) Stress-linked redox signalling: What's in pROS-pect. *Plant, Cell & Environment*
Fridovich I (1989) Superoxide dismutases: an adaptation to a paramagnetic gas. J Biol Chem 264(14):7761–7764
Fu Y, Li H, Yang Z (2002) The ROP2 GTPase controls the formation of cortical fine F-actin and the early phase of directional cell expansion during Arabidopsis organogenesis. Plant Cell 14(4): 777–794
Gapper C, Dolan L (2006) Control of plant development by reactive oxygen species. Plant Physiol 141(2):341–345
Gardner PR, Fridovich I (1991) Superoxide sensitivity of the Escherichia coli 6-phosphogluconate dehydratase. J Biol Chem 266(3):1478–1483
Geisler DA, Sampathkumar A, Mutwil M, Persson S (2008) Laying down the bricks: logistic aspects of cell wall biosynthesis. Curr Opin Plant Biol 11(6):647–652
Girotti AW (2001) Photosensitized oxidation of membrane lipids: reaction pathways, cytotoxic effects, and cytoprotective mechanisms. J Photochem Photobiol B Biol 63(1–3):103–113
Gonzalez N, Vanhaeren H, Inzé D (2012) Leaf size control: complex coordination of cell division and expansion. Trends Plant Sci 17(6):332–340
Halliwell B (2006) Reactive species and antioxidants. Redox biology is a fundamental theme of aerobic life. Plant Physiol 141(2):312–322
Halliwell B, Gutteridge JM (2015) Free radicals in biology and medicine. Oxford university press, USA
Han C, Liu Q, Yang Y (2009) Short-term effects of experimental warming and enhanced ultraviolet-B radiation on photosynthesis and antioxidant defense of Picea asperata seedlings. Plant Growth Regul 58(2):153–162
Hernández JA, Ferrer MA, Jiménez A, Barceló AR, Sevilla F (2001) Antioxidant systems and O2.–/H2O2 production in the apoplast of pea leaves. Its relation with salt-induced necrotic lesions in minor veins. Plant Physiol 127(3):817–831
Holländer-Czytko H, Grabowski J, Sandorf I, Weckermann K, Weiler EW (2005) Tocopherol content and activities of tyrosine aminotransferase and cystine lyase in Arabidopsis under stress conditions. J Plant Physiol 162(7):767–770
Holt NE, Zigmantas D, Valkunas L, Li XP, Niyogi KK, Fleming GR (2005) Carotenoid cation formation and the regulation of photosynthetic light harvesting. Science 307(5708):433–436
Horling F, Lamkemeyer P, Konig J, Finkemeier I, Kandlbinder A, Baier M, Dietz KJ (2003) Divergent light-, ascorbate-, and oxidative stress-dependent regulation of expression of the peroxiredoxin gene family in Arabidopsis. Plant Physiol 131(1):317–325

Hu WH, Song XS, Shi K, Xia XJ, Zhou YH, Yu JQ (2008) Changes in electron transport, superoxide dismutase and ascorbate peroxidase isoenzymes in chloroplasts and mitochondria of cucumber leaves as influenced by chilling. Photosynthetica 46(4):581–588

Jithesh MN, Prashanth SR, Sivaprakash KR, Parida AK (2006) Antioxidative response mechanisms in halophytes: their role in stress defence. J Genet 85(3):237–254

Karuppanapandian T, Moon JC, Kim C, Manoharan K, Kim W (2011) Reactive oxygen species in plants: their generation, signal transduction, and scavenging mechanisms. Aust J Crop Sci 5(6): 709–725

Kaya H, Nakajima R, Iwano M, Kanaoka MM, Kimura S, Takeda S et al (2014) Ca2+−activated reactive oxygen species production by Arabidopsis RbohH and RbohJ is essential for proper pollen tube tip growth. Plant Cell 26(3):1069–1080

Keleş Y, Öncel I (2002) Response of antioxidative defence system to temperature and water stress combinations in wheat seedlings. Plant Sci 163(4):783–790

Koppenol WH (2001) The Haber-Weiss cycle–70 years later. Redox Rep 6(4):229–234

Kravchik M, Bernstein N (2013) Effects of salinity on the transcriptome of growing maize leaf cells point at cell-age specificity in the involvement of the antioxidative response in cell growth restriction. BMC Genomics 14(1):1–13

Kwak JM, Mori IC, Pei ZM, Leonhardt N, Torres MA, Dangl JL et al (2003) NADPH oxidase AtrbohD and AtrbohF genes function in ROS-dependent ABA signaling in Arabidopsis. EMBO J 22(11):2623–2633

Loggini B, Scartazza A, Brugnoli E, Navari-Izzo F (1999) Antioxidative defense system, pigment composition, and photosynthetic efficiency in two wheat cultivars subjected to drought. Plant Physiol 119(3):1091–1100

MacAdam JW, Grabber JH (2002) Relationship of growth cessation with the formation of diferulate cross-links and p-coumaroylated lignins in tall fescue leaf blades. Planta 215(5):785–793

Maheshwari R, Dubey RS (2009) Nickel-induced oxidative stress and the role of antioxidant defence in rice seedlings. Plant Growth Regul 59(1):37–49

Marino D, Dunand C, Puppo A, Pauly N (2012) A burst of plant NADPH oxidases. Trends Plant Sci 17(1):9–15

Meriga B, Reddy BK, Rao KR, Reddy LA, Kishor PK (2004) Aluminium-induced production of oxygen radicals, lipid peroxidation and DNA damage in seedlings of rice (Oryza sativa). J Plant Physiol 161(1):63–68

Miller GAD, Suzuki N, Ciftci-Yilmaz SULTAN, Mittler RON (2010) Reactive oxygen species homeostasis and signalling during drought and salinity stresses. Plant Cell Environ 33(4): 453–467

Mishra S, Jha AB, Dubey RS (2011) Arsenite treatment induces oxidative stress, upregulates antioxidant system, and causes phytochelatin synthesis in rice seedlings. Protoplasma 248(3): 565–577

Mittler R (2002) Oxidative stress, antioxidants and stress tolerance. Trends Plant Sci 7(9):405–410

Mohammed A, Omar A, Christine B, Ibrahim A, Sarra A, Al Ossam Mohammed A (2004) Pathological roles of reactive oxygen species and their defence mechanisms. Saudi Pharm J 12(1):1–18

Møller IM (2001) Plant mitochondria and oxidative stress: electron transport, NADPH turnover, and metabolism of reactive oxygen species. Annu Rev Plant Biol 52(1):561–591

Møller IM, Jensen PE, Hansson A (2007) Oxidative modifications to cellular components in plants. Annu Rev Plant Biol 58:459–481

Munné-Bosch S (2005) The role of α-tocopherol in plant stress tolerance. J Plant Physiol 162(7): 743–748

Murphy MP (2009) How mitochondria produce reactive oxygen species. Biochem J 417(1):1–13

Navrot N, Rouhier N, Gelhaye E, Jacquot JP (2007) Reactive oxygen species generation and antioxidant systems in plant mitochondria. Physiol Plant 129(1):185–195

Noctor G, De Paepe R, Foyer CH (2007) Mitochondrial redox biology and homeostasis in plants. Trends Plant Sci 12(3):125–134

Noctor G, Foyer CH (1998) Ascorbate and glutathione: keeping active oxygen under control. Annu Rev Plant Biol 49(1):249–279

Noctor G, Veljovic-Jovanovic S, Foyer CH (2000) Peroxide processing in photosynthesis: antioxidant coupling and redox signalling. Philos Trans R Soc Lond Ser B Biol Sci 355(1402): 1465–1475

Noctor G, Veljovic-Jovanovic SONJA, Driscoll S, Novitskaya L, Foyer CH (2002) Drought and oxidative load in the leaves of C3 plants: a predominant role for photorespiration? Ann Bot 89(7):841–850

Palma JM, Corpas FJ, del Río LA (2009) Proteome of plant peroxisomes: new perspectives on the role of these organelles in cell biology. Proteomics 9(9):2301–2312

Pang CH, Wang BS (2008) Oxidative stress and salt tolerance in plants. In *Progress in botany* (pp. 231–245). Springer, Berlin, Heidelberg

Pfannschmidt T (2003) Chloroplast redox signals: how photosynthesis controls its own genes. Trends Plant Sci 8(1):33–41

Pinto E, Sigaud-kutner TC, Leitao MA, Okamoto OK, Morse D, Colepicolo P (2003) Heavy metal–induced oxidative stress in algae 1. J Phycol 39(6):1008–1018

Polyn S, Willems A, De Veylder L (2015) Cell cycle entry, maintenance, and exit during plant development. Curr Opin Plant Biol 23:1–7

Purvis AC (1997) Role of the alternative oxidase in limiting superoxide production by plant mitochondria. Physiol Plant 100(1):165–170

Racchi ML, Bagnoli F, Balla I, Danti S (2001) Differential activity of catalase and superoxide dismutase in seedlings and in vitro micropropagated oak (Quercus robur L.). Plant Cell Rep 20(2):169–174

Rasmusson AG, Geisler DA, Møller IM (2008) The multiplicity of dehydrogenases in the electron transport chain of plant mitochondria. Mitochondrion 8(1):47–60

Ratnayaka HH, Molin WT, Sterling TM (2003) Physiological and antioxidant responses of cotton and spurred anoda under interference and mild drought. J Exp Bot 54(391):2293–2305

Rebeiz CA, Montazer-Zouhoor A, Mayasich JM, Tripathy BC, Wu SM, Rebeiz CC, Friedmann HC (1988) Photodynamic herbicides. Recent developments and molecular basis of selectivity. Crit Rev Plant Sci 6(4):385–436

Rhoads DM, Umbach AL, Subbaiah CC, Siedow JN (2006) Mitochondrial reactive oxygen species. Contribution to oxidative stress and interorganellar signaling. Plant Physiol 141(2):357–366

del Río LA, Sandalio LM, Corpas FJ, Palma JM, Barroso JB (2006) Reactive oxygen species and reactive nitrogen species in peroxisomes. Production, scavenging, and role in cell signaling. Plant Physiol 141(2):330–335

Rodríguez AA, Córdoba AR, Ortega L, Taleisnik E (2004) Decreased reactive oxygen species concentration in the elongation zone contributes to the reduction in maize leaf growth under salinity. J Exp Bot 55(401):1383–1390

Romero-Puertas MC, Palma JM, Gómez M, Del Rio LA, Sandalio LM (2002) Cadmium causes the oxidative modification of proteins in pea plants. Plant Cell Environ 25(5):677–686

Rubio MC, González EM, Minchin FR, Webb KJ, Arrese-Igor C, Ramos J, Becana M (2002) Effects of water stress on antioxidant enzymes of leaves and nodules of transgenic alfalfa overexpressing superoxide dismutases. Physiol Plant 115(4):531–540

Saed-Moucheshi A, Shekoofa A, Pessarakli M (2014) Reactive oxygen species (ROS) generation and detoxifying in plants. J Plant Nutr 37(10):1573–1585

Sagi M, Fluhr R (2006) Production of reactive oxygen species by plant NADPH oxidases. Plant Physiol 141:336–340

Scandalios JG (1993) Oxygen stress and superoxide dismutases. Plant Physiol 101(1):7

Scandalios JG (2002) Oxidative stress responses-what have genome-scale studies taught us? Genome Biol 3(7):1–6

Schippers JH, Foyer CH, van Dongen JT (2016) Redox regulation in shoot growth, SAM maintenance and flowering. Curr Opin Plant Biol 29:121–128

Schopfer P (2006) Biomechanics of plant growth. Am J Bot 93(10):1415–1425

Shah K, Kumar RG, Verma S, Dubey RS (2001) Effect of cadmium on lipid peroxidation, superoxide anion generation and activities of antioxidant enzymes in growing rice seedlings. Plant Sci 161(6):1135–1144

Shao H, Chu L (2005) Plant molecular biology in China: opportunities and challenges. Plant Mol Biol Report 23(4):345–358

Shao HB, Chu LY, Lu ZH, Kang CM (2008) Primary antioxidant free radical scavenging and redox signaling pathways in higher plant cells. Int J Biol Sci 4(1):8

Sharma P, Dubey RS (2005) Drought induces oxidative stress and enhances the activities of antioxidant enzymes in growing rice seedlings. Plant Growth Regul 46(3):209–221

Sharma P, Jha AB, Dubey RS, Pessarakli M (2012) Reactive oxygen species, oxidative damage, and antioxidative defense mechanism in plants under stressful conditions. J Bot 2012:217037. https://doi.org/10.1155/2012/217037

Shi YC, Fu YP, Liu WQ (2012) NADPH oxidase in plasma membrane is involved in stomatal closure induced by dehydroascorbate. Plant Physiol Biochem 51:26–30

Sies H (1993) Strategies of antioxidant defense. Eur J Biochem 215(2):213–219

Šimonovičová M, Huttová J, Mistrík I, Široká B, Tamás L (2004) Peroxidase mediated hydrogen peroxide production in barley roots grown under stress conditions. Plant Growth Regul 44(3): 267–275

Skelly MJ, Loake GJ (2013) Synthesis of redox-active molecules and their signaling functions during the expression of plant disease resistance. Antioxid Redox Signal 19(9):990–997

Smirnoff N (2000) Ascorbic acid: metabolism and functions of a multi-facetted molecule. Curr Opin Plant Biol 3(3):229–235

Smirnoff N, Running JA, & Gatzek S (2004) Ascorbate biosynthesis: a diversity of pathways. *Vitamin C: its functions and biochemistry in animals and plants*, 7–29

Srivastava S, Dubey RS (2011) Manganese-excess induces oxidative stress, lowers the pool of antioxidants and elevates activities of key antioxidative enzymes in rice seedlings. Plant Growth Regul 64(1):1–16

Stohs SJ, Bagchi D (1995) Oxidative mechanisms in the toxicity of metal ions. Free Radic Biol Med 18(2):321–336

Tanou G, Molassiotis A, Diamantidis G (2009a) Induction of reactive oxygen species and necrotic death-like destruction in strawberry leaves by salinity. Environ Exp Bot 65(2–3):270–281

Tanou G, Molassiotis A, Diamantidis G (2009b) Hydrogen peroxide-and nitric oxide-induced systemic antioxidant prime-like activity under NaCl-stress and stress-free conditions in citrus plants. J Plant Physiol 166(17):1904–1913

Torres MA, Dangl JL, Jones JD (2002) Arabidopsis gp91phox homologues AtrbohD and AtrbohF are required for accumulation of reactive oxygen intermediates in the plant defense response. Proc Natl Acad Sci 99(1):517–522

Torres-Franklin ML, Contour-Ansel D, Zuily-Fodil Y, Pham-Thi AT (2008) Molecular cloning of glutathione reductase cDNAs and analysis of GR gene expression in cowpea and common bean leaves during recovery from moderate drought stress. J Plant Physiol 165(5):514–521

Triantaphylides C, Krischke M, Hoeberichts FA, Ksas B, Gresser G, Havaux M et al (2008) Singlet oxygen is the major reactive oxygen species involved in photooxidative damage to plants. Plant Physiol 148(2):960–968

Tripathy BC, Oelmüller R (2012) Reactive oxygen species generation and signaling in plants. Plant Signal Behav 7(12):1621–1633

Tsuboi H, Kouda K, Takeuchi H, Takigawa M, Masamoto Y, Takeuchi M, Ochi H (1998) 8-hydroxydeoxyguanosine in urine as an index of oxidative damage to DNA in the evaluation of atopic dermatitis. Br J Dermatol 138(6):1033–1035

Türkan I, Bor M, Özdemir F, Koca H (2005) Differential responses of lipid peroxidation and antioxidants in the leaves of drought-tolerant P. acutifolius Gray and drought-sensitive P. vulgaris L. subjected to polyethylene glycol mediated water stress. Plant Sci 168(1):223–231

Turrens JF (2003) Mitochondrial formation of reactive oxygen species. J Physiol 552(2):335–344

Wheeler GL, Jones MA, Smirnoff N (1998) The biosynthetic pathway of vitamin C in higher plants. Nature 393(6683):365–369

Willekens H, Chamnongpol S, Davey M, Schraudner M, Langebartels C, Van Montagu M et al (1997) Catalase is a sink for H2O2 and is indispensable for stress defence in C3 plants. EMBO J 16(16):4806–4816

Wojtaszek P (1997) Oxidative burst: an early plant response to pathogen infection. Biochem J 322(3):681–692

Yamauchi Y, Furutera A, Seki K, Toyoda Y, Tanaka K, Sugimoto Y (2008) Malondialdehyde generated from peroxidized linolenic acid causes protein modification in heat-stressed plants. Plant Physiol Biochem 46(8–9):786–793

Yoshie Y, Goto K, Takai R, Iwano M, Takayama S, Isogai A, Che FS (2005) Function of the rice gp91phox homologs OsrbohA and OsrbohE genes in ROS-dependent plant immune responses. Plant Biotechnology 22(2):127–135

Zhang HX, Hodson JN, Williams JP, Blumwald E (2001) Engineering salt-tolerant Brassica plants: characterization of yield and seed oil quality in transgenic plants with increased vacuolar sodium accumulation. Proc Natl Acad Sci 98(22):12832–12836

Zhou H, Finkemeier I, Guan W, Tossounian MA, Wei B, Young D et al (2018) Oxidative stress-triggered interactions between the succinyl-and acetyl-proteomes of rice leaves. Plant Cell Environ 41(5):1139–1153

Reactive Oxygen Species: Role in Senescence and Signal Transduction

7

Yamshi Arif, Anayat Rasool Mir, and Shamsul Hayat

Abstract

Reactive oxygen species (ROS) play a dual role in the plant kingdom. Previously, they were recognized as an inevitable toxic byproduct of cellular metabolism but accepted as a central regulator of important signaling events in cells. ROS participates in several biological processes like seed germination, cell division and differentiation, pollen tube development, growth of root hair, and programmed cell death. Moreover, they participate in several redox processes; conversely, at the same time, ROS induces oxidative stress, thereby prompting a physiological or programmed cell death. The homeostasis between ROS generation and eradication is managed, and the signals conveyed to the cells to initiate senescence in plant cells or organs are yet to be revealed. The current chapter focuses on how ROS contributes to senescence and redox signaling in plants besides sustaining cellular proliferation and physiological function.

Keywords

Cellular metabolism · Biological processes · Programmed cell death · Senescence

Y. Arif · A. R. Mir · S. Hayat (✉)
Faculty of Life Sciences, Department of Botany, Aligarh Muslim University, Aligarh, India

M. Faizan et al. (eds.), *Reactive Oxygen Species*,
https://doi.org/10.1007/978-981-19-9794-5_7

7.1 Introduction

Biotic and abiotic stresses resulted in the loss of productivity and yield, causing global food security (Gan and Amasino 1997; Lee et al. 2012). The continuous rise in the stresses caused a catastrophic impact on global warming and climate alteration, assessing the importance of plant tolerance against stresses (Prochazkova et al. 2001). ROS plays a central role in stress sensing, combining various stress response signaling networks to activate plant defense mechanisms (Mittler et al. 2022). Understanding the role of ROS in plant development, senescence, and stress response will allow us to enhance plant tolerance and increase metabolic and developmental processes (Mittler et al. 2022).

IJMS Senescence in plants is a complex detrimental process that causes the death of a whole plant or a single organ (Jajić et al. 2015). It is regulated by internal factors like age, developmental process, plant hormone level, and environmental stimuli such as photoperiod, stress, shading, and wounding (Gan and Amasino 1997). The synthesis of ROS is the response of plant cells under senescence and abiotic stresses (Lee et al. 2012). ROS are the product of aerobic energy metabolism in plants under biotic and environmental stresses (Silva et al. 2010; Choudhury et al. 2013). Under a typical environment, the generation of ROS in cells is sustained at levels by enzymatic antioxidants. This balance is maintained by reducing the antioxidants or the enhanced accumulation of ROS, causing oxidative stress and damaging cellular macromolecules and membrane that enhance lipid peroxidation (Lushchak 2011). ROS-enhanced oxidative stress limits agricultural yield (Ashraf 2009), and plants have evolved several mechanisms that protect themselves from adverse abiotic stress conditions, like tolerance processes that involve ROS generation along with antioxidant defense machinery and the activation of signal transduction (Jajić et al. 2015).

7.2 ROS Generation and Removal in Plants

ROS is present in ionic and molecular forms in plants. Ionic states of ROS consist of superoxide anions ($O2^{\cdot-}$) and hydroxyl radicals ($^{\bullet}OH$), whereas molecular records of ROS consist of singlet oxygen ($^1O_2$) and hydrogen peroxide (H_2O_2) (Apel and Hirt 2004; Mittler et al. 2004). Every type of ROS produced in the plant has a different oxidative capacity and affects genes that regulate plants' physio-biochemical processes (Zhang et al. 2017). $^1O_2$ is delivered inside photosystem II of chloroplast and has robust oxidizability (Jajić et al. 2015). The life span of $^1O_2$ is concise and highly unstable in a cell; it negatively impacts photosynthesis. $O_2^{\cdot-}$ acts as a precursor of several ROS because of its instability and oxidizing solid and reducing properties. The average concentration of $O2^{\cdot-}$ in plant stem cells causes stability (Zeng et al. 2017). A high concentration of O2·causes excessive ROS levels, which leads to senescence or cell death (Gill and Tuteja 2010). In rice plants, different organs such as stems and roots are the primary sites for O2·- generation; it regulates rice adaptation in an aquatic environment (Yamauchi et al. 2017). $O2^{\cdot-}$ is generated in electron transport chains during photosynthesis, respiratory electron transport chains inside mitochondria, and membrane-dependent NADPH oxidase system that reacts with H^+ to form oxygen molecules; or reacts with antioxidant enzyme superoxide dismutase (SOD) that includes H_2O_2 (Bose et al. 2014; Mhamdi and Van Breusegem 2018). Among ROS, H_2O_2 is regarded as a potent redox molecule because of its specific physical and chemical properties, such as its stability inside plant cells and fast and reversible oxidation of proteins (Mittler 2017; Mhamdi and Van Breusegem 2018). H_2O_2 is transported via aquaporins present in the cell membrane. It participates in cell signaling regulation and causes long-distance oxidative damage (Bienert et al. 2007; Wudick et al. 2015). H_2O_2 plays a significant role in cell differentiation, cell wall formation, apoptosis, and senescence in plants (Kärkönen and Kuchitsu 2015; Schippers et al. 2016; Waszczak et al. 2016; Ribeiro et al. 2017; Zeng et al. 2017). Furthermore, H_2O_2 interconnection with plant hormones regulates plant metabolism, senescence, and stress responses (Haung et al. 2019).$^{\bullet}OH$ is formed when oxygen molecule bonds in H_2O_2 cleaves. $^{\bullet}OH$ is active and most reactive ROS as it acts nearby its site of production. Thus, $^{\bullet}OH$ reacts with membranes and causes cell wall loosening by oxidizing cell wall polysaccharides, and can also enhance the breakage of DNA (Kärkönen and Kuchitsu 2015). Under normal environmental conditions, ROS levels are scavenged by several antioxidant defense machineries. The equilibrium between ROS production and removal can be disturbed by several biotic and ecological factors (Jajić et al. 2015). These disturbances may lead to excessive ROS production inside cells, damaging cellular structures. Altogether, plants mitigate the higher amount of ROS by maintaining cellular redox homeostasis (Mir et al., 2022). Thus, the increased ROS levels are detoxified by ROS- scavenging systems (Mir et al., 2022).

ROS-detoxifying machinery is classified into two types enzymatic and nonenzymatic antioxidant defense systems, which work synergistically to neutralize free radicals (Apel and Hirt 2004). The enzymatic antioxidant includes superoxide dismutase (SOD), catalase (CAT), glutathione peroxidase (GPX), and ascorbate peroxidase (APX) (Mir et al. 2022). In the rice, several genes that participate in ROS removal exhibits tissue/organ-specific expression profile (Gechev et al. 2006). However, ROS role in homeostasis and modulation of gene expression is not clear. SOD converts $\cdot OH$ to H_2O_2, and this H_2O_2 is converted into H_2O and O_2 via CAT and POX (Mittler 2017). The nonenzymatic antioxidant machinery, such as glutathione, flavonoids, and ascorbic acid (AsA), removes toxic $^{\bullet}OH$ and $^1O_2$ (Gechev et al. 2006). When the ROS level exceeds beyond threshold level or above the status of the scavenging system, then the cell enters an oxidative state leading to a specific stressful environment and cell damage, causing cell and plant senescence (Fujita et al. 2006). When the ROS concentration is low inside a cell, ROS acts as a secondary messenger, and it participates in cell division, differentiation, organogenesis, and cell maintenance under several biotic and abiotic stresses (Zeng et al. 2017). Therefore, it is essential to maintain ROS levels inside the cell in the right concentration for plant health. Additionally, changes in ROS level are parts of average plant growth and development, but they should not exceed the threshold concentration inside a cell (Mittler 2017).

7.3 ROS Detection

The primary issue with the precise determination of the ROS role during senescence and abiotic stress is the continuous production of ROS (Hideg et al. 2002; Laloi et al. 2007). Thus, a limited amount of non-invasive and specific methods can be used for determination. For instance, in a plant that suffers from moderate stress, H_2O_2, $^1O_2$, and $O_2^{\bullet-}$ are released continuously in the plant cell, making it difficult to find their discrete specific roles (Laloi et al. 2007). This is not easy to evaluate significant differences in the generation of ROS when senescence is applied artificially or when the plant ages naturally (Springer et al. 2015). The effect of gene expression is different when H_2O_2 is used exogenously and when produced endogenously (Golemiec et al. 2014). Several reports suggested that various stressor in plants alters the gene expression patterns (Rizhsky et al. 2004; Mittler et al. 2022).

There are several assays that can be applied for the detection and evaluation of ROS in plant tissues, such as colorimetric measurement of $O_2^{\bullet-}$ with XTT and fluorometric measurement of H_2O_2 with Amplex Red (Zhou et al. 1997; Frahry and Schopfer 2001). These methods are easy to use, but they lack specificity. Ample Red is unstable, and it can get autooxidized and produce H_2O_2 and $O^{\bullet}{}_2^{-}$, while XTT is reduced by short-chain sugars (Gomes et al. 2005; Dikalov et al. 2007). Higher specificity for ROS detection can be obtained using the spin trapping technique via electron paramagnetic resonance (EPR) spectroscopy. Spin trapping techniques use the reaction carried out between free radicals and nitroso or nitrone compounds to form a stable spin adduct (Swartz et al. 2007). The spin adduct gives a specific EPR spectrum property of the particular radical that is trapped. Under normal conditions, the ROS flux produced inside the cells is maintained in low concentration via antioxidant defense machinery, which makes ROS detection difficult (Swartz et al. 2007). For effective detection of ROS, the imbalance between ROS generation and decay should be created (Huang et al. 2019). The sample illumination can obtain this with visible light in the presence of a suitable spin trap or by adding a spin trap after illumination (Swartz et al. 2007). The limitation of spin trapping is that this method does not provide information about specific sites of ROS production in tissues due to solvent incompatibility with living tissue (Hideg et al. 1994, 2011). Non-invasive, in vivo determinations of ROS can be obtained by using fluorescent probes that are combined with confocal laser scanning microscopy or fluorescence microscopy (Sandalio et al. 2008). The advantage of CLSM technique is that it can evaluate the intracellular location of ROS by using different fluorescent probes for specific organelles (Rodríguez-Serrano et al. 2006). 2′,7′-Dichlorofluorescein diacetate (DCF-DA) is a dye used to detect ROS in plant tissues, specifically H_2O_2, but it can also react with other peroxides (Tarpey et al. 2004). Another probe used for the detection of $O^{\bullet}{}_2^{-}$ is Dihydroethidium (DHE) (Corpas et al. 2006). The evaluation of singlet oxygen species with singlet oxygen sensor green (SOSG) reagent is particular to $^1O_2$ without showing specificity to hydroxyl radicals or superoxide (Flors et al. 2006). The monitoring of ROS inside living cells can be obtained with a genetically encoded probe like reduction-oxidation sensitive green fluorescent protein (roGFPs) and HyPer (Lukyanov and Belousov 2014). HyPer shows its sensitivity towards

H_2O_2 but is insensitive to other oxidants. Thus, HyPer does not cause more ROS production (Lukyanov and Belousov 2014). On the other hand, roGFP is used for the detection of H_2O_2 and also for detecting thiol redox state (Hernández-Barrera et al. 2013). But this probe has certain disadvantages, including the necessity of antioxidants and pH (Lukyanov and Belousov 2014).

7.4 Role of ROS in Senescence

7.4.1 Superoxide Anions

In plants, $O^{\bullet}{}_2{}^-$ is produced in different cell compartments such as peroxisome, chloroplasts, mitochondria, apoplast, and plasma membrane (Rhoads et al. 2006; Rosenwasser et al. 2011; Sandalio et al. 2013). The primary source superoxide anion in the chloroplast is carried out by Mehler reactions, in which O_2 is reduced by electrons from the photosynthetic electron transport chain (Allen and Hall 1973). $O^{\bullet}{}_2{}^-$ produced is converted into H_2O_2, mainly via the action of CuZn-SOD (Asada 2006). Thus, SOD evaluates the $O^{\bullet}{}_2{}^-$ life in cells and its association with biochemical reactions. The $O^{\bullet}{}_2{}^-$ is a less reactive, short-lived ROS having a half-life of 2–4 μs, as it does not cross the chloroplast membrane (Dat et al. 2000).$O^{\bullet}{}_2{}^-$ is not generated in the chloroplast (Sandalio et al. 1988). In peroxisomes, $O^{\bullet}{}_2{}^-$ is produced in two different ways in the peroxisomal matrix by the action of xanthine oxidase and via the help of the electron transport chain (ETC) in a peroxisomal membrane (Del Río and Donaldson 1995). The peroxisome is an essential source of signaling molecules. They detoxify toxic H_2O_2 and $O^{\bullet}{}_2{}^-$via the presence of antioxidants found in organelles (Del Río and Donaldson 1995; Corpas et al. 2006). NADPH oxidases (NOX) are an essential source of $O^{\bullet}{}_2{}^-$ in plants, it is also known as respiratory burst oxidase homologs (Rbohs), that catalyzes the production of $O^{\bullet}{}_2{}^-$ (Sagi and Fluhr 2006; Kaur et al. 2014). Rbohs play a crucial role in several physio-biochemical processes, like ROS signaling (Kwak et al. 2003; Torres and Dangl 2005). ($O^{\bullet}{}_2{}^-$ is generated in the cytosol by the action of aldehyde oxidase and xanthine dehydrogenase (Yesbergenova et al. 2005; Zarepour et al. 2010). It was reported that increased generation of $O^{\bullet}{}_2{}^-$ causes senescence (McRae and Thompson 1983; Pastori and del Río 1997). Moreover, signaling plays a crucial role in ROS production and its conversion of $O^{\bullet}{}_2{}^-$ to H_2O_2 (McRae and Thompson 1983). High synthesis of ROS is destructive to the cell due to oxidative modifications of cellular compartments, leading to senescence (Van Breusegem and Dat 2006). It was reported that under the high temperature, large quantities of ROS such as of O•2– to H2O2 were produced and accumulated in the leaves of cucumber, which causes senescence and reduces protein and chlorophyll content, and increases lipid peroxidation (Scarpeci et al. 2008). In *Arabidopsis thaliana* plants, the signaling role of $O^{\bullet}{}_2{}^-$ was detected, and the plant was exposed to methyl viologen, a superoxide anion propagator under a light. The production of $O^{\bullet}{}_2{}^-$ in the absence of H_2O_2 accumulation reported nuclear-encoded genes that are likely to be precise for $O^{\bullet}{}_2{}^-$ associated signal transduction (Scarpeci et al. 2008). Several reports suggested that the upregulation of genes that participates during abiotic stress responses, such as WRKY6, plays a key role in the senescence and defense-amelioration process (Robatzek and Somssich 2001). In barley, production of $O^{•}_2{}^{-}$ was observed during physiological processes such as the development and senescence of secondary leaves via the help of the EPR method combined with DMPO as a spin trap (Jajić et al. 2015). Thus, the synthesis of $O^{•}_2{}^{-}$ increases during the developmental process and reaches its maximum level after the beginning of senescence. Additionally, $O^{•}_2{}^{-}$ generation started to reduce till the senescence get over (Jajić et al. 2014). This causes a rise in membrane fluidity, which is also responsible for the increase in ROS production (Jajić et al. 2014). The surge in $O^{•}_2{}^{-}$ level was detected in the interveinal region of senescing tobacco leaves, and it was also present in minor veins of mature and old senescent leaves. However, it was not found in the central veins (Niewiadomska et al. 2009). It was found that superoxide content plays a vital role in the downregulation of genes associated with photosynthesis. $O^{•}_2{}^{-}$ acts as a signaling molecule found in *Medicago* leaves where ROS generation was reduced with diphenyleneiodonium (DPI). Increased synthesis of $O^{•}_2{}^{-}$ was observed in pea plants under Cd stress (Rodríguez-Serrano et al. 2006). Cd exposure causes oxidative stress, which is due to disorders in antioxidant defense machinery and reduction in NO level. In lupine roots, NO ameliorate the toxic effect of Cd, as it promotes antioxidant machinery to prevent oxidative damage produced by $O^{•}_2{}^{-}$ (Kopyra and Gwóźdź 2003). Thus, $O^{•}_2{}^{-}$ contributes to senescence during biotic and abiotic stresses (Huang et al. 2019).

7.4.2 Hydrogen Peroxide (H_2O_2)

H_2O_2 plays a key role in plants during the stress response; it acts as the signaling molecule that facilitates the physiological process (Quan et al. 2008). It participates in the regulation of senescence, mitigation against the pathogenic attack, and abiotic stress mitigation (Peng et al. 2005). Additionally, it can affect the expression of several genes (Yun et al. 2010). H2O2 is synthesized in plants with the help of two possible pathways like $O^{•}_2{}^{-}$ dismutation with the use of SOD with the help of oxidases like amino and oxalate oxidases (Hu et al. 2003). The H_2O_2 level is kept at the control level by the network of enzymatic and nonenzymatic antioxidants that reduces the accumulation of H_2O_2 (Foyer and Noctor 2005). The equilibrium between SODs and H_2O_2-detoxifying enzymes in cells is important in determining the H_2O_2 steady level (Mittler et al. 2004).

In contrast with other ROS, H_2O_2 is considered the most stable and least reactive ROS, and it can easily cross the plasma membrane. Thus it acts as a signaling molecule (Yang and Poovaiah 2002; Quan et al. 2008). H_2O_2 plays a key role in plants as a signaling molecule and serves as the abiotic and biotic stress regulator (Quan et al. 2008). H_2O_2 in higher concentrations induces plant cell death and contributes to cell degradation during senescence (Dat et al. 2000; Zimmermann et al. 2006). H_2O_2 functions in a dose-dependent manner. A low concentration (600 μM) increases the vase life of a cut flower, whereas a high concentration shows a negative effect (Liao et al. 2012). H_2O_2 works in a concentration-dependent manner in wax apples. Low concentration induces growth, yield, and quality (Khandaker et al. 2012). H_2O_2 serves as an essential signaling molecule during senescence, where it promotes senescence in different plant species (Bieker et al. 2012). It was observed that an increase in H_2O_2 concentration during the bolting and flowering process is due to a decrease in ascorbate peroxidase activity at the same time (Zimmermann et al. 2006). This causes an increase in the expression of transcription factors (TFs) and senescence-related genes. Therefore, expression analysis reveals that H_2O_2 application increased 14 salt-triggered senescence-associated genes and 15 senescence-linked NAC genes (Balazadeh et al. 2010). Thus, some salt-induced senescence is part of H_2O_2 signaling via NAC transcription factors. H_2O_2 signaling increases the expression of NAC TFs ORS1, JUB1, and ATAF1, which considerably induces senescence (Garapati et al. 2015). Overexpression of ORS1 increases the expression of genes associated with senescence; in contrast, expression of JUB1 inhibits senescence, diminishes cellular H_2O_2 level, and grows plants' abiotic stress tolerance (Jajić et al. 2015). Overexpression of ATAF1 facilitates the senescence process, promoting expression of TF ORE1 and reducing the maintenance of chloroplast TF GLK1 (Jajić et al. 2015). In barley, H_2O_2 was produced during the developmental and senescence process (Jajić et al. 2015). H_2O_2 plays a key role during the senescence process in two different aspects: it serves as a signaling molecule in the initiation of senescence and degradation of molecules at later phases of senescence (Huang et al. 2019).

Several reports suggested that H_2O_2 can be linked with other signaling molecules, such as abscisic acid (ABA) and ethylene, that play an essential role in plant development and senescence (Jubany-Marí et al. 2009; Chen et al. 2012). Ethephon application causes an increase in H_2O_2 level, which causes an increase in the expression of catalase (CAT). The removal of H_2O_2 caused by exogenous glutathione levels reduces ethephon-associated effects (Chen et al. 2012). The association between ABA, H_2O_2, and ascorbic acid was investigated in Mediterranean shrubs during drought. Thus, it was observed that drought stress caused ABA and H_2O_2 interaction that increases ascorbic acid content, maintaining the stability and integrity of plants during stress (Gao et al. 2010; Ishibashi et al. 2011; Zhang et al. 2011).

7.4.3 Singlet Oxygen

Singlet oxygen is highly reactive, excited molecular oxygen that is produced in a reaction between O_2 and the chlorophyll triplet state (Krieger-Liszkay 2005). The generation of $^1O_2$ is not escorted by the transfer of an electron to O_2, but one of the unpaired electrons is transferred to a higher energy orbital (Gill and Tuteja 2010). Under normal conditions, $^1O_2$ is produced during photosynthesis by the activation of photosensitizers like chlorophyll and its precursors (Krieger-Liszkay 2005). $^1O_2$ is produced during senescence and under different environmental stresses (Springer et al. 2015). Like other ROS, $^1O_2$ also has dual roles. It acts as an oxidizing agent which reacts with several biological molecules and causes cell death and senescence

(Op den Camp et al. 2003). It also serves as the signaling molecule that expresses different genes (Kim et al. 2008). In conditional fluorescent (flu) Arabidopsis mutant, photosensitizer protochlorophyllide was accumulated in the dark and produced $^1O_2$ after the light transfer (Fischer et al. 2007).

Additionally, after light illumination, several different nuclear genes are activated in the flu mutant. They are different from superoxide anion and hydrogen peroxide. This suggests that $^1O_2$ does not act as the toxin but as the signaling molecule that activates various stress response signal pathways (Op den Camp et al. 2003). It was reported that $^1O_2$ induces senescence and also enhances tolerance against stress. It also increases tolerance in Arabidopsis plants through the action of β-cyclocitral (Ramel et al. 2012). β-cyclocitral is the derivative of β-carotene that is generated in high light and increases expression of several genes produced by $^1O_2$ (Ramel et al. 2012). The npq1lut2 mutant significantly accumulates $^1O_2$ due to the loss of zeaxanthin and lutein that participates in the scavenging of free radicals of chlorophyll and oxygen (Dall'Osto et al. 2006, 2010). High light caused $^1O_2$ responses in Arabidopsis cell suspension culture (ACSC) having functional chloroplast (González-Pérez et al. 2011). Thus, remarkable changes in the transcript expression upregulated $^1O_2$ caused plant tolerance against stress (Huang et al. 2019).

When it comes to senescence, the role of $^1O_2$ is limited and scarce ROS synthesis increases during senescence (Prochazkova et al. 2001; Lee et al. 2012). Therefore, the increase in the $^1O_2$ is detected simultaneously with the other ROS. It makes it difficult to isolate the role of $^1O_2$. It was observed that $^1O_2$ is the main reason of senescence-associated oxidative stress in the chloroplast (Munné-Bosch et al. 2001). Therefore, this was concluded that on the basis of degradation of α-tocopherol and β-carotene under drought condition plants, which suggests the increased formation of single oxygen (Munné-Bosch et al. 2001). Additionally, in recent investigation a mass production of $^1O_2$ was evaluated in the early stages of hormone-applied barley but then reduced, while in senescence plant continuous synthesis of $^1O_2$ was observed (Springer et al. 2015). Additionally, increase in $^1O_2$ synthesis, senescing plants also contained β-carotene oxidative breakdown products like β-cyclocistral, which act as important messenger in $^1O_2$ signaling during stress (Ramel et al. 2012). In barley leaves, it was observed that during its development and senescence continuous production of $^1O_2$ (Jajić et al. 2015).$^1O_2$ leads to increase in production of lipid peroxidation (LPO) which causes cell death and senescence (Zhang et al. 2010). In Arabidopsis, $^1O_2$ caused 80% production of LPO which ultimately causes senescence (Triantaphylidès et al. 2008). Moreover, LPO causes increase in production of free radicals which induces senescence (Arora et al. 2002). This causes increase in lipoxygenase activity, which promotes LPO and forms $^1O_2$, causing cell death and senescence (Triantaphylidès et al. 2008).

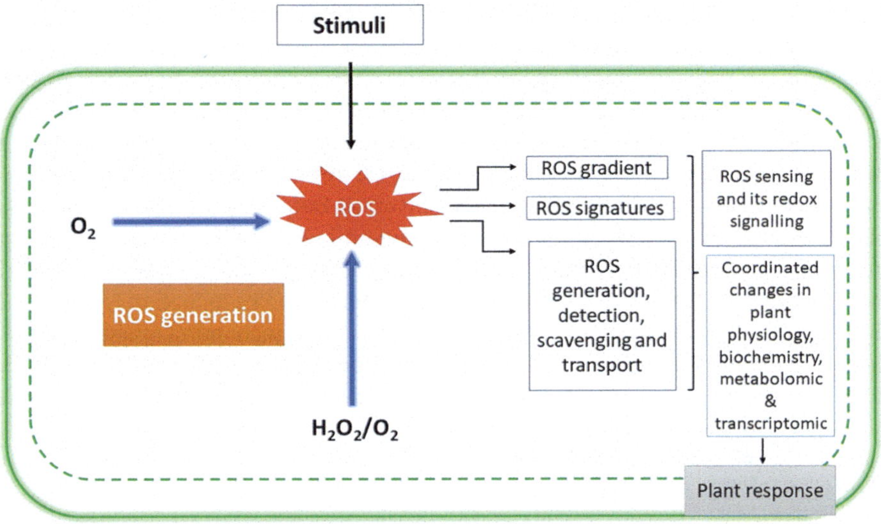

Fig. 7.1 ROS generation and homeostasis in plants

7.5 ROS Signal Transduction in Plants

Plants regularly sense and determine the concentration of ROS and regulate their gene expression and enzyme activities to adapt in changing environmental condition (Fig. 7.1; Asada 2006). ROS signaling is regulated by two opposing processes for synthesis and scavenging. There are three different mechanism of sensing ROS inside plant cells they are receptor proteins, redox-sensing TFs and inhibition of phosphatases by ROS ((Mittler 2002; Mittler et al. 2004; Apel and Hirt 2004; Miller et al. 2008). ROS can be identified due to their effect on several metabolic processes and their redox potential of specific cellular proteins. These can be membrane linked or soluble and can be found in different cellular compartments (Mittler et al. 2004)). There are several developmental, senescence or environmental signals that feed into ROS signal transduction and disturb ROS homeostasis in cellular compartment specific or cell-specific manner (Mittler et al. 2004). Disturbed ROS levels are alleged by several enzymes, proteins, or receptors and regulate several developmental, senescence and defense pathways (Huang et al. 2019). ROS generation is carried out by several important life processes such as photosynthesis and respiration, also by various enzymes and proteins like xanthine oxidase, NADPH oxidase and amine oxidase (Mittler et al. 2004). Contrary to this, ROS detoxification is carried by several ROS-scavenging enzymes and several antioxidants like CAT, POX, SOD, and APX (Mittler et al. 2004). The duration, intensity, and localization of the several ROS signals inside the cells are detected by crosstalk between these opposite forces such as ROS detoxification and production, and these signal transduction decoding will evaluate several cellular responses to environmental stress, that alters several physio-biochemical processes (Mittler 2017). These processes require a strong regulation and involve amplification and feedback inhibition mode. Additionally, regulation of intensity and time of different ROS signals, the ROS-detoxifying processes are required for maintaining a low-level baseline of ROS on which there are several other signals (Mittler 2017). ROS act as the diverse signaling molecule that regulates several biotic and abiotic stresses. Stresses disrupt metabolic homeostasis of cell, causing increased synthesis of ROS (Mittler et al. 2004). Alteration in plant metabolic pathways causes evolution and utilization of ROS as the signaling molecule that participates plant growth, senescence, defense, and hormonal signal transduction (Mittler 2002). During pathogenic attack, ROS production is increased that triggers R-gene associated signaling pathways that increase synthesis of ROS via help of NADPH oxidases, that are localized on plasma membrane. Thus, ROS activates several signal transduction pathways (Torres et al. 2006). Abiotic stress produces osmotic stress which is sensed by several signaling pathways causes increase in production of ROS via NADPH oxidase and ROS-generating signals (Miller et al. 2008).

In Arabidopsis, mutants lacking cytosolic ascorbate peroxidase 1 (apx1) are used in ROS signal transduction. APX1 in the cytosol serves as the buffer that controls ROS level that is transported to nuclei and activate expression of gene (Rizhsky et al. 2004; Pnueli et al. 2003; Davletova et al. 2005; Suzuki et al. 2005, 2008; Mittler 2006; Ciftci-Yilmaz et al. 2007). In the absence of APX1, ROS accumulates inside cytosol, which is sensed by several redox-response TFs, like heat shock TFs, that facilitates signal transduction cascade of several TFs that include members of zinc finger protein ZAT family and TFs of WRKY family (Davletova et al. 2005). This signal transduction pathway includes MBF1c acts as transcriptional coactivator and NADPH oxidase (RbohD) amplifies ROS signal transduction (Ciftci-Yilmaz et al. 2007). ROS sources such as mitochondria or apoplast generate specific signaling stimuli that include cytosolic ROS- detoxifying mechanisms like GPX, thioredoxin, and peroxiredoxin (Mittler 2017). The ROS signals produced in different cellular compartments reach nuclei and activate expression of several genes that promotes response to perceived stimuli (Rizhsky et al. 2004; Pnueli et al. 2003; Davletova et al. 2005; Suzuki et al. 2005, 2008; Mittler 2006; Ciftci-Yilmaz et al. 2007).

The several detoxifying and generating enzymes encoded by ROS genes are found in several subcellular and cellular compartments. It was also found that more than one enzyme activity specific to one ROS can be found in different organelles (Mittler et al. 2004). ROS metabolism in specific cellular compartments can alter ROS signal transduction and homeostasis of an adjacent cellular compartment or it reaches to nuclei and activate expression of gene (Mittler 2017). Antioxidants ascorbic acid and glutathione transporters play key role in evaluating the specific concentration of such compounds and their redox potential in different cellular organelles (Mittler et al. 2004). Vacuole plays main role in ROS signaling network, because of its larger cellular volume it controls ROS metabolism in plants. In Arabidopsis, mode of coordination in between various components of ROS exclusion network in plants is multifaceted and complex (Mittler 2017). In Arabidopsis, during light stress there is increase of cytosolic and not of chloroplastic

ROS defense enzyme, even though most of the ROS is synthesized in chloroplasts and peroxisome (Mittler et al. 2004). The cytosolic ROS-scavenging signaling pathways require chloroplast protection (Davletova et al. 2005). It was observed that different cellular pathways are activated inside the cells in response to induced ROS synthesis inside cytosol or peroxisome. Activation of both cytosolic and peroxisomal signaling pathways caused synthesis of new signal that is entirely different from that activated by the individual cytosolic or peroxisomal signals (Mittler 2017).

NADPH oxidase plays essential role in signal transduction of ROS and plant defense mechanism during abiotic stress (Torres and Dangl 2005). They synthesize $^1O_2$ by oxidizing NADPH and electron transfer to O_2 (Torres and Dangl 2005). The Arabidopsis genome consists of ten major NADPH oxidase genes, which consist of almost cytosolic 300 amino acid amino-terminal extension having two EF- that binds with calcium ions at least on one phosphorylation site (Mittler 2017). It was observed that NADPH oxidase activation involves phosphorylation of serine N-terminal via calcium-dependent protein kinase (CDPK) and its interaction with Rho-like GTPase (ROP). NADPH oxidase phosphorylation and its binding with calcium combines its activation, as it functions as calcium sensor (Ogasawara et al. 2008; Takeda et al. 2008). The NADPH oxidase or ROP interaction modulated by the calcium binding to two EF-hand motifs at the oxidase terminus (Wong et al. 2007). NADPH oxidase activation caused increased production of O2·- which is converted into H_2O_2, inside apoplastic spaces (Mittler 2017). ROS produced signal that can reach the nuclei, that activates expression of genes in response to external stimuli (Mittler 2017). In root tips, interaction between pH, ROS, and calcium signaling induces cell elongation which further improves plant physiology and biochemistry (Monshausen et al. 2007; Van Breusegem et al. 2008).

7.6 ROS-Induced Redox Signaling

Organelle to nucleus communication, typically induced retrograde signaling that alters the expression of nuclear genes that encodes organelle proteins which have structural and metabolic functions (Huang et al. 2019). Organelles such as chloroplast, mitochondria, and peroxisomes synthesize ROS via photosynthesis, respiration and photorespiration (Foyer and Noctor 2003). Increased ROS synthesize, earlier considered as a precursor of oxidative stress but now it causes oxidant signaling altering several proteins via redox-associated, posttranslational modifications that regulate molecular master switches (Mittler 2017). Coordination of expression of genes among the genomes of plant cells supports the integration of cell signaling pathways and uses retrograde signal transduction pathway in the center of cellular reaction to certain stimulus (Chan et al. 2016). Among many ROS synthesized, H_2O_2 act as the signaling molecule that crosses plasma membrane, it have major role cell to cell signal transduction that optimize functioning of several cells (Bienert et al. 2006, 2007; Mubarakshina et al. 2010). It was reported that ROS-damaged molecules are the participants in the signaling pathway (Huang et al.

2019). Thus, disturbance in the ROS synthesis or detoxification in plants causes impact on the ROS-associated redox signaling, then causing oxidative stress that increases cell death (Mittler 2017).

The synchronization between chloroplast and the nucleus is important and needs a continual stream of communication between the organelles, also called the chloroplast-to-nucleus retrograde signaling (Barajas-López et al. 2013). The chloroplast is the main site for ROS synthesis under stress and during development of plastid, causing alteration in the gene expression of nucleus via the redox-dependent retrograde signal transduction (Pesaresi et al. 2006; Woodson and Chory 2008). The chloroplast-to-nucleus retrograde signal transduction pathways are activated by three different factors like intermediate of tetrapyrrole biosynthesis, damage of organellar transcription and translation process and alteration in the redox potential in chloroplast and ROS accumulation (Galvez-Valdivieso and Mullineaux 2010).

The respiratory process in the mitochondria synthesizes ROS during electron transport chain. The regulation of such mitochondrial processes is essential for cell homeostasis that requires sensing and specific signal transduction information towards nucleus and transcriptional machinery reprograming for acclimation responses (Welchen et al. 2014). This type of control of nuclear gene expression from mitochondria-to-nucleus retrograde signaling is exception and important in recent years. In plants, the mitochondria-to-nucleus retrograde signal transduction process is increased by the expression of nuclear genes such as alternative oxidase 1 (AOX1) (Vanlerberghe 2013). The determination of mitochondrial stress-linked genes is visualized as the typical motif present in their promoter called mitochondrial dysfunction motif (MDM), that is seen by TFs of family NAM, NAC, WRKY, and ATAF1/2 families (De Clercq et al. 2013).

It was observed that organelle ROS and redox regulatory networks reveal that signals are connected physiologically, spatially, metabolically, and temporally (Huang et al. 2019). ROS generated in other subcellular compartments are used to transport and travel around signals that facilitate reactions outside cellular organelles. The transmission or transfer of redox signal from many organelles to the cytosol, includes multiple singnaling pathways (Dietz et al. 2016). Signal transmission of redox signaling from several other organelles inside cytosol, requires active transport of ROS (Dietz et al. 2016). Signal transmission can be achieved by cytosolic streaming (Tominaga and Ito 2015). Cytosol is important for regulating metabolic process, posttranslational modification and controls protein synthesis (Moore et al. 2016).

7.7 Conclusions

The determination of ROS in plant biology started with the focus on the ROS-detoxifying synthesis mechanism in the chloroplast. ROS signaling completely depends upon the production, scavenging and its role in plant physio-biochemistry. This chapter gives better understanding of ROS in plants. It is very well documented the role of different ROS in senescence. Organelles such as chloroplast, mitochondria, and peroxisomes synthesize ROS via photosynthesis, respiration, and photorespiration, and participates in redox and retrograde signaling. $O_2^{\cdot -}$, H_2O_2, and $^1O_2$ act as the signaling molecule that plays important role in the plant development, senescence, and signaling.

ROS plays important role in signaling and senescence. Therefore, unraveling the contribution of different genes, TFs, and proteins need to be explored. More studies are needed portrait the stress-induced ROS signal transduction of cells and linking plant metabolic, cellular, genomic, transcriptional and proteomic networks are needed to be fully explored.

References

Allen JF, Hall DO (1973) Superoxide reduction as a mechanism of ascorbate-stimulated oxygen uptake by isolated chloroplasts. Biochem Biophys Res Commun 52(3):856–862

Apel K, Hirt H (2004) Reactive oxygen species: metabolism, oxidative stress, and signal transduction. Annu Rev Plant Biol 55:373–399

Arora A, Sairam RK, Srivastava GC (2002) Oxidative stress and antioxidative system in plants. Curr Sci 82:1227–1238

Asada K (2006) Production and scavenging of reactive oxygen species in chloroplasts and their functions. Plant Physiol 141(2):391–396

Ashraf M (2009) Biotechnological approach of improving plant salt tolerance using antioxidants as markers. Biotechnol Adv 27(1):84–93

Balazadeh S, Wu A, Mueller-Roeber B (2010) Salt-triggered expression of the ANAC092-dependent senescence regulon in Arabidopsis thaliana. Plant Signal Behav 5(6):733–735

Barajas-López JDD, Blanco NE, Strand Å (2013) Plastid-to-nucleus communication, signals 960 controlling the running of the plant cell. Biochim Biophys Acta 1833(2):425–437

Bieker S, Riester L, Stahl M, Franzaring J, Zentgraf U (2012) Senescence-specific alteration of hydrogen peroxide levels in Arabidopsis thaliana and oilseed rape spring variety Brassica napus L. cv. Mozart J Integr Plant Biol 54(8):540–554

Bienert GP, Møller AL, Kristiansen KA, Schulz A, Møller IM, Schjoerring JK, Jahn TP (2007) Specific aquaporins facilitate the diffusion of hydrogen peroxide across membranes. J Biol Chem 282(2):1183–1192

Bienert GP, Schjoerring JK, Jahn TP (2006) Membrane transport of hydrogen peroxide. Biochim Biophys Acta 1758(8):994–1003

Bose J, Rodrigo-Moreno A, Shabala S (2014) ROS homeostasis in halophytes in the context of salinity stress tolerance. J Exp Bot 65(5):1241–1257

Chan KX, Phua SY, Crisp P, McQuinn R, Pogson BJ (2016) Learning the languages of the 1003 chloroplast: retrograde signaling and beyond. Annu Rev Plant Biol 67:25–53

Chen HJ, Wu SD, Huang GJ, Shen CY, Afiyanti M, Li WJ, Lin YH (2012) Expression of a cloned sweet potato catalase SPCAT1 alleviates ethephon-mediated leaf senescence and H2O2 elevation. J Plant Physiol 169(1):86–97

Choudhury S, Panda P, Sahoo L, Panda SK (2013) Reactive oxygen species signaling in plants under abiotic stress. Plant Signal Behav 8(4):e23681

Ciftci-Yilmaz S, Morsy MR, Song L, Coutu A, Krizek BA, Lewis MW, Warren D, Cushman J, Connolly EL et al (2007) The EAR-motif of the Cys2/His2-type zinc finger protein Zat7 plays a key role in the defense response of Arabidopsis to salinity stress. J Biol Chem 282(12): 9260–9268

Corpas FJ, Fernández-Ocaña A, Carreras A, Valderrama R, Luque F, Esteban FJ, Rodríguez-Serrano M, Chaki M, Pedrajas JR et al (2006) The expression of different superoxide dismutase forms is cell-type dependent in olive (Olea europaea L.) leaves. Plant Cell Physiol 47(7): 984–994

Dall'Osto L, Cazzaniga S, Havaux M, Bassi R (2010) Enhanced photoprotection by protein-bound vs. free xanthophyll pools: a comparative analysis of chlorophyll b and xanthophyll biosynthesis mutants. Mol Plant 3(3):576–593

Dall'Osto L, Lico C, Alric J, Giuliano G, Havaux M, Bassi R (2006) Lutein is needed for efficient chlorophyll triplet quenching in the major LHCII antenna complex of higher plants and effective photoprotection in vivo under strong light. BMC Plant Biol 6(32):32

Dat J, Vandenabeele S, Vranová E, Van Montagu M, Inzé D, Van Breusegem F (2000) Dual action of the active oxygen species during plant stress responses. Cell Mol Life Sci 57(5):779–795

Davletova S, Rizhsky L, Liang H, Shengqiang Z, Oliver DJ, Coutu J, Shulaev V, Schlauch K, Mittler R (2005) Cytosolic ascorbate peroxidase 1 is a central component of the reactive oxygen gene network of Arabidopsis. Plant Cell 17(1):268–281

De Clercq I, Vermeirssen V, Van Aken O, Vandepoele K, Murcha MW, Law SR, Inzé A, Ng S, Ivanova A et al (2013) The membrane-bound NAC transcription factor ANAC013 functions in mitochondrial retrograde regulation of the oxidative stress response in Arabidopsis. Plant Cell 25(9):3472–3490

Del Río LA, Donaldson RP (1995) Production of superoxide radicals in glyoxysomal membranes from castor bean endosperm. J Plant Physiol 146(3):283–287

Dietz KJ, Turkan I, Krieger-Liszkay A (2016) Redox- and reactive oxygen species-dependent signaling into and out of the photosynthesizing chloroplast. Plant Physiol 171(3):1541–1550

Dikalov S, Griendling KK, Harrison DG (2007) Measurement of reactive oxygen species in cardiovascular studies. Hypertension 49(4):717–727

Fischer BB, Krieger-Liszkay A, Hideg E, Snyrychová I, Wiesendanger M, Eggen RI (2007) Role of singlet oxygen in chloroplast to nucleus retrograde signaling in Chlamydomonas reinhardtii. FEBS Lett 581(29):5555–5560

Flors C, Fryer MJ, Waring J, Reeder B, Bechtold U, Mullineaux PM, Nonell S, Wilson MT, Baker NR (2006) Imaging the production of singlet oxygen in vivo using a new fluorescent sensor, singlet oxygen Sensor Green®. J Exp Bot 57(8):1725–1734

Foyer CH, Noctor G (2003) Redox sensing and signalling associated with reactive oxygen in chloroplasts, peroxisomes and mitochondria. Physiol Plant 119(3):355–364

Foyer CH, Noctor G (2005) Redox homeostasis and antioxidant signaling: a metabolic interface between stress perception and physiological responses. Plant Cell 17(7):1866–1875

Frahry G, Schopfer P (2001) NADH-stimulated, cyanide-resistant superoxide production in maize coleoptiles analyzed with a tetrazolium-based assay. Planta 212(2):175–183

Fujita M, Fujita Y, Noutoshi Y, Takahashi F, Narusaka Y, Yamaguchi-Shinozaki K, Shinozaki K (2006) Crosstalk between abiotic and biotic stress responses: a current view from the points of convergence in the stress signaling networks. Curr Opin Plant Biol 9(4):436–442

Galvez-Valdivieso G, Mullineaux PM (2010) The role of reactive oxygen species in signalling from chloroplasts to the nucleus. Physiol Plant 138(4):430–439

Gan S, Amasino RM (1997) Making sense of senescence (molecular genetic regulation and manipulation of leaf senescence). Plant Physiol 113(2):313–319

Gao Y, Guo Y-K, Lin S-H, Fang Y-Y, Bai J-G (2010) Hydrogen peroxide pretreatment alters the activity of antioxidant enzymes and protects chloroplast ultrastructure in heat-stressed cucumber leaves. Sci Hortic 126(1):20–26

Garapati P, Xue GP, Munné-Bosch S, Balazadeh S (2015) Transcription factor ATAF1 in Arabidopsis promotes senescence by direct regulation of key chloroplast maintenance and senescence transcriptional cascades. Plant Physiol 168(3):1122–1139

Gechev TS, Van Breusegem F, Stone JM, Denev I, Laloi C (2006) Reactive oxygen species as signals that modulate plant stress responses and programmed cell death. BioEssays 28(11): 1091–1101

Gill SS, Tuteja N (2010) Reactive oxygen species and antioxidant machinery in abiotic stress tolerance in crop plants. Plant Physiol Biochem 48(12):909–930

Golemiec E, Tokarz K, Wielanek M, Niewiadomska E (2014) A dissection of the effects of ethylene, H2O2 and high irradiance on antioxidants and several genes associated with stress and senescence in tobacco leaves. J Plant Physiol 171(3–4):269–275
Gomes A, Fernandes E, Lima JL (2005) Fluorescence probes used for detection of reactive oxygen species. J Biochem Biophys Methods 65(2–3):45–80
González-Pérez S, Gutiérrez J, García-García F, Osuna D, Dopazo J, Lorenzo Ó, Revuelta JL, Arellano JB (2011) Early transcriptional defense responses in Arabidopsis cell suspension culture under high-light conditions. Plant Physiol 156(3):1439–1456
Hernández-Barrera A, Quinto C, Johnson EA, Wu HM, Cheung AY, Cárdenas L (2013) Using hyper as a molecular probe to visualize hydrogen peroxide in living plant cells: a method with virtually unlimited potential in plant biology. Methods Enzymol 527:275–290
Hideg E, Barta C, Kálai T, Vass I, Hideg K, Asada K (2002) Detection of singlet oxygen and superoxide with fluorescent sensors in leaves under stress by photoinhibition or UV radiation. Plant Cell Physiol 43(10):1154–1164
Hideg E, Kálai T, Hideg K (2011) Direct detection of free radicals and reactive oxygen species in thylakoids. Methods Mol Biol 684:187–200
Hideg E, Spetea C, Vass I (1994) Singlet oxygen production in thylakoid membranes during photoinhibition as detected by epr spectroscopy. Photosynth Res 39(2):191–199
Hu X, Bidney DL, Yalpani N, Duvick JP, Crasta O, Folkerts O, Lu G (2003) Overexpression of a gene encoding hydrogen peroxide-generating oxalate oxidase evokes defense responses in sunflower. Plant Physiol 133(1):170–181
Huang H, Ullah F, Zhou D-X, Yi M, Zhao Y (2019) Mechanisms of ROS regulation of plant development and stress responses. Front Plant Sci 10:800. https://doi.org/10.3389/fpls.2019.00800
Ishibashi Y, Yamaguchi H, Yuasa T, Iwaya-Inoue M, Arima S, Zheng SH (2011) Hydrogen peroxide spraying alleviates drought stress in soybean plants. J Plant Physiol 168(13):1562–1567
Jajić I, Sarna T, Szewczyk G, Strzałka K (2015) Changes in production of reactive oxygen species in illuminated thylakoids isolated during development and senescence of barley. J Plant Physiol 184:49–56
Jajić I, Wiśniewska-Becker A, Sarna T, Jemioła-Rzemińska M, Strzałka K (2014) EPR spin labeling measurements of thylakoid membrane fluidity during barley leaf senescence. J Plant Physiol 171(12):1046–1053
Jubany-Marí T, Munné-Bosch S, López-Carbonell M, Alegre L (2009) Hydrogen peroxide is involved in the acclimation of the Mediterranean shrub, Cistus albidus L., to summer drought. J Exp Bot 60(1):107–120
Kärkönen A, Kuchitsu K (2015) Reactive oxygen species in cell wall metabolism and development in plants. Phytochemistry 112:22–32
Kaur G, Sharma A, Guruprasad K, Pati PK (2014) Versatile roles of plant NADPH oxidases and emerging concepts. Biotechnol Adv 32(3):551–563
Khandaker MM, Boyce AN, Osman N (2012) The influence of hydrogen peroxide on the growth, development and quality of wax apple (Syzygium samarangense, [Blume] Merrill & L.M. Perry var. jambu madu) fruits. Plant Physiol Biochem 53:101–110
Kim C, Meskauskiene R, Apel K, Laloi C (2008) No single way to understand singlet oxygen signalling in plants. EMBO Rep 9(5):435–439
Kopyra M, Gwóźdź EA (2003) Nitric oxide stimulates seed germination and counteracts the inhibitory effect of heavy metals and salinity on root growth of Lupinus luteus. Plant Physiol Biochem 41(11–12):1011–1017
Krieger-Liszkay A (2005) Singlet oxygen production in photosynthesis. J Exp Bot 56(411):337–346
Kwak JM, Mori IC, Pei ZM, Leonhardt N, Torres MA, Dangl JL, Bloom RE, Bodde S, Jones JD et al (2003) NADPH oxidase AtrbohD and AtrbohF genes function in ROS-dependent ABA signaling in Arabidopsis. EMBO J 22(11):2623–2633

Laloi C, Stachowiak M, Pers-Kamczyc E, Warzych E, Murgia I, Apel K (2007) Cross-talk between singlet oxygen- and hydrogen peroxide-dependent signaling of stress responses in Arabidopsis thaliana. Proc Natl Acad Sci U S A 104(2):672–677

Lee S, Seo PJ, Lee HJ, Park CM (2012) A NAC transcription factor NTL4 promotes reactive oxygen species production during drought-induced leaf senescence in Arabidopsis. Plant J 70(5):831–844

Liao WB, Zhang ML, Huang GB, Yu JH (2012) Hydrogen peroxide in the vase solution increases vase life and keeping quality of cut oriental× trumpet hybrid lily 'Manissa'. Sci Hortic 139:32–38

Lukyanov KA, Belousov VV (2014) Genetically encoded fluorescent redox sensors. Biochim Biophys Acta 1840(2):745–756

Lushchak VI (2011) Adaptive response to oxidative stress: bacteria, fungi, plants and animals. Comp Biochem Physiol C Toxicol Pharmacol: CBP 153(2):175–190

McRae DG, Thompson JE (1983) Senescence-dependent changes in superoxide anion production by illuminated chloroplasts from bean leaves. Planta 158(3):185–193

Mhamdi A, Van Breusegem F (2018) Reactive oxygen species in plant development. Development 145:15. Available from p.dev:164376

Miller G, Shulaev V, Mittler R (2008) Reactive oxygen signaling and abiotic stress. Physiol Plant 133(3):481–489

Mir AR, Alam P, Hayat S (2022) Perspective of melatonin-mediated stress resilience and Cu remediation efficiency of Brassica juncea in Cu-contaminated soils. Front Plant Sci 13

Mittler R (2002) Oxidative stress, antioxidants and stress tolerance. Trends Plant Sci 7(9):405–410

Mittler R (2006) Abiotic stress, the field environment and stress combination. Trends Plant Sci 11(1):15–19

Mittler R (2017) ROS are good. Trends Plant Sci 22(1):11–19

Mittler R, Vanderauwera S, Gollery M, Van Breusegem F (2004) Reactive oxygen gene network of plants. Trends Plant Sci 9(10):490–498

Mittler R, Zandalinas SI, Fichman Y, Van Breusegem F (2022) Reactive oxygen species signalling in plant stress responses. Nat Rev Mol Cell Biol 23(10):663–679

Monshausen GB, Bibikova TN, Messerli MA, Shi C, Gilroy S (2007) Oscillations in extracellular pH and reactive oxygen species modulate tip growth of Arabidopsis root hairs. Proc Natl Acad Sci U S A 104(52):20996–21001

Moore M, Gossmann N, Dietz KJ (2016) Redox regulation of cytosolic translation in plants. Trends Plant Sci 21(5):388–397

Mubarakshina MM, Ivanov BN, Naydov IA, Hillier W, Badger MR et al (2010) Production and diffusion of chloroplastic H_2O_2 and its implication to signalling. J Exp Bot 61(1306):3577–3587

Munné-Bosch S, Jubany-Marí T, Alegre L (2001) Drought-induced senescence is characterized by a loss of antioxidant defences in chloroplasts. Plant Cell Environ 24(12):1319–1327

Niewiadomska E, Polzien L, Desel C, Rozpadek P, Miszalski Z, Krupinska K (2009) Spatial patterns of senescence and development-dependent distribution of reactive oxygen species in tobacco (Nicotiana tabacum) leaves. J Plant Physiol 166(10):1057–1068

Ogasawara Y, Kaya H, Hiraoka G, Yumoto F, Kimura S, Kadota Y, Hishinuma H, Senzaki E, Yamagoe S et al (2008) Synergistic activation of the Arabidopsis NADPH oxidase AtrbohD by Ca2+ and phosphorylation. J Biol Chem 283(14):8885–8892

Op den Camp RG, Przybyla D, Ochsenbein C, Laloi C, Kim C, Danon A, Wagner D, Hideg E, Göbel C et al (2003) Rapid induction of distinct stress responses after the release of singlet oxygen in Arabidopsis. Plant Cell 15(10):2320–2332

Pastori GM, del Río LA (1997) Natural senescence of pea leaves (an activated oxygen-mediated function for peroxisomes). Plant Physiol 113(2):411–418

Peng LT, Jiang YM, Yang SZ, Pan SY (2005) Accelerated senescence of fresh-cut Chinese water chestnut tissues in relation to hydrogen peroxide accumulation. J Plant Physiol Mol Biol 31:527–532

Pesaresi P, Masiero S, Eubel H, Braun HP, Bhushan S, Glaser E, Salamini F, Leister D (2006) Nuclear 1345 photosynthetic gene expression is synergistically modulated by rates of protein synthesis in 1346 chloroplasts and mitochondria. Plant Cell 18(4):970–991

Pnueli L, Hongjian L, Mittler R (2003) Growth suppression, abnormal guard cell response, and augmented induction of heat shock proteins in cytosolic ascorbate peroxidase (Apx1)—deficient Arabidopsis plants. Plant J 34:187–203

Prochazkova D, Sairam RK, Srivastava GC, Singh DV (2001) Oxidative stress and antioxidant activity as the basis of senescence in maize leaves. Plant Sci 161(4):765–771

Quan LJ, Zhang B, Shi WW, Li HY (2008) Hydrogen peroxide in plants: a versatile molecule of the reactive oxygen species network. J Integr Plant Biol 50(1):2–18

Ramel F, Birtic S, Ginies C, Soubigou-Taconnat L, Triantaphylidès C, Havaux M (2012) Carotenoid oxidation products are stress signals that mediate gene responses to singlet oxygen in plants. Proc Natl Acad Sci U S A 109(14):5535–5540

Rhoads DM, Umbach AL, Subbaiah CC, Siedow JN (2006) Mitochondrial reactive oxygen species. Contribution to oxidative stress and interorganellar signaling. Plant Physiol 141(2):357–366

Ribeiro CW, Korbes AP, Garighan JA, Jardim-Messeder D, Carvalho FEL, Sousa RHV, Caverzan A, Teixeira FK, Silveira JAG et al (2017) Rice peroxisomal ascorbate peroxidase knockdown affects ROS signaling and triggers early leaf senescence. Plant Sci 263:55–65

Rizhsky L, Davletova S, Liang H, Mittler R (2004) The zinc-finger protein Zat12 is required for cytosolic ascorbate peroxidase 1 expression during oxidative stress in Arabidopsis. J Biol Chem 279(12):11736–11743

Robatzek S, Somssich IE (2001) A new member of the Arabidopsis WRKY transcription factor family, AtWRKY6, is associated with both senescence- and defence-related processes. Plant J 28(2):123–133

Rodríguez-Serrano MARÍA, Romero-Puertas MC, Zabalza ANA, Corpas FJ, Gómez M, Del Río LA, Sandalio LM (2006) Cadmium effect on oxidative metabolism of pea (Pisum sativum L.) roots. Imaging of reactive oxygen species and nitric oxide accumulation in vivo. Plant Cell Environ 29(8):1532–1544

Rosenwasser S, Rot I, Sollner E, Meyer AJ, Smith Y, Leviatan N, Fluhr R, Friedman H (2011) Organelles contribute differentially to reactive oxygen species-related events during extended darkness. Plant Physiol 156(1):185–201

Sagi M, Fluhr R (2006) Production of reactive oxygen species by plant NADPH oxidases. Plant Physiol 141(2):336–340

Sandalio LM, Férnandez VM, Rupérez FL, del Río LA (1988) Superoxide free radicals are produced in glyoxysomes. Plant Physiol 87:1–4

Sandalio LM, Rodríguez-Serrano M, Romero-Puertas MC, del Río LA (2008) Imaging of reactive oxygen species and nitric oxide in vivo in plant tissues. Methods Enzymol 440:397–409

Sandalio LM, Rodríguez-Serrano M, Romero-Puertas MC, del Río LAD (2013) Role of peroxisomes as a source of reactive oxygen species (ROS) signaling molecules. Subcell Biochem 69:231–255

Scarpeci TE, Zanor MI, Carrillo N, Mueller-Roeber B, Valle EM (2008) Generation of superoxide anion in chloroplasts of Arabidopsis thaliana during active photosynthesis: a focus on rapidly induced genes. Plant Mol Biol 66(4):361–378

Schippers JH, Foyer CH, van Dongen JT (2016) Redox regulation in shoot growth, SAM maintenance and flowering. Curr Opin Plant Biol 29:121–128

Silva EN, Ferreira-Silva SL, Fontenele Ade V, Ribeiro RV, Viégas RA, Silveira JAG (2010) Photosynthetic changes and protective mechanisms against oxidative damage subjected to isolated and combined drought and heat stresses in Jatropha curcas plants. J Plant Physiol 167(14):1157–1164

Springer A, Acker G, Bartsch S, Bauerschmitt H, Reinbothe S, Reinbothe C (2015) Differences in gene expression between natural and artificially induced leaf senescence in barley. J Plant Physiol 176:180–191. https://doi.org/10.1016/j.jplph.2015.01.004

Suzuki N, Bajad S, Shuman J, Shulaev V, Mittler R (2008) The transcriptional co-activator MBF1c is a key regulator of thermotolerance in Arabidopsis thaliana. J Biol Chem 283(14):9269–9275
Suzuki N, Rizhsky L, Liang H, Shuman J, Shulaev V, Mittler R (2005) Enhanced tolerance to environmental stress in transgenic plants expressing the transcriptional coactivator multiprotein bridging factor 1c. Plant Physiol 139(3):1313–1322
Swartz HM, Khan N, Khramtsov VV (2007) Use of electron paramagnetic resonance spectroscopy to evaluate the redox state in vivo. Antioxid Redox Signal 9(10):1757–1771
Takeda S, Gapper C, Kaya H, Bell E, Kuchitsu K, Dolan L (2008) Local positive feedback regulation determines cell shape in root hair cells. Science 319(5867):1241–1244
Tarpey MM, Wink DA, Grisham MB (2004) Methods for detection of reactive metabolites of oxygen and nitrogen: in vitro and in vivo considerations. Am J Physiol Regul Integr Comp Physiol 286(3):R431–R444
Tominaga M, Ito K (2015) The molecular mechanism and physiological role of cytoplasmic 1462 streaming. Curr Opin Plant Biol 27:104–110
Torres MA, Dangl JL (2005) Functions of the respiratory burst oxidase in biotic interactions, abiotic stress and development. Curr Opin Plant Biol 8(4):397–403
Torres MA, Jones JD, Dangl JL (2006) Reactive oxygen species signaling in response to pathogens. Plant Physiol 141(2):373–378
Triantaphylidès C, Krischke M, Hoeberichts FA, Ksas B, Gresser G, Havaux M, Van Breusegem F, Mueller MJ (2008) Singlet oxygen is the major reactive oxygen species involved in photooxidative damage to plants. Plant Physiol 148(2):960–968
Van Breusegem F, Bailey-Serres J, Mittler R (2008) Unraveling the tapestry of networks involving reactive oxygen species in plants. Plant Physiol 147(3):978–984
Van Breusegem F, Dat JF (2006) Reactive oxygen species in plant cell death. Plant Physiol 141(2): 384–390
Vanlerberghe GC (2013) Alternative oxidase: a mitochondrial respiratory pathway to maintain metabolic and Signaling homeostasis during abiotic and biotic stress in plants. Int J Mol Sci 14(4):6805–6847
Waszczak C, Kerchev PI, Mühlenbock P, Hoeberichts FA, Van Der Kelen K, Mhamdi A, Willems P, Denecker J, Kumpf RP et al (2016) SHORT-ROOT deficiency alleviates the cell death phenotype of the Arabidopsis catalase2 mutant under photorespiration-promoting conditions. Plant Cell 28(8):1844–1859
Welchen E, García L, Mansilla N, Gonzalez DH (2014) Coordination of plant 1501 mitochondrial biogenesis: keeping pace with cellular requirements. Front Plant Sci 4(551):1502
Wong HL, Pinontoan R, Hayashi K, Tabata R, Yaeno T, Hasegawa K, Kojima C, Yoshioka H, Iba K et al (2007) Regulation of rice NADPH oxidase by binding of Rac GTPase to its N-terminal extension. Plant Cell 19(12):4022–4034
Woodson JD, Chory J (2008) Coordination of gene expression between organellar and nuclear 1508 genomes. Nat Rev Genet 9(5):383–395
Wudick MM, Li X, Valentini V, Geldner N, Chory J, Lin J, Maurel C, Luu DT (2015) Subcellular redistribution of root aquaporins induced by hydrogen peroxide. Mol Plant 8(7):1103–1114
Yamauchi T, Yoshioka M, Fukazawa A, Mori H, Nishizawa NK, Tsutsumi N, Yoshioka H, Nakazono M (2017) An NADPH oxidase RBOH functions in rice roots during lysigenous aerenchyma formation under oxygen-deficient conditions. Plant Cell 29(4):775–790
Yang TPBW, Poovaiah BW (2002) Hydrogen peroxide homeostasis: activation of plant catalase by calcium/calmodulin. Proc Natl Acad Sci U S A 99(6):4097–4102
Yesbergenova Z, Yang G, Oron E, Soffer D, Fluhr R, Sagi M (2005) The plant Mo-hydroxylases aldehyde oxidase and xanthine dehydrogenase have distinct reactive oxygen species signatures and are induced by drought and abscisic acid. Plant J 42(6):862–876
Yun KY, Park MR, Mohanty B, Herath V, Xu F, Mauleon R, Wijaya E, Bajic VB, Bruskiewich R et al (2010) Transcriptional regulatory network triggered by oxidative signals configures the early response mechanisms of japonica rice to chilling stress. BMC Plant Biol 10(1):16

Zarepour M, Kaspari K, Stagge S, Rethmeier R, Mendel RR, Bittner F (2010) Xanthine dehydrogenase AtXDH1 from Arabidopsis thaliana is a potent producer of superoxide anions via its NADH oxidase activity. Plant Mol Biol 72(3):301–310

Zeng J, Dong Z, Wu H, Tian Z, Zhao Z (2017) Redox regulation of plant stem cell fate. EMBO J 36(19):2844–2855

Zhang G, Zhang M, Zhao Z, Ren Y, Li Q, Wang W (2017) Wheat TaPUB1 modulates plant drought stress resistance by improving antioxidant capability. Sci Rep 7(1):7549

Zhang MP, Zhang CJ, Yu GH, Jiang YZ, Strasser RJ, Yuan ZY, Yang XS, Chen GX (2010) Changes in chloroplast ultrastructure, fatty acid components of thylakoid membrane and chlorophyll a fluorescence transient in flag leaves of a super-high-yield hybrid rice and its parents during the reproductive stage. J Plant Pysiol 167(4):277–285. 121

Zhang X-L, Jia X-F, Yu B, Gao Y, Bai J-G (2011) Exogenous hydrogen peroxide influences antioxidant enzyme activity and lipid peroxidation in cucumber leaves at low light. Sci Hortic 129(4):656–662

Zhou M, Diwu Z, Panchuk-Voloshina N, Haugland RP (1997) A stable nonfluorescent derivative of resorufin for the fluorometric determination of trace hydrogen peroxide: applications in detecting the activity of phagocyte NADPH oxidase and other oxidases. Anal Biochem 253(2):162–168

Zimmermann P, Heinlein C, Orendi G, Zentgraf U (2006) Senescence-specific regulation of catalases in Arabidopsis thaliana (L.) Heynh. Plant Cell Environ 29(6):1049–1060

Hazardous Phytotoxic Nature of Reactive Oxygen Species in Agriculture

8

Khushbu Sharma, Priyanka Devi, Prasann Kumar, Abhijit Dey, and Padmanabh Dwivedi

Abstract

During the course of normal metabolic activities in plants such as photosynthesis, respiration, and cellular metabolism, several reactive oxygen species are produced (hydroxyl peroxide and superoxide anion). The presence of these reactive oxygen species can cause plants to suffer from oxidative stress, meaning they suffer from an unhealthful environment. Because of the accumulation of lipids in the cell membrane, increased amounts of ROS are produced endogenously within the cell. This adversely affects the production of soluble proteins and sugars, as well as reduces the absorption of elements throughout the cell membrane. During stress, the whole production process of ROS is regulated by a signal protein that acts as a signal. These oxidative stresses are very common in agriculture crops because of some natural and anthropogenic factors such as the accumulation of heavy metal ions in the rhizosphere. From the rhizosphere, these ions get transported all over the plant and cause damage to the normal metabolic functioning of the plant under stress. In this instance, there are not enough energy molecules available for the plant to continue photosynthesis, which further results in inadequate yields. The plant experiences some physiological changes as a result of oxidative stress in order to scavenge the effects of oxidative stress. There are several issues we address in this chapter, including the mechanisms by which

K. Sharma · P. Devi · P. Kumar (✉)
Department of Agronomy, School of Agriculture, Lovely Professional University, Phagwara, India
e-mail: prasann.21784@lpu.co.in

A. Dey
Department of Life Sciences, Presidency University, Kolkata, India

P. Dwivedi (✉)
Department of Plant Physiology, Institute of Agricultural Sciences, Banaras Hindu University, Varanasi, India

M. Faizan et al. (eds.), *Reactive Oxygen Species*,
https://doi.org/10.1007/978-981-19-9794-5_8

reactive oxygen species are introduced into crops, generation of free radicals, mitochondrial oxidative stress, signaling, and the phytotoxic nature of these in agriculture crops. The chapter also deals with the scavenging mechanisms that plants have evolved to protect themselves from reactive oxygen species damage at the enzymatic and molecular levels.

Keywords

Agriculture · Biotic · Phytotoxic · Reactive oxygen species · Rhizosphere · Toxicity

Abbreviations

ABA	Abscisic acid
ADP	Adenosine di-phosphate
Ca^{2+}	Calcium
Cd	Cadmium
DHAR	Dehydroascorbate reductase
DNA	Deoxyribonucleic acid
GPX	Glutathione peroxidase
GSH	Glutathione
GST	Glutathione-S-transferases
H_2O_2	Hydrogen peroxide
Hg	Mercury
MDA	Malondialdehyde
MDHAR	Mono-dehydroascorbate reductase
Na^+	Sodium
NADPH	Nicotinamide Adenine Dinucleotide Phosphate
02	Oxygen
Pb	Lead
PPO	Polyphenol oxidase
ROS	Reactive oxygen species

8.1 Introduction

There is a steady increase in abiotic stresses in the environment these days such as heavy metal stress, salinity stress, drought stress, heat stress, and others. As a result of these types of stresses in the environment, crop production is greatly affected. Under stressed conditions, plants undergo a variety of physiological changes as a result of the disruption in normal metabolic activities such as photorespiration, photosynthesis, etc. In the cell, the plant undergoes oxidative stress, such as the generation of reactive oxygen species (ROS), such as hydrogen peroxide, hydroperoxyl, and hydroxyl, among others. This will unveil the important role that oxygen plays in electron transport by taking up the final waste products from respiration, all molecules with low energy and free ions of hydrogen, but during stress situations, whole physiological cycles become disrupted, and the low energy molecules do not get accepted by oxygen, leading to the generation of Reactive Oxygen species, free radical ions of oxygen (Mishra et al. 2011). There is, therefore, a disturbance in the electron transport system in the mitochondrial membrane that occurs as a result of damaged mitochondria that generate ROS, and over-accumulation of these ROS ions in the cellular system of cells that leads to toxicity, increase in lipid concentration on the cell membrane, low chlorophyll content, ultimately resulting in low photosynthesis, total protein concentration, etc. (Evans et al. 2004). During the process of mitigating stress in plants, several enzymes get activated, such as catalase, peroxidase, amylase, superoxide dismutase in order to maintain equilibrium within the cell; these enzymes enable some reduction reactions to regulate the concentration of ROS (Mishra et al. 2011). It has been demonstrated that both enzymatic and nonenzymatic activities are involved in maintaining the optimal level of ROS within a cell. Normal growth and development are impossible without the ability to regulate the ROS levels in the body. In the case of plants, the defense mechanism adopted in order to scavenge the ROS produced will be managed by the cells themselves (Meriga et al. 2004). The cell is ultimately killed as a result of all disturbances in it, such as photobleaching and excess ROS production. The activation of signal proteins during stress conditions results in transporting them to the nuclei of the cells where all the enzymatic reactions begin. The ROS are always the first to respond during times of stress because environmental factors always get the first response.

8.2 Chemistry of Reactive Oxygen Species

Approximately 20.95 percent of the oxygen found in the atmosphere is found in the atmosphere, which puts it third behind hydrogen and helium. There are many compounds that are found in the earth's crust. One of the most common is calcium carbonate. In addition to the number of unpaired electrons present in spin, the reactivity of oxygen is also influenced by the number of unpaired electrons within each spin. The mitochondria and chloroplasts of plant cells are the two most significant sites for the production of ROS. As a result of the chlorophyll pigment's ability to absorb light quantum in the chloroplast while photosynthesis is taking place, the pigment gains energy from the triplet state. This time period is very short (3.1–3.9 *s) and a molecule of oxygen is transported outside the chloroplast and diffuses to the nuclei through the plasma membrane to communicate with them (Meriga et al. 2004). Further, the oxygen molecule that is released at the end of the electron transport chain is also activated by plasma membrane oxidation. It will combine with hydrogen peroxide during the process. It should be noted that there are some species of ROS that are not radical such as hydrogen peroxide, ozone, hydroxide, etc. Other groups of no radical are also found in plants like carbonyls, hypochlorous acids, and many more. There are many types of reactive oxygen species which can be classified into different categories such as reactive oxygen intermediates, carbonyl, hypobromous acid, carbonate, hypoiodous acid, etc. Further, ROS species are a consequence of incomplete reduction processes.

Whenever a plant is subjected to abiotic stress, it undergoes oxidative stress, which results in the evolution of ROS species. There is a partnership between the redox potential and the enzymatic activity that maintains the equilibrium in stress situations. In contrast, a group of several enzymes found in plant cells plays an active role in maintaining cell metabolism during times of stress. As mentioned before, in plant cells, there is a critical level of ROS that must be sustained in order for the cell to perform its metabolic activities. ROS production, as well as ROS scavenging mechanisms in cells, were both controlled by certain signal proteins present in the cell. Consequently, ROS is itself considered to be a signaling molecule that is able to control the physiological activities of cells (Evans et al. 2004). However, when ROS are produced within the cell, the result is oxidative damage to the cell. There is a strong correlation between the proper regulation of ROS concentration and concomitant processes (phenomena affected by natural processes). It is believed that during abiotic stress conditions, the equilibrium of ROS generation and scavenging is disrupted, and the accumulation of ROS leads to overproduction of ROS. All the metabolic functions of the cell are also disturbed. This leads to photobleaching of chlorophyll, accumulation of lipid peroxidase activity in the cell, decrease in protein content, and increase of the membrane injury index. Several enzymes within a cell such as catalase, lipid peroxidase, and superoxide dismutase, are stimulated to mitigate the effects of stress.

8.3 Oxidative Stress under Abiotic Stress

There are both natural and anthropogenic factors that play a role in causing abiotic stress to occur in plants. Fields are normally where plants are grown, and the major source of energy for plants is the soil and the water that makes up the soil. But changing weather conditions and contamination of soil and water create a great threat to plants, as well as humans. Moreover, the waste produced from industries gets imported directly into rivers, which results in eutrophication of the water and excessive concentration of heavy metals in the soil and water. A condition called heavy metal stress could nicely describe this condition (Mishra et al. 2011). The presence of heavy metal stress is not the only source of ROS in cells. There are several other stresses such as drought stress, salinity stress, heat stress, all of which contribute to the formation of ROS species inside the cell. We already knew that ROS is produced at a number of locations such as the apoplast, chloroplast, mitochondria, plasma membrane, and peroxisomes as the major sites of ROS generation. One of the major disturbances resulting from abiotic stresses is seen in photosynthesis as a result of a reduction in carbon dioxide and oxygen concentration. In contrast, there is a significant variation in the production of ROS, even within species, which reflects differences in morphology, genetic resilience and the degree to which stress is revealed.

8.3.1 Oxidative Stress under Salinity

Plants are impacted by salinity stress by changing the rhizosphere as well as the plants themselves. This takes place due to ion toxicity in the soil, and the plant receives insufficient amounts of water which results in nutritional deficiencies resulting in excessive ROS production (Evans et al. 2004). It has been observed that the major impact of salinity stress has been observed in the young leaves of the plants. Researchers found that the rice grown under salinity stress exhibited higher levels of ROS and electrolyte leakage, as well as higher levels of lipid peroxidase activity at the end of the growing season than in the control.

8.4 Overview of Plant Antioxidant Defense System

The formation of ROS can be directly or indirectly slowed down by a small group of antioxidants. Additionally, there are some antioxidant enzymes that are also active in regulating the oxidative stress response in plants, since their lower molecular weight allows them to be transported into the cell with ease. In addition to this, a host of nonenzymatic antioxidants that provide a defense mechanism for our cells are also natural antioxidants such as alkaloids, phenolic compounds, flavonoids, and alkaloids. In order to inhibit overproduction of ROS, the equilibrium of ROS production gets signaled during induction of abiotic stresses, and the non-protein amino acids act as scavengers, including catalase, superoxide dismutase, peroxidase, polyphenol oxidase (PPO), APX, MDHAR, DHAR, GR, GPX, GST, TRX, and PRX.

8.5 Revisiting ROS Signaling in Plant Defense

As a result of the disturbance of different metabolic functions and physiological disorders, excessive amounts of ROS are generated under conditions of abiotic stress. It should be noted that the antioxidant defense pathways such as the AsA-GSH pathway require a lot of energy in the form of NADPH. Once this energy is depleted, these pathways are incapable of preventing ROS toxicity (Evans et al. 2004). It was only at the end of the twentieth century and the beginning of the twenty-first century that ROS (especially H_2O_2) began to receive significant attention as a mechanism of plant responses to stresses. H_2O_2 has been identified by some scientists as a signaling molecule, which results in adaptation processes and confers tolerance to a variety of biotic and abiotic stresses. During stress, the chloroplast may generate ROS that may divert electrons from the photosynthetic machinery, preventing the overloading of the antenna and resulting damage. During stress, mitochondria may also be protected by ROS. Depending on the source of ROS generation, it might be that H_2O_2 could be used to trigger downstream signaling cascades through the activation of Ca^{2+} and MAPK pathway through cell wall peroxidase. Further, plant hormones such as ethylene (ET) and abscisic acid (ABA) are known to play a dual role in response to stress, i.e., they interact with ROS to achieve stress tolerance and also affect the stress response over a wide range of conditions. Under abiotic stress conditions, ROS can not only transmit signals and communicate with hormones, but can also regulate metabolic fluxes. This enables them to control plant acclimation processes involving redox reactions affecting transcription and translation of stress acclimation proteins and enzymes, ultimately protecting plant cells from damage. The production and degradation of H_2O_2 are further mediated by NO and Ca^{2+} signaling pathways. These pathways indirectly control plant growth and development, as well as cellular and physiological responses to adverse abiotic conditions. We have listed some key reports dealing with the effects of H_2O_2 treatment under different abiotic stress conditions, since endogenous H_2O_2 increases tolerance to abiotic stress. Exogenous application of H_2O_2 is also gaining increasing attention, especially as recent studies have shown its efficacy in increasing tolerance to abiotic stress. In addition, ROS collaborates with RNS, RSS, and RCS as a signal transduction pathway that takes place under stress conditions. Because cellular antioxidant levels can affect ROS generation and signaling, antioxidant levels may also influence ROS generation. In the meantime, RSS is involved in the production, perception, and further signaling of ROS and RNS, while RCS is involved in the downstream perception and mediation of ROS under different stress conditions. Plant cells contain both enzymatic and nonenzymatic antioxidants that are catalyzed in a variety of places like cystol, mitochondria, thylakoids, peroxisomes, chloroplasts, and vacuoles. In order for superoxide dismutase to perform its activity, free molecules of oxygen ions must be converted into hydrogen peroxide, which is diffused into water molecules by the action of a second enzyme called catalase.

8.6 The Hazardous Effects of ROS

During incomplete reduction of molecular oxygen in the cell, reactive oxygen species are formed. These highly active molecules contain oxygen, as a result of incomplete reduction of molecular oxygen. There are many free radical species known which can be synthesized from free radicals or compounds. ROS are important by-products of oxygen metabolism, and their concentration in the cell is determined by the balance of their cellular synthesis and clearance by a number of antioxidant substances and enzymes. In the decade following their discovery in biological materials, ROS were assumed to produce only harmful effects. They had been linked to a variety of diseases. When the production of ROS exceeds the level of antioxidant defense mechanism in the cell, oxidative stress results, causing damage to nucleic acids, proteins, and lipids. The ROS have potentially deleterious influence on cancer development, neurotoxicity, cholesterol, diabetes, and aging that has been linked to innumerable factors. The existence of ROS in cells, however, has been proposed in the recent past that oxidative stress production may have been evolutionarily selected to serve certain important functions in cells.

8.7 Oxidative Damage as a Biomarker of Ageing

In order to regulate themselves, free radicals tend to take electrons from other molecules, which makes them very dangerous in cells where they may damage macromolecules like DNA, proteins, and lipids (Pham-Huy et al. 2008). DNA undergoes a change from guanine to 8-oxoguanine when it is exposed to ROS in order to mate with cytosine and adenine. As a result of this mutation, there can be double-strand breaks (DSBs) in DNA that cause genomic instability and result in double-strand breaks in the DNA (Kregel and Zhang 2007). In some cases, proteins can be damaged when ROS oxidize amino acid side chains and backbones, particularly in residues of cysteine- and methionine-containing thiols. As a result, structural changes can occur in the proteins, which may result in a loss of functionality (Sohal 2002) or can be used to manipulate the signaling capabilities of ROS. It is also important to note that when ROS is present in the environment, lipids undergo lipid peroxidation, which leads to damage of the cell membrane and the production of reactive by-products that can further damage the cell (Mylonas and Kouretas 1999). Researchers have found that DNA damage (Moskalev et al. 2013), protein carbonization (Jha and Rizvi 2011), lipid peroxidation (Massudi et al. 2012), and mitochondrial DNA damage are all associated with aging (Park and Larsson 2011). Thus, for a variety of specific variables, there have been proposed biomarkers of aging (Liguori et al. 2018). Plant ROS are crucial signaling molecules that manage normal development and stress responses as well as cause permanent DNA damage and cell death for plants.

8.8 ROS and Cell Damage

In vivo, cells are subject to a delicate balance between the formations of ROS and the antioxidant defense systems that maintain cell integrity. As well as elevated levels of oxygenation, inflammation, and hypoperfusion (a process that creates excessive ROS), there can also be deficiencies or restrictions on antioxidant defences that can disrupt this equilibrium. The mechanism of ROS-induced cell death has been proposed to be a combination of mechanisms. A higher concentration of ROS can damage proteins, lipids, and nucleic acids directly, leading to cell death. It could be argued that protein oxidation and nitrosylation (carbonylation, nitration, and nitrotyrosine production, for instance) involve significant cellular damage and can seriously compromise both enzyme function and growth factor production

(Stadtman and Levine 2000). As a result of lipoperoxidation, ceramide is released and sphingomyelinase is activated, triggering cell death as well as apoptosis (Fruhwirth and Hermetter 2008). An important component of necrosis and apoptosis lies in the oxidation of nucleic acid, together with premature ageng and DNA breakage (Auten et al. 2002). It is dependent on the severity and consequences of the alteration and the damage that the cells can heal.

8.9 ROS Damage to Biomolecules

The management of ROS secretion and disposal is essential for the prevention of oxidative stress. In order to be considered for "oxidative stress," a cell must accumulate a sufficient amount of ROS to overcome the cell's defense mechanisms (Meriga et al. 2004). When the balance between ROS generation and scavenging breakdowns is due to a variety of stress factors such as salt, drought, strong light, metal toxicity, viruses, and many more, it causes damage to tissues. There are a wide variety of biomolecules that are harmed by a buildup of ROS which include lipids, proteins, and DNA. It has been shown that these changes can affect intrinsic membrane characteristics such as fluidity, ion transport, enzyme activity loss, protein cross-linking, protein synthesis suppression, DNA damage, and so on, all of which can contribute to cell death (Fig. 8.1).

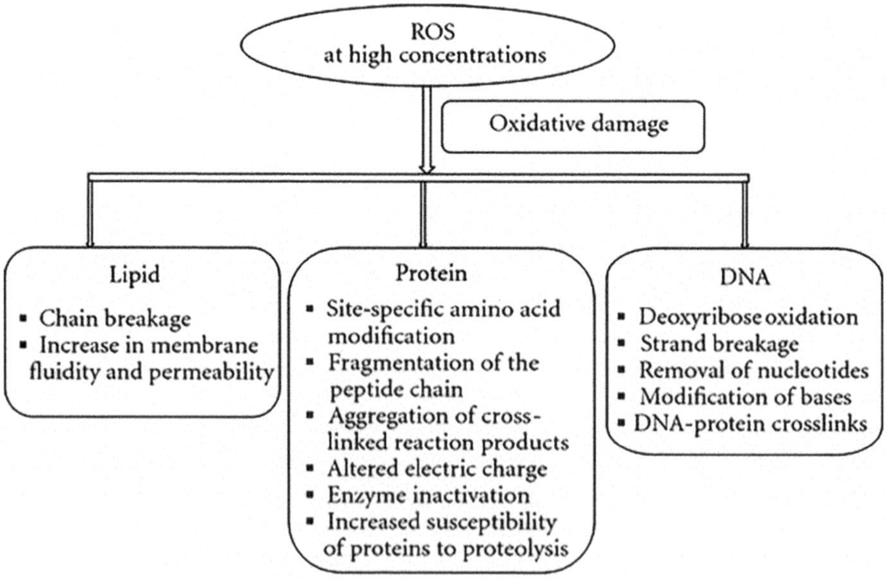

Fig. 8.1 ROS-induced oxidative damage to lipids, proteins, and DNA

8.10 Oxidative Damage to Lipids

There is a threshold level above which the proportion of ROS exceeds, which leads to an increase in lipid peroxidation in both the cellular and organelle membranes, which negatively affects normal cellular function. As a consequence of lipid peroxidation, oxidative stress is exacerbated because lipid-derived radicals are formed, which can react with proteins and DNA to cause damage. It is often used to detect ROS-mediated damage to cell membranes under stress conditions by measuring the amount of lipid peroxidation in the membrane, and thus determining the severity of that damage. There is growing evidence that plants growing in stressful environments experience oxidation (that is, degradation) of their body fats (Mishra et al. 2011). Due to an increased level of lipid peroxidation and ROS, as a result of an increase in lipid peroxidation, a high level of ROS is produced resulting in increased lipid peroxidation. As a result of the peroxidation of unsaturated fatty acids in the phospholipids, malondialdehyde (MDA) is one of the end products, which can damage the cell membrane (Halliwell and Gutteridge 2015). There are two typical sites of ROS assault on phospholipid molecules, namely the unsaturated (double) bond between two carbon atoms and the ester connection between glycerol and the fatty acid. In addition to phospholipids, membrane phospholipids also contain polyunsaturated fatty acids (PUFAs) that are particularly vulnerable to ROS attacks. Several studies have shown that the peroxidation of polyunsaturated fatty acids causes the release of ROS that increase cell function and permeability.

8.11 ROS Damage to Proteins

There are a number of ways in which ROS)can alter proteins in a direct or indirect manner depending on their reaction to the ROS. Among the many examples of direct changes in a protein's function are nitrogenation, carbonylation, disulfide bond formation, and glutathionylation. There is an indirect way to alter proteins by coupling with the breakdown products of fatty acid peroxidation (Yamauchi et al. 2008). Enhanced susceptibility to protein proteolysis is caused by excessive ROS generation because it causes amino acid fragmentation, fragmentation of peptide chains, aggregation of cross-linked reaction products, and a change in electric charge. In tissues that have been damaged by oxidative stress, carbonated proteins are present in higher concentrations, which is considered a measure of the amount of protein oxidation (Romero-Puertas et al. 2002). As a result of different adverse environmental conditions, plants show increased levels of protein modification (Romero-Puertas et al. 2002). There is a wide variation in the resistance of amino acids within a peptide to ROS damage. There is a high risk of ROS causing damage to thiol groups and amino acids containing sulfur. As a result of the activation of oxygen, H-atom can be removed from cysteine residues, forming a thiol radical which can then form a disulfide bridge when it crosses with another thiol radical. It has been shown that the depletion of the active thiol groups in proteins, including Cd, Pb, and Hg has been caused by the toxicity of heavy metals, such as Cd, Pb, and

Hg (Stohs and Bagchi 1995). It is also possible to form methionine sulphoxide derivatives by adding oxygen to a methionine molecule (Brot and Weissbach 1982). Upon exposure to ROS, tyrosine forms rapid crosslinks that give rise to tyrosine products, which can then bind to the protein site (Davies 1987). It is an irreversible process in which ROS react with iron and sulfide, resulting in the activation of different enzymes (Gardner and Fridovich 1991). An iron ion (Fe) forms a divalent bond with a protein at the binding site, where it reacts with the protein to form an OH^- ion, which results in the oxidation of residues of amino acids surrounding the protein (Stadtman 1986).

8.12 ROS Hazard to DNA

The ROS which produce damage to DNA play an important role. There are several types of DNA in plants such as those that are found in the nucleus, mitochondria, and chloroplasts that experience oxidative damage. The DNA of a cell is the genetic material of the cell, and any damage to it can cause drastic changes in the proteins encoded in the DNA, thus leading to problems or the complete inactivation of the genes encoded in the DNA. The oxidative attack on DNA is associated with a variety of consequences including deoxyribose oxidation, strand breakage, nucleotide loss, variation in the organic bases of nucleotides, and changes in DNA-protein crosslinks. It is also possible for alterations in one nucleotide to cause conflict with the nucleotide in the other nucleotide, which may result in mutations in the other strand. The DNA degradation rate of plants under different conditions of environmental stress such as salinity and metal toxicity has been found to increase significantly with these stressors (Meriga et al. 2004). The sugar moiety of DNA, as well as the base moieties, are both very sensitive to oxidation by ROS. The most common oxidative stress on DNA bases is the addition of hydroxyl radicals to double bonds, while the most common cause of sugar degradation is the abstraction of hydrogen from deoxyribose. There are many types of purine and pyrimidine bases that react with the hydroxyl radical in nature, including the deoxyribose backbone (Halliwell and Gutteridge 2015). It has been shown that OH can produce a variety of compounds from DNA bases, the most common of which are hydroxymethyl urea, urea, thymine glycol, thymine, and adenine ring-opened products, and saturated products. As a result, the most prevalent by-product is 8-Hydroxyguanine. It has been shown that H_2O_2 and O_2 do not react with bases at all, whereas O_2 interacts exclusively with guanine (Dizdaroglu 1993). As a result of ROS-induced DNA damage, mutagenic changes may also occur. In particular, mutations caused by selective alteration of the G: C regions are indicative of oxidative DNA damage. Indirectly, ROS attack DNA bases through the production of reactive products produced by ROS attacking other macromolecules like lipids. The ROS attack on DNA sugars results in the breaking of single-strand DNA strands. It is believed that ROS are responsible for the loss of a hydrogen atom from deoxyribose's C4 position, resulting in the formation of the radical deoxyribose, which then reacts to cause DNA strand breaks (Evans et al. 2004). In healthy conditions, neither H_2O_2

nor O_2• can break strands in vitro causing strand breaks. The DNA and protein crosslinks are produced when OH oxidizes DNA and proteins linked to it. As previously mentioned, DNA-protein chains are difficult to repair and can be fatal if replication or transcription occurs before repair. It has been demonstrated that mitochondrial and chloroplast DNA lack protective proteins, histones, and are located near ROS-producing systems, which make them more vulnerable to oxidative damage than nuclear DNA (Richter 1992). Despite the fact that damaged DNA has a repair system, excessive alterations caused by ROS can lead to irreversible damage to the DNA, which can have detrimental effects on the cell.

8.13 Conclusion

In response to stress, a signal protein that acts as a signal affects the whole production process of ROS in the body. In other words, there is an alteration in the electron transport system that is caused by damaged mitochondria that are capable of generating ROS. The over-accumulation of ROS ions in the cell membrane leads to potential toxicity, an increase in the level of lipids on the membrane, a drop in chlorophyll content, ultimately resulting in a reduction in the amount of photosynthesis, total protein concentration, and so forth. When a plant is exposed to abiotic stress, it becomes oxidative stressed, resulting in the emergence of ROS species in the plant. It has been proven through studies that when stress is applied, there exists a partnership between redox potential and enzymatic activity that maintains the equilibrium of the system. During times of stress, on the other hand, a group of several enzymes present in plant cells plays an active role in maintaining cell metabolism. For a plant cell to function correctly, it needs a certain level of ROS. In order for the cell to perform its metabolic functions, this degree of ROS must be sustained. The production of ROS, as well as the mechanisms that scavenge ROS, are both controlled by certain signal proteins involved in the cell.

Acknowledgments All the authors acknowledge thanks to Lovely Professional University, India and Banaras Hindu University, India for providing possible assistance.

Conflict of Interest We wish to confirm that there are no known conflicts of interest associated with this publication that could have influenced its outcome.

References

Auten RL, Whorton MH, Nicholas Mason S (2002) Blocking neutrophil influx reduces DNA damage in hyperoxia-exposed newborn rat lung. Am J Respir Cell Mol Biol 26(4):391–397

Brot N, Weissbach H (1982) The biochemistry of methionine sulfoxide residues in proteins. Trends Biochem Sci 7(4):137–139

Davies KJ (1987) Protein damage and degradation by oxygen radicals. I. General aspects. J Biol Chem 262(20):9895–9901

Dizdaroglu M (1993) Chemistry of free radical damage to DNA and nucleoproteins. DNA Free Radicals:19–39
Evans MD, Dizdaroglu M, Cooke MS (2004) Oxidative DNA damage and disease: induction, repair and significance. Mutat Res Rev Mutat Res 567(1):1–61
Fruhwirth GO, Hermetter A (2008) Mediation of apoptosis by oxidized phospholipids. Lipids Health Dis:351–367
Gardner PR, Fridovich I (1991) Superoxide sensitivity of the Escherichia coli 6-phosphogluconate dehydratase. J Biol Chem 266(3):1478–1483
Halliwell B, Gutteridge JM (2015) Free radicals in biology and medicine. Oxford university press, USA
Jha R, Rizvi SI (2011) Carbonyl formation in erythrocyte membrane proteins during aging in humans. Biomed Pap Med Fac Univ Palacky Olomouc Czech Repub 155(1):39–42
Kregel KC, Zhang HJ (2007) An integrated view of oxidative stress in aging: basic mechanisms, functional effects, and pathological considerations. Am J Phys Regul Integr Comp Phys 292(1): R18–R36
Liguori I, Russo G, Curcio F, Bulli G, Aran L, Della-Morte D et al (2018) Oxidative stress, aging, and diseases. Clin Interv Aging 13:757
Massudi H, Grant R, Braidy N, Guest J, Farnsworth B, Guillemin GJ (2012) Age-associated changes in oxidative stress and NAD+ metabolism in human tissue. PLoS One 7(7):e42357
Meriga B, Reddy BK, Rao KR, Reddy LA, Kishor PK (2004) Aluminium-induced production of oxygen radicals, lipid peroxidation and DNA damage in seedlings of rice (Oryza sativa). J Plant Physiol 161(1):63–68
Mishra S, Jha AB, Dubey RS (2011) Arsenite treatment induces oxidative stress, upregulates antioxidant system, and causes phytochelatin synthesis in rice seedlings. Protoplasma 248(3): 565 577
Moskalev AA, Shaposhnikov MV, Plyusnina EN, Zhavoronkov A, Budovsky A, Yanai H, Fraifeld VE (2013) The role of DNA damage and repair in aging through the prism of Koch-like criteria. Ageing Res Rev 12(2):661–684
Mylonas C, Kouretas D (1999) Lipid peroxidation and tissue damage. In Vivo (Athens, Greece) 13(3):295–309
Park CB, Larsson NG (2011) Mitochondrial DNA mutations in disease and aging. J Cell Biol 193(5):809–818
Pham-Huy LA, He H, Pham-Huy C (2008) Free radicals, antioxidants in disease and health. Int J Biomed Sci 4(2):89
Richter C (1992) Reactive oxygen and DNA damage in mitochondria. Mutat Res/DNAging 275(3–6):249–255
Romero-Puertas MC, Palma JM, Gómez M, Del Rio LA, Sandalio LM (2002) Cadmium causes the oxidative modification of proteins in pea plants. Plant Cell Environ 25(5):677–686
Sohal RS (2002) Role of oxidative stress and protein oxidation in the aging process. Free Radic Biol Med 33:37–44
Stadtman ER (1986) Oxidation of proteins by mixed-function oxidation systems: implication in protein turnover, ageing and neutrophil function. Trends Biochem Sci 11(1):11–12
Stadtman ER, Levine RL (2000) Protein oxidation. Ann N Y Acad Sci 899:191–208
Stohs SJ, Bagchi D (1995) Oxidative mechanisms in the toxicity of metal ions. Free Radic Biol Med 18(2):321–336
Yamauchi Y, Furutera A, Seki K, Toyoda Y, Tanaka K, Sugimoto Y (2008) Malondialdehyde generated from peroxidized linolenic acid causes protein modification in heat-stressed plants. Plant Physiol Biochem 46(8–9):786–793

Hormonal Response in Plants Influenced by Reactive Oxygen Species

9

Huma Arshad and Ghazala Mustafa

Abstract

Reactive oxygen species are being produced because of various cellular metabolic activities. Environmental stresses result in an excessive production of ROS causing an oxidative burst. The ROS accumulation may lead to apoptosis. On the other hand, ROS also act as secondary messengers in a variety of cellular mechanisms, like growth, development, and responses to biotic and abiotic stresses in plants. ROS and plant hormones are intrinsically interwoven and form an integrated network of signaling pathways that regulates various metabolic processes from seed germination to stress tolerance acquisition. The interaction of ROS with abscisic acid and gibberellic acid controls the seed dormancy and programmed cell death in the aleuronic cell layer during seed germination and is well-documented. ROS are also involved in auxin signaling, gravitropism, and stomatal opening and closing. ROS regulate the antioxidant enzyme activity by inducing the expression of associated genes by mediating the ABA signaling pathways. In this chapter we will discuss the complex nature of the interplay of ROS and phytohormones in detail. In the first section, the role of ROS in the regulation of various phytohormonal pathways will be discussed individually. Afterward, the molecular nature of these interaction and the cellular signaling pathways involved in this interaction will be reviewed.

Keywords

Signaling pathway · Phytohormones · Stomatal opening · Gibberellic acid

H. Arshad · G. Mustafa (✉)
Department of Plant Sciences, Quaid-i-Azam University, Islamabad, Pakistan
e-mail: mghazala@qau.edu.pk

M. Faizan et al. (eds.), *Reactive Oxygen Species*,
https://doi.org/10.1007/978-981-19-9794-5_9

9.1 Introduction

A variety of plant cellular responses such as growth and development, hormone signaling, gravitropism, and programmed cell death (PCD) are mediated by ROS. ROS such as H_2O_2, $O_2^{\bullet-}$, and $OH^{\bullet}$ are generated and utilized during various aerobic metabolic pathways. These are majorly produced during photosynthesis (chloroplast), photorespiration (peroxisomes), and by the electron transport (mitochondria). Although plants have an efficient system of antioxidants which rapidly removes these ROS, H_2O_2 is accumulated in chloroplast and mitochondria under biotic and abiotic stresses (Mustafa and Komatsu 2016). Biotic or abiotic stress causes metabolic shock in plants which activates membrane-associated NADPH oxidases and peroxidases resulting in an oxidative burst (Mustafa and Komatsu 2016). ROS production changes the cellular redox state which acts as signals for various developmental processes (Mustafa et al. 2022). The ROS accumulation is a stress response which in turn regulates the gene expression. However, an excessive accumulation can also trigger PCD (Li et al. 2020). ROS are also produced during hormone-mediated regulation of plant development (Vivancos et al. 2010). ROS are also involved in hormone signaling, as secondary messengers, which regulates plant growth, development, and stress tolerance. Phytohormones activate NADPH oxidases resulting in ROS production (Mustafa and Komatsu 2017). On the other hand, H_2O_2 regulates phytohormone biosynthesis by interacting with their precursors. Further efforts are still required to elucidate the complex interplay of ROS-generating/scavenging mechanisms and their interaction with plant hormones. In this chapter the interaction between ROS and hormone signaling is discussed first, followed by ROS and hormonal cross talk and the molecular nature of the ROS-phytohormone interaction.

9.2 ROS-Mediated Phytohormonal Responses

9.2.1 Auxins

Auxin is responsible for the regulation of various key aspects of plant development. The role of auxin in cell division, elongation, and differentiation is well established. ROS are closely associated with this auxin-mediated regulation of plant development. ROS directly mediate the auxin-induced cell elongation and cell wall loosening (Li et al. 2020). Auxin-regulated root gravitropism is also mediated by ROS.

Auxin activates a Rho-like small GTPase, i.e., RAC/ROPs, which regulates auxin signaling, gene expression, and transport (Li et al. 2020). The RAC/ROPs interact with the N-terminal region of NADPH oxidase in a Ca^{2+}-dependent manner and directly regulate its function (Wong et al. 2007). ROS are accumulated under stress conditions and downregulate the expression of auxin signaling and transport-associated genes. ROS also cause oxidative degradation and inactivation of auxin, altering auxin signaling (Peer et al. 2013). This ROS-mediated downregulation of auxin signaling involves specific mitogen-activated protein kinase cascades like

MEKK1-MPK4 and ANP1-MPK3/6 (Nakagami et al. 2006). Oxidative stress also induces the Trp-dependent biosynthesis of auxin by upregulating the gene expression (Han et al. 2013). In addition, glutathione homeostasis is also involved in the ROS-mediated regulation of auxin signaling (Bashandy et al. 2010).

ROS also affect root curvature and tropism by inhibiting cell growth. *N*-1-naphthylphthalamic acid (auxin transport inhibitor) does not affect the H_2O_2-induced root curvature, suggesting that in auxin signaling and gravitropism, ROS plays a role downstream to auxin transport (Li et al. 2020). On the other hand, auxin activates RSL4 which upregulates the expression of NADPH oxidases and class III peroxidases. These oxidases and peroxidases catalyze ROS and control ROS-mediated polar root hairs (Mangano et al. 2017). The growth of root apical meristems requires a delicate balance of ROS, phytohormones, and other signaling molecules. The ROS and auxin interact antagonistically for the regulation of meristematic growth (Huang et al. 2019).

9.2.2 Gibberellins

Gibberellins (GAs) also play a key role in plant growth and developmental regulation. GA functioning involves the regulation of nuclear growth-repressing DELLA proteins. DELLA proteins negatively regulate the GA signaling. DELLA proteins are accumulated under stress and suppress plant growth for the chance of plant survival. DELLA proteins regulate plant growth and stress responses in response to environmental stimuli and attenuate the developmental programs (Yang et al. 2012). GA signaling regulates cellular redox homeostasis and mediates stress tolerance. DELLA proteins also upregulate the expression of a subset of antioxidant-associated genes which regulates the ROS level by enhancing the ROS-scavenging capacity (Achard et al. 2008). Reduced level of DELLA causes ROS accumulation and enhanced root growth, whereas reduced levels of GAs cause increased abundance of DELLA proteins leading to reduced ROS level, root length reduction, and a decreased sensitivity to diphenyleneiodonium (NADPH oxidase inhibitor) (Achard et al. 2008).

ROS and GA cross talk also regulates various events of seed germination. Aleurone cells remobilize endosperm reserves by secreting hydrolytic enzymes for the growth of the embryo. ROS are also involved in the programmed cell death of aleurone cells (Kacprzyk et al. 2022). NADPH oxidases are inactivated during the dicotyledonous seed germination causing increased ROS generation (Leymarie et al. 2012). ROS also upregulate the biosynthesis and signaling of GAs during germination. These findings suggest that during the early events of seed germination a complex network of cross talk exists between GA and ROS signaling in various cells.

9.2.3 Abscisic Acid

ABA is a phytohormone which mainly controls plant responses against abiotic stresses. It is mainly involved in promoting dormancy and inhibiting root development and germination (Xiong et al. 2002). ABA signal perception network involves many components. The pyrabactin resistance protein (PYR) or PYR-like proteins are involved in the perception of ABA binding. The binding of ABA to the PYR1/PYL complex induces a signaling cascade which inhibits protein phosphatase type 2Cs (PP2Cs) (Postiglione and Muday 2020). The PP2Cs are responsible for the protein dephosphorylation and negative regulation of ABA signaling.

In guard cells, NADPH oxidase-dependent ROS production regulates the ABA-mediated stomatal closure. ABA-mediated ROS production activates Ca^{2+} channels causing an increase in cytosolic Ca^{2+} concentration which promotes ABA-induced stomatal closure in guard cells (Sharipova et al. 2021). The PP2C proteins regulate OST1 which is a member of the SnRK2 family. OST1 acts upstream to ROS signaling and mediates ABA-induced closure of the stomata. In *Arabidopsis*, OST1 also mediates phosphorylation of the RBOHF subunit of NADPH oxidase, thus regulating ROS generation (Postiglione and Muday 2020). The ABA-mediated stress tolerance in seedlings is also regulated by ROS. ROS accumulation as well as the expression of antioxidant enzyme-associated genes is upregulated in response to ABA treatment which results in higher antioxidant activity (Pilarska et al. 2021). The NADPH oxidase forms a positive feedback loop with the MAPK cascades in which the MAPK cascades act upstream as well as downstream to the ROS production (Xing et al. 2008). In *Arabidopsis*, H_2O_2 mediated the catabolism of ABA during seed imbibition (Liu et al. 2010a, b), whereas in barley seeds, the activity and expression of an ABA-associated kinase, i.e., PKABA, were found to be suppressed by H_2O_2. PKABA regulates the α-amylase expression by regulating the GA-dependent Myb transcription factor (GAmyb) (Ishibashi et al. 2012). Therefore, it can be concluded that during seed imbibitions, H_2O_2 acts antagonistic to ABA signaling. ABA-mediated ROS accumulation caused by the activation of NADPH oxidase also inhibits the primary and lateral root growth (Duan et al. 2019; Jiao et al. 2013). Complex network of ROS-ABA interactions regulates the root architecture.

9.2.4 Ethylene

Ethylene (ET) is a an important phytohormone which is involved in the regulation of plant growth, stress response, and senescence. Several ethylene-induced responses in plants are mediated by ROS. In *Arabidopsis*, H_2O_2 produced by RbohF mediates the ethylene-induced stomatal closure (Bright et al. 2006). Ethylene promotes tolerance to potassium deficiency by promoting ROS production (Pilarska et al. 2021). ET also mediates the effects of ozone by suppressing the closure of stomata (Wilkinson and Davies 2009). ET along with NADPH oxidases regulates the salinity stress responses in *Arabidopsis* (Pilarska et al. 2021). ET activates the

MKK9-MPK3/6 signaling cascade by inhibiting the CTR1 (Yoo et al., 2008). ERF6 (ethylene response factor 6) which transcriptionally regulates the ROS-associated genes is phosphorylated by MPK6 (Wang et al. 2013). ET mediated MAPK cascades transcriptionally regulate the ROS-associated RBOH genes. ET and ROS may also mediate cell death as a response to various environmental stimuli. Moreover, genes associated with ET biosynthesis are also regulated by ROS, thereby facilitating the ET signaling cascade. ET and ROS are also involved in root nodule initiation and function in a semiaquatic legume. They were found to act downstream of the Nod factor response (D'Haeze et al. 2003).

9.2.5 Brassinosteroids

Brassinosteroids (BRs) are steroid-based phytohormones which regulate plant growth, development, as well as stress response (Kour et al. 2021). BRs also stimulate ROS production to improve stress tolerance in plants (Xia et al. 2011). Xia et al. 2009, suggested the role of ROS in the BR signaling pathway as well as other hormonal signaling pathways. Low BR levels cause stomatal opening due to transient production of ROS in the epidermal strips which changes the redox state of the cell, whereas high levels of BR cause stomatal closing and ABA biosynthesis due to prolonged accumulation of ROS (Xia et al. 2015). BRs inhibit the ROS-mediated lipid degradation under metal stress and stimulate the antioxidant enzyme activity (Soares et al. 2016). BRs are actively involved in the heavy metal stress amelioration in plants by ROS scavenging and upregulating the activities of superoxide dismutase (SOD), peroxidase (POD), ascorbate peroxidase (APX), catalase (CAT), glutathione reductase (GR), glutathione-S-transferase (GST), and guaiacol peroxidase (Bücker-Neto et al. 2017). The ROS-mediated BR signaling also actively regulates the heat stress tolerance. The BZR1 (a BR regulator) directly binds to the promoter of *RBOH1* which triggers tapetal cell degradation and programmed cell death (PCD) by mediating the ROS production (Yan et al. 2020). BRs also regulate the concentration of calcium ions (Ca^{2+}) and ROS in the cytosol, thereby reducing the length of the hypocotyl in dark-grown seedlings (Zhao et al. 2013). BRs also induce ROS production mediated by the NADPH oxidase in the apoplast (Nie et al. 2013).

9.2.6 Jasmonic Acid

Jasmonic acid (JA) is also a stress-related hormone of higher plants (Wang et al. 2020). In abiotic stress, JA is usually involved in physiological and molecular responses (Karpets et al. 2014). Endogenous JA maintains ROS homeostasis and enhances salt tolerance in tomato (Abouelsaad and Renault 2018). In higher plants JA and its derivatives also regulate the plant response against wounding and environmental stresses (Xia et al. 2015). JA biosynthesis during the early phases of wound response is mainly regulated by ROS production, Ca^{2+} gradient and activation of MAPK cascades, whereas ROS also act as a secondary messenger influencing JA response. JA also mediates ROS-dependent signaling pathway which causes stomatal closure (Gomes et al. 2014). Devireddy et al. (2018) showed that stomatal closure in response to light stress was coordinated by rapid systemic signals in leaves within the canopy of the *Arabidopsis* plant, including the ones not directly exposed to light. This response was mediated by JA, ABA, and a rapid increase in ROS production. Leaf senescence is positively regulated by JA and H_2O_2 signaling. Zhang et al. (2020) reported the role of ROS (H_2O_2) downstream to JA in promoting leaf senescence. They indicated that JA stimulated the H_2O_2 accumulation to promote leaf senescence, while higher levels of H_2O_2 may result in lipid peroxidation which in turn upregulates the biosynthesis of JA. These findings suggest a positive feedback regulatory loop between H_2O_2 and JA during leaf senescence (Sabater and Martín 2013).

9.2.7 Salicylic Acid

Salicylic acid (SA) is a phenol-based plant hormone which plays a crucial role in plant defense signaling. It regulates a number of plant processes such as growth, development, photosynthesis, flowering, and the activity of antioxidant enzymes in response to changing environmental conditions. SA is capable of playing a dual role in ROS accumulation as well as scavenging depending upon the different environmental conditions, such as drought, chilling stress, high light intensity, and pathogenic attack (Saleem et al. 2021). The role of SA in maintaining ROS homeostasis has been recently explored (Mahmoud et al. 2021; Poór 2020). ROS produced due the catalytic activity of peroxidases mediate the SA-induced stomatal closure (Miura and Tada 2014; Miura et al. 2013). SA at lower concentrations facilitates abiotic stress tolerance, whereas higher concentrations upregulate the ROS production causing oxidative burst which leads to cell death (Poór 2020; Miura and Tada 2014). SA is also a key regulating molecule of plant stress response against high temperature. Heat stress causes elevation in the SA levels which protects the photosystem II (PSII) complex from the harmful effects of ROS, thus increasing photosynthetic capacity and imparting heat stress tolerance (Duan et al. 2019). ROS are constantly being generated by different compartments in mitochondria at every stage of the cell cycle, and their detoxification is crucial in order to protect the cellular machinery from their toxic effects. SA mediates a number of these cellular detoxification strategies by scavenging and limiting the ROS in mitochondria in a concentration- and time-dependent manner (Poór 2020).

The ROS forms a feedback loop with various plant hormones during the regulation of plant developmental processes and stress response. ROS accumulated under stressful conditions not only affect the biosynthesis of these plant hormones but also attenuate their signaling pathways. These altered pathways are responsible for the alterations in the plant metabolic events (Fig. 9.1). These phytohormonal pathways are also capable of regulating ROS concentration. The MAPK cascade and NADPH oxidases play a central role in this regulation.

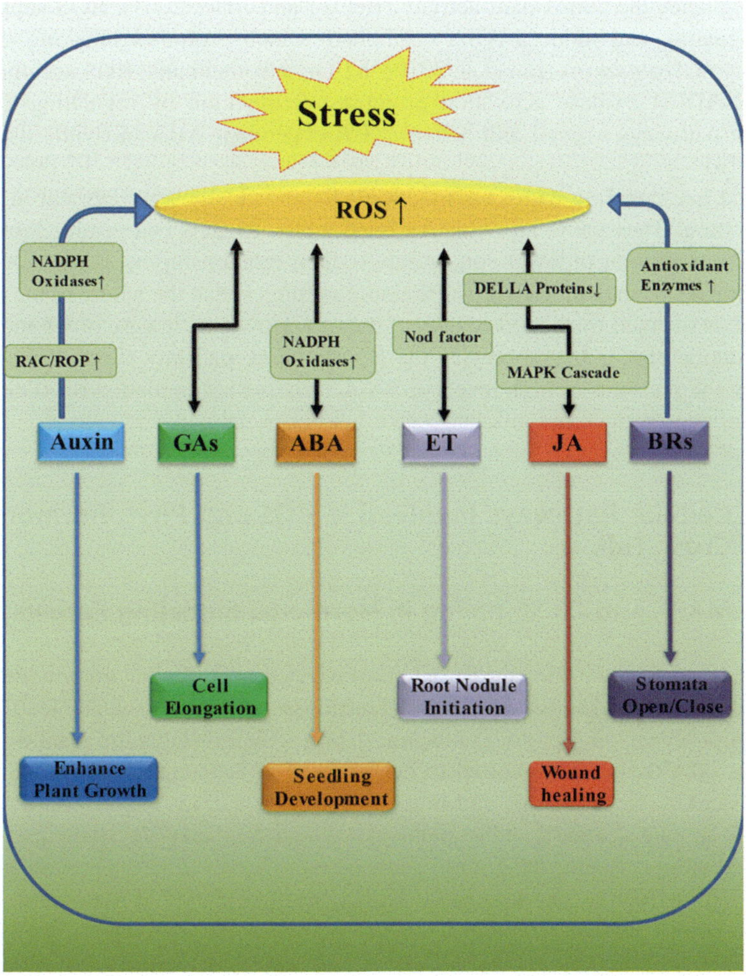

Fig. 9.1 The schematic representation of ROS-mediated hormonal regulation of the key metabolic processes in plants. ROS are accumulated under stressful environmental conditions which results in the altered regulation of plant developmental processes

9.3 Cross Talk between Hormonal Signaling Pathways

The above discussion highlights the role of ROS in regulating different plant hormonal responses individually. However, these pathways do not operate in isolation. They involve a complex network of interactions and extensive metabolic cross talk among various hormonal signaling pathways.

In aleurone cells, GA promotes ROS accumulation which induces ROS-mediated PCD, whereas ABA downregulates GA-induced oxidant production in these cells and upregulates the antioxidant activity (Bethke and Jones 2001). ROS suppresses ABA signaling and induces ABA catabolism which promotes breaking of seed dormancy in *Arabidopsis* (Liu et al. 2010a, b). On the other hand, ROS accumulation due to NADPH oxidase activation promotes germination by regulating the GA signaling pathways as well and activates BR-dependent ABA biosynthesis (Zhou et al. 2014).

JA and SA signaling pathways function antagonistically when present in higher concentration. They cause an oxidative burst which triggers cell death (Zhou et al. 2014). When present in lower concentration, their relationship becomes synergistic to one another (Han et al. 2013). These findings suggest that the nature of the JA-SA cross talk is dictated by the redox state of the cell. However, in case of ET induction due to abiotic stress, JA suppresses the SA signaling pathway (Leon-Reyes et al. 2009). In the presence of high levels of SA, ET causes accumulation of ROS which further influences the cross talk between other hormonal pathways.

9.4 Cellular Pathways Involved in ROS and Phytohormonal Cross Talk

9.4.1 MAPK and Ca Signaling in Hormonal Signaling Pathways

MAPK cascades play a crucial role in almost every aspect of plant growth, development, senescence, and stress signaling and response as they are responsible for signal transduction events and protein phosphorylation which dictates the plant hormone signaling. MAPK is well reported to be involved in jasmonic acid, salicylic acid, abscisic acid, auxin, ethylene, and brassinosteroid signaling. There is an extensive cross-talk among these signaling pathways which regulates the dynamics of the whole signaling network. ABA and ET are actively involved in the protein stability mediated by MAPK as well as the overall control of the MAPK cascade. In oat AsMAP1 was negatively regulated by GA (Zhou et al. 2014). In contrast, GA upregulated the gene expression of *PsMAPK3* in unpollinated pea ovary (Leymarie et al. 2012). The overexpression of GhMKK4 gene in *N. benthamiana* has been reported to increase the sensitivity of the plant to GA and ABA (Li et al. 2017).

The molecular mechanism of MAPK regulation requires further elucidation. The expression of MAPK-associated components, assembly of protein complexes, and the subcellular localization of the MAPK cascade are highly coordinated (Bigeard and Hirt 2018). The MAPK, MAPKK, and MAPKKKs are generally localized in the nucleus and cytoplasm. The active form of MAPKKK18 is found in the nucleus, while the inactive isoform is localized in the cytoplasm, suggesting that MAPKKK18 activity and concentration are tightly controlled and specific to target organelle (Mitula et al. 2015). The localization of MAPK is associated with its enzymatic activity. MPK3, MPK4, and MPK6 are regulated by MPK1 protein phosphatase (Bartels et al. 2009), whereas MKP2 regulates the MPK3 and MPK6 activity (Lumbreras et al. 2010).

ABA treatment inhibited the activities of MPK3 and MPK6 kinase via AP2C1 and PP2C5 phosphatases (Brock et al. 2010). ABI1 protein phosphatase also regulates the MPK6 and MAPKKK18 which in turn negatively regulated the ABA signaling (Mitula et al. 2015). The cross talk between jasmonic acid, ethylene, salicylic acid, and auxin responses is regulated by AtMPK4. The ET and BR signaling are precisely regulated by AtMKK4/5-MPK6. Dephosphorylation of specific protein phosphatases regulates the MAPK inactivation, which forms a negative feedback loop controlling various hormonal responses.

9.4.2 Role of Phytohormones in ROS-Dependent Cell Death

Programmed cell death is a pathway for the removal of damaged or infected cells which is genetically regulated. Plant hormones also play a key role in PCD. JA, SA, and ET regulate various developmental processes including senescence. ET and SA are actively involved in PCD regulation, where ethylene is primarily involved in developmental PCD whereas SA contributes to pathogen-induced cell death (Zhou et al. 2014).

PCD is a tightly regulated process which involves the upregulation of genes associated with several metabolic processes such as transport, ROS homeostasis, ABA and ET signaling, DNA/protein binding, binding of metal ions, carotene metabolism, lipid catabolism, glutamine synthase 2, caspase activity, and autophagy, whereas genes associated with photosynthesis, carotenoid and chlorophyll biosynthesis, cytokinin signaling, amino acid metabolism, and glutamine synthase 1 are downregulated (Zhou et al. 2014). Expression profiles of ROS-scavenging enzymes such as peroxidase, APX, SOD, CAT, and GPX as well as the nonenzymatic antioxidants, i.e., AsA and GSH, were also found to be affected in response to PCD in *Triticum aestivum* (Liu et al., 2018). These findings conclude that plant hormones along with various metabolites play key roles in regulating PCD.

9.5 ROS and Phytohormone Integration during Systemic Signaling

Systemic acquired resistance (SAR) is a long-distance signaling response due to pathogen exposure. The establishment of SAR requires SA, JA, as well as ROS (Durrant and Dong 2004). Pathogen recognition triggers cell death and a localized oxidative burst which leads to a systemic oxidative burst. This causes changes in the redox state of the local as well as systemic tissue cells and is responsible for initiating the SAR. ROS mediate SAR in systemic leaves by facilitating the de novo synthesis of SA (Xia et al. 2011). The RBOHD-mediated ROS production is crucial for swift systemic signaling in response to wounding (Miller et al. 2009). ROS are actively accumulated in local as well as systemic tissues which upregulate the pathways of JA synthesis. ROS also influence the JA response to wounding by acting as secondary messengers (Orozco-Cárdenas et al. 2001). The molecular basis of this relationship between accumulation of ROS and JA synthesis upregulation still remains unclear. ABA and ET are well-studied mediators of long-distance signaling in response to several stresses (Jackson 2008). The relative balance of these plant hormones dictates the transcriptional regulation in response to stresses.

9.6 Conclusion and Future Prospects

The above discussion highlights the effect of ROS in plant hormone signaling and response against various environmental stimuli. ROS produced during various metabolic activities regulate a number of plant metabolic activities, stress response, and tolerance. The regulation of membrane-bound NADPH oxidase is crucial for the cross talk between ROS and plant hormones and in systemic signaling as well. ROS signaling, MAPK cascades, and Ca^{2+} signals form a complex feedback loop system that regulates hormonal signaling. The RAC/ROP proteins regulate NADPH oxidase activity in auxin signaling pathways. Similar mechanisms may also exist in case of other hormones which are still to be explored. The dynamics of Ca^{2+} signaling modulates the NADPH oxidase activity in response to stress. The ROS-phytohormone cross talk involves a complex interaction network which is still not completely understood. The spatial temporal regulation of ROS production is still to be explored, whereas a vast information gap also exists regarding the identification of proteins involved in the cross talk and regulation of different hormone signaling pathways.

References

Abouelsaad I, Renault S (2018) Enhanced oxidative stress in the jasmonic acid-deficient tomato mutant def-1 exposed to NaCl stress. J Plant Physiol 226:136–144

Achard P, Renou JP, Berthomé R, Harberd NP, Genschik P (2008) Plant DELLAs restrain growth and promote survival of adversity by reducing the levels of reactive oxygen species. Curr Biol 18:656–660

Bartels S, Anderson JC, González Besteiro MA, Carreri A, Hirt H, Buchala A et al (2009) MAP KINASE PHOSPHATASE1 and PROTEIN TYROSINE PHOSPHATASE1 are repressors of salicylic acid synthesis and SNC1-mediated responses in *Arabidopsis*. Plant Cell 21:2884–2897. https://doi.org/10.1105/tpc.109.067678

Bashandy T, Guilleminot J, Vernoux T, Caparros-Ruiz D, Ljung K, Meyer Y, Reichheld JP (2010) Interplay between the NADP-linked thioredoxin and glutathione systems in Arabidopsis auxin signaling. Plant Cell 22:376–391

Bethke PC, Jones RL (2001) Cell death of barley aleurone protoplasts is mediated by reactive oxygen species. Plant J 25:19–29

Bigeard J, Hirt H (2018) Nuclear signaling of plant MAPKs. Front Plant Sci 9:469. https://doi.org/10.3389/fpls.2018.00469

Bright J, Desikan R, Hancock JT, Weir IS, Neill SJ (2006) ABA induced NO generation and stomatal closure in Arabidopsis are dependent on H_2O_2 synthesis. Plant J 45:113–122

Brock AK, Willmann R, Kolb D, Grefen L, Lajunen HM, Bethke G et al (2010) The Arabidopsis mitogen-activated protein kinase phosphatase PP2C5 affects seed germination, stomatal aperture, and abscisic acid-inducible gene expression. Plant Physiol 153:1098–1111. https://doi.org/10.1104/pp.110.156109

Bücker-Neto L, Paiva ALS, Machado RD, Arenhart RA, Margis-Pinheiro M (2017) Interactions between plant hormones and heavy metals responses. Genet Mol Biol 40:373–386

Devireddy AR, Zandalinas SI, Gómez-Cadenas A, Blumwald E, Mittler R (2018) Coordinating the overall stomatal response of plants: rapid leaf-to-leaf communication during light stress. Sci Signal 11(518):eaam9514

D'Haeze W, Rycke RD, Mathis R, Goormachtig S, Pagnotta S, Verplancke C, Capoen W, Holsters M (2003) Reactive oxygen species and ethylene play a positive role in lateral root base nodulation of a semiaquatic legume. Protoc Natl Acad Sci USA 100:11789–11794

Duan J, Lee KP, Dogra V, Zhang S, Liu K, Caceres-Moreno C et al (2019) Impaired PSII proteostasis promotes retrograde signaling via salicylic acid. Plant Physiol 180(4):2182–2197

Durrant WE, Dong X (2004) Systemic acquired resistance. Annu Rev Phytopathol 42:185–209

Gomes MP, Smedbol É, Carneiro MMLC, Garcia QS, & Juneau P (2014) Reactive oxygen species and plant hormones. In *Oxidative damage to plants* (pp. 65–88). Academic Press

Han Y, Mhamdi A, Chaouch S, Noctor G (2013) Regulation of basal and oxidative stress-triggered jasmonic acid-related gene expression by glutathione. Plant Cell Environ 36:1135–1146

Huang H, Ullah F, Zhou D-X, Yi M, Zhao Y (2019) Mechanisms of ROS regulation of plant development and stress responses. Front Plant Sci 10:800. https://doi.org/10.3389/fpls.2019.00800

Ishibashi Y, Tawaratsumida T, Kondo K, Kasa S, Sakamoto M, Aoki N, Zheng SH, Yuasa T, Iwaya-Inoue M (2012) Reactive oxygen species are involved in gibberellin/abscisic acid signalling in barley aleurone cells. Plant Physiol 158:1705–1714

Jackson MB (2008) Ethylene-promoted elongation: an adaptation to submergence stress. Ann Bot 101:229–248

Jiao Y, Sun L, Song Y, Wang L, Liu L, Zhang L, Liu B, Li N, Miao C, Hao F (2013) AtrbohD and AtrbohF positively regulate abscisic acid-inhibited primary root growth by affecting Ca^{2+} signalling and auxin response of roots in Arabidopsis. J Exp Bot 64:4183–4192

Kacprzyk J, Burke R, Schwarze J, McCabe PF (2022) Plant programmed cell death meets auxin signalling. FEBS J 289:1731–1745. https://doi.org/10.1111/febs.16210

Karpets YV, Kolupaev YE, Lugovaya AA, Oboznyi AI (2014) Effect of jasmonic acid on the pro–/antioxidant system of wheat coleoptiles as related to hyperthermia tolerance. Russ J Plant Physiol 61(3):339–346

Kour J, Kohli SK, Khanna K, Bakshi P, Sharma P, Singh AD, ... & Sharma A (2021) Brassinosteroid signaling, crosstalk, and physiological functions in plants under heavy metal stress. *Frontiers in Plant Science, 12*

Leon-Reyes A, Spoel SH, Lange ESD, Abe H, Kobayashi M, Tsuda S, Millenaar FF, Welschen RA, Ritsema T, Pieterse CM (2009) Ethylene modulates the role of NONEXPRESSOR OF PATHOGENESISRELATED GENES1 in cross talk between salicylate and jasmonate signaling. Plant Physiol 149:1797–1809

Leymarie J, Vitkauskaité G, Hoang HH, Gendreau E, Chazoule V, Meimoun P, Corbineau F, El-Maarouf-Bouteau H, Bailly C (2012) Role of reactive oxygen species in the regulation of Arabidopsis seed dormancy. Plant Cell Physiol 53:96–106

Li A, Lakshmanan P, He W, Tan H, Liu L, Liu H, Liu J, Huang D, Chen Z (2020) Transcriptome profiling provides molecular insights into auxin-induced adventitious root formation in sugarcane (*Saccharum* spp. interspecific hybrids) microshoots. Plants (Basel, Switzerland) 9(8):931. https://doi.org/10.3390/plants9080931

Li Y, Cai H, Liu P, Wang C, Gao H, Wu C et al (2017) Arabidopsis *MAPKKK18* positively regulates drought stress resistance via downstream *MAPKK3*. Biochem Biophys Res Commun 484:292–297. https://doi.org/10.1016/j.bbrc.2017.01.104

Liu Y, Li X, Tan H, Liu M, Zhao X, Wang J (2010a) Molecular characterization of RsMPK2, a C1 subgroup mitogen-activated protein kinase in the desert plant Reaumuria soongorica. Plant Physiol Biochem 48:836–844. https://doi.org/10.1016/j.plaphy.2010.07.001

Liu Y, Ye N, Liu R, Chen M, Zhang J (2010b) H2O2 mediates the regulation of ABA catabolism and GA biosynthesis in Arabidopsis seed dormancy and germination. J Exp Bot 61:2979–2990
Liu Z, Shi X, Li S, Hu G, Zhang L, Song X (2018) Tapetal-Delayed Programmed Cell Death (PCD) and oxidative stress-induced male sterility of Aegilops uniaristata cytoplasm in wheat. Int J Mol Sci 19(6):1708. MDPI AG. https://doi.org/10.3390/ijms19061708
Lumbreras V, Vilela B, Irar S, Solé M, Capellades M, Valls M et al (2010) MAPK phosphatase MKP2 mediates disease responses in Arabidopsis and functionally interacts with MPK3 and MPK6. Plant J 63:1017–1030. https://doi.org/10.1111/j.1365-313X.2010.04297.x
Mahmoud LM, Vincent CI, Grosser JW, Dutt M (2021) The response of salt-stressed Valencia sweet orange (*Citrus sinensis*) to salicylic acid and methyl jasmonate treatments. Plant Physiol Rep 26(1):137–151
Mangano S, Denita-Juarez SP, Choi HS, Marzol E, Hwang Y, Ranocha P, Velasquez SM, Borassi C, Barberini ML, Aptekmann AA, Muschietti JP, Nadra AD, Dunand C, Cho HT, Estevez JM (2017) Molecular link between auxin and ROS-mediated polar growth. Proc Natl Acad Sci U S A 114(20):5289–5294. https://doi.org/10.1073/pnas.1701536114
Miller G, Schlauch K, Tam R, Cortes D, Torres MA, Shulaev V, Dangl JL, Mittler R (2009) The plant NADPH oxidase RBOHD mediates rapid systemic signaling in response to diverse stimuli. Sci Signal 2:ra45
Mitula F, Tajdel M, Ciesla A, Kasprowicz-Maluski A, Kulik A, Babula-Skowronska D et al (2015) Arabidopsis ABA-activated kinase MAPKKK18 is regulated by protein phosphatase 2C ABI1 and the ubiquitin-proteasome pathway. Plant Cell Physiol 56:2351–2367
Miura K, Okamoto H, Okuma E, Shiba H, Kamada H, Hasegawa PM, Murata Y (2013) SIZ1 deficiency causes reduced stomatal aperture and enhanced drought tolerance via controlling salicylic acid-induced accumulation of reactive oxygen species in Arabidopsis. Plant J 73(1): 91–104
Miura K, Tada Y (2014) Regulation of water, salinity, and cold stress responses by salicylic acid. Front Plant Sci 5:4
Mustafa G, Farooq A, Riaz Z, Hasan M, Amna (2022) Nano-proteomics of stress tolerance in crop plants. In: Faizan M, Hayat S, Yu F (eds) Sustainable Agriculture Reviews 53. Sustainable Agriculture Reviews, vol 53. Springer, Cham. https://doi.org/10.1007/978-3-030-86876-5_15
Mustafa G, Komatsu S (2016) Toxicity of heavy metals and metal-containing nanoparticles on plants. Proteins Proteomics 1864(8):932–944
Mustafa G, Komatsu S (2017) Nanoparticles mediated soybean response mechanism at morphological, physiological, and proteomic level. Curr Proteomics 14:1. https://doi.org/10.2174/1570164613666161128145103
Nakagami H, Soukupová H, Schikora A, Zárský V, Hirt H (2006) A mitogen-activated protein kinase kinase kinase mediates reactive oxygen species homeostasis in Arabidopsis. J Biol Chem 281:38697–38704
Nie WF, Wang MM, Xia XJ, Zhou YH, Shi K, Chen Z, Yu JQ (2013) Silencing of tomato RBOH1 and MPK2 abolishes brassinosteroid-induced H_2O_2 generation and stress tolerance. Plant Cell Environ 36(4):789–803
Orozco-Cárdenas ML, Narváez-Vásquez J, Ryan CA (2001) Hydrogen peroxide acts as a second messenger for the induction of defense genes in tomato plants in response to wounding, systemin, and methyl jasmonate. Plant Cell 13:179–191
Peer WA, Cheng Y, Murphy AS (2013) Evidence of oxidative attenuation of auxin signalling. J Exp Bot 64:2629–2639
Pilarska M, Bartels D, Niewiadomska E (2021) Differential regulation of NAPDH oxidases in salt-tolerant *Eutrema salsugineum* and salt-sensitive *Arabidopsis thaliana*. Int J Mol Sci 22(19): 10341. https://doi.org/10.3390/ijms221910341
Poór P (2020) Effects of salicylic acid on the metabolism of mitochondrial reactive oxygen species in plants. Biomol Ther 10(2):341
Postiglione AE, Muday GK (2020) The role of ROS homeostasis in ABA-induced guard cell Signaling. Front Plant Sci 11:968. https://doi.org/10.3389/fpls.2020.00968

Sabater B, Martín M (2013) Hypothesis: increase of the ratio singlet oxygen plus superoxide radical to hydrogen peroxide changes stress defense response to programmed leaf death. Front Plant Sci 4:479

Saleem M, Fariduddin Q, Castroverde CDM (2021) Salicylic acid: a key regulator of redox signaling and plant immunity. Plant Physiol Biochem 168:381–397

Sharipova G, Ivanov R, Veselov D, Akhiyarova G, Shishova M, Nuzhnaya T, Kudoyarova G (2021) Involvement of reactive oxygen species in ABA-induced increase in hydraulic conductivity and aquaporin abundance. Int J Mol Sci 22:9144. https://doi.org/10.3390/ijms22179144

Soares C, de Sousa A, Pinto A, Azenha M, Teixeira J, Azevedo RA, Fidalgo F (2016) Effect of 24-epibrassinolide on ROS content, antioxidant system, lipid peroxidation and Ni uptake in Solanum nigrum L. under Ni stress. Environ Exp Bot 122:115–125

Vivancos PD, Dong Y, Ziegler K, Markovic J, Pallardó FV, Pellny TK, Verrier PJ, Foyer CH (2010) Recruitment of glutathione into the nucleus during cell proliferation adjusts whole-cell redox homeostasis in Arabidopsis thaliana and lowers the oxidative defence shield. Plant J 64: 825–838

Wang J, Song L, Gong X, Xu J, Li M (2020) Functions of Jasmonic acid in plant regulation and response to abiotic stress. Int J Mol Sci 21(4):1446. https://doi.org/10.3390/ijms21041446

Wang PC, Du YY, Zhao XL, Miao YC, Song CP (2013) The MPK6- ERF6-ROS-responsive cis-acting element7/GCC box complex modulates oxidative gene transcription and the oxidative response in Arabidopsis. Plant Physiol 161:1392–1408

Wilkinson S, Davies WJ (2009) Ozone suppresses soil drying- and abscisic acid (ABA)-induced stomatal closure via an ethylene-dependent mechanism. Plant Cell Environ 32:949–959

Wong HL, Pinontoan R, Hayashi K, Tabata R, Yaeno T, Hasegawa K, Kojima C, Yoshioka H, Iba K, Kawasaki T (2007) Regulation of rice NADPH oxidase by binding of Rac GTPase to its N-terminal extension. Plant Cell 19:4022–4034

Xia XJ, Wang YJ, Zhou YH, Tao Y, Mao WH, Shi K et al (2009) Reactive oxygen species are involved in brassinosteroid-induced stress tolerance in cucumbers. Plant Physiol 150(2): 801–814

Xia XJ, Zhou YH, Ding J, Shi K, Asami T, Chen Z, Yu JQ (2011) Induction of systemic stress tolerance by brassinosteroid in Cucumis sativus. New Phytol 191:706–720

Xia XJ, Zhou YH, Shi K, Zhou J, Foyer CH, Yu JQ (2015) Interplay between reactive oxygen species and hormones in the control of plant development and stress tolerance. J Exp Bot 66(10):2839–2856

Xing Y, Jia W, Zhang J (2008) AtMKK1 mediates ABA-induced CAT1 expression and H_2O_2 production via AtMPK6-coupled signaling in Arabidopsis. Plant J 54:440–451

Xiong L, Schumaker KS, Zhu JK (2002) Cell signaling during cold, drought, and salt stress. Plant Cell 14:S165–S183

Yan MY, Xie DL, Cao JJ, Xia XJ, Shi K, Zhou YH et al (2020) Brassinosteroid-mediated reactive oxygen species are essential for tapetum degradation and pollen fertility in tomato. Plant J 102(5):931–947

Yang DL, Yao J, Mei CS, Tong XH, Zeng LJ, Li Q, Xiao LT, Sun TP, Li J, Deng XW (2012) Plant hormone jasmonate prioritizes defense over growth by interfering with gibberellin signaling cascade. Proc Natl Acad Sci USA 109:1192–1200

Yoo SD, Cho YH, Tena G, Xiong Y, Sheen J (2008) Dual control of nuclear EIN3 by bifurcate MAPK cascades in C2H4 signalling. Nature 451(7180):789–795. https://doi.org/10.1038/nature06543

Zhang Y, Ji TT, Li TT, Tian YY, Wang LF, Liu WC (2020) Jasmonic acid promotes leaf senescence through MYC2-mediated repression of CATALASE2 expression in Arabidopsis. Plant Sci 299: 110604

Zhao Y, Qi Z, Berkowitz GA (2013) Teaching an old hormone new tricks: cytosolic Ca^{2+} elevation involvement in plant brassinosteroid signal transduction cascades. Plant Physiol 163(2): 555–565

Zhou J, Wang J, Li X, Xia XJ, Zhou YH, Shi K, Chen Z, Yu JQ (2014) H_2O_2 mediates the crosstalk of brassinosteroid and abscisic acid in tomato responses to heat and oxidative stresses. J Exp Bot 65:4371–4383

The Dual Role of Reactive Oxygen Species as Signals that Influence Plant Stress Tolerance and Programmed Cell Death

10

Mohd Soban Ali, Asif Hussain Hajam, Mohammad Suhel, Sheo Mohan Prasad, and Gausiya Bashri

Abstract

Reactive oxygen species (ROS) are molecules and radicals which are continuously made by the redox cascades of aerobic metabolism in plants. They may even act as a signal by altering the redox equilibrium of the cell, which can change how specific proteins work or how genes are expressed and respond to biotic and abiotic stress. The network of redox signals controls metabolism to regulate how energy is produced and utilized at every stage of plant growth by interfering with the main signaling molecules (hormones) to adapt to environmental changes. In normal conditions, plants have many ways to fight ROS, but when they are stressed, ROS starts building up and causes oxidative stress. This duality can only be achieved by strictly regulating ROS generation and consumption by the antioxidant defense system within cells. Indeed, ROS are involved in a wide variety of redox-governing actions within cells, which are necessary for maintaining cellular homeostasis. However, its excess synthesis has been linked to oxidative stress, a potentially harmful process that damages biomolecules and cellular structures and contributes to programmed cell death (PCD). The chapter provides a concise view of the dual role of ROS signaling in plant acclamatory defense processes and PCD.

Keywords

Antioxidants · Oxidative stress · Signaling · Tolerance

M. S. Ali · A. H. Hajam · G. Bashri (✉)
Department of Botany, Faculty of Life Sciences, Aligarh Muslim University, Aligarh, India

M. Suhel · S. M. Prasad
Ranjan Plant Physiology and Biochemistry Laboratory, Department of Botany, University of Allahabad, Prayagraj, India

M. Faizan et al. (eds.), *Reactive Oxygen Species*,
https://doi.org/10.1007/978-981-19-9794-5_10

10.1 Introduction

Life on earth originated in a reducing environmental condition, and after that, the evolution of photosynthetic organisms (cyanobacteria) changed the environmental conditions from reducing to oxidizing by the liberation of O_2 in the atmosphere as a byproduct of photosynthesis (Bendall et al. 2008). Since that time, reactive oxygen species (ROS) have been an unavoidable component of aerobic metabolism. The formation of ROS is typically a consequence of normal aerobic metabolism, including mostly membrane-linked electron transport mechanisms and redox cascades, and is exacerbated by unfavorable environmental conditions (Das and Roychoudhury 2014). Under unfavorable environmental conditions such as heavy metals, drought, excessive salt concentrations, high and low temperatures, exposure to high levels of ultraviolet radiation or ozone, and pathogen infections, the health of plants may be negatively impacted (Raza et al. 2019). All of these unfavorable situations have one thing in common: increased ROS generation within various subcellular compartments of the plant cell (Bhattacharjee 2005).

All cellular compartments, like cell wall, plasma membrane, chloroplast, mitochondria, peroxisomes, glyoximes, and apoplast, produce ROS as byproducts (Podolyan et al. 2019; Breygina and Klimenko 2020). In most cases, oxygenic organisms' electron transport mechanisms generate ROS because O_2 is a powerful electron acceptor. Unlike molecular O_2, ROS, which includes hydrogen peroxide (H_2O_2), singlet oxygen ($^1O_2$), superoxide radical ($O_2^{\bullet-}$), and hydroxyl radical ($^{\bullet}OH$), are derivatives of O_2 that have been partly reduced or activated. These ROS are highly reactive and damaging, and they are capable of initiating oxidative damage to macromolecules such as proteins, DNA, and lipids (Gill and Tuteja 2010). Besides the toxic nature of ROS, it also acts as a signaling molecule and has been reported to be involved in regulating growth and developmental processes such as seed germination, growth of root hairs and leaf, stomatal movement, root gravitropism, and programmed cell death (PCD) (Mittler 2017; Huang et al. 2019). The concentration of ROS is tightly controlled by the antioxidant defense system to balance the cellular level of ROS in order to avoid their toxicity to the cellular components of plants (Gull et al. 2019). This is carried out by a complicated signaling network that detects environmental challenges that significantly govern and control the spatiotemporal titer of ROS in plants by changing the processes responsible for creating and scavenging ROS (Foyer and Noctor 2005; Xie et al. 2014). In point of fact, it is reasonable to suppose that plants could own the systems that regulate the concentration of ROS in accordance with the requirements of the cells. There are increasing numbers of studies showing the toxic nature of ROS that depends on their concentrations, cross talk with cellular components, the status of defense systems, and environmental conditions (Xie et al. 2014; Huang et al. 2019). However, when the balance between ROS and the antioxidant defense system is upset, it results in altered cellular redox homeostasis, leading to various physiological difficulties, which are collectively known as "oxidative stress" (Apel and Hirt 2004). Extremely high concentrations of reactive oxygen species represent a severe risk of causing DNA damage and improper timing of PCD (Xie et al. 2014). This chapter will highlight the findings of a few research conducted over the last decade that expanded our knowledge of ROS and its dual function in plant tolerance and programmed cell death.

10.2 Site of ROS Generation in Plants

In plants, environmental stresses, such as extreme hot or cold temperatures, intense light, salinity, drought, heavy metals, air pollutants, pests, and disease attacks, lead to the overgeneration of ROS. Several cellular structures generate ROS, including the mitochondria (Huang et al. 2016), chloroplast (Dietz et al. 2016), and peroxisomes (Sandalio and Romero-Puertas 2015). There are two forms of reactive oxygen species: O_2-centered radicals like superoxide radical ($O_2^{\bullet -}$), alkoxyl radical ($RO^{\bullet}$), peroxyl radical ($ROO^{\bullet}$), and hydroxyl radical ($^{\bullet}OH$) and O_2-centered non-radicals like hydrogen peroxide (H_2O_2) and singlet oxygen ($^1O_2$) (Mittler et al. 2004; Gill and Tuteja 2010). Approximately 1% of the oxygen taken in by plants is utilized to make ROS in subcellular sites such as peroxisomes, chloroplasts, and mitochondria (Sandalio and Romero-Puertas 2015; Dietz et al. 2016; Huang et al. 2016). Electron transport systems carrying oxygenic organisms generally produce ROS because O_2 is a powerful electron acceptor. Molecular O_2 is reduced in four phases, yielding a number of O_2 radical species. In the first step of O_2 reduction, superoxide radical ($O_2^{\bullet -}$) is made, which has a half-life of 2–4 μs and does not spread easily (Halliwell 2006). When O_2 is further reduced, it produces H_2O_2, a molecule with a relatively long lifetime (1 ms). Hydrogen peroxide is made in the absence of an enzyme when $O_2^{\bullet -}$ is dismutated to H_2O_2 under low-pH circumstances or more often in the presence of enzyme superoxide dismutase (SOD) (Das and Roychoudhury 2014). Further, in the presence of metal catalysts, $^{\bullet}OH$ radical is generated in the form of Haber-Weiss or Fenton-type reactions. Hydroxyl radical is very reactive because it reacts with almost every molecule in the body and often has an effect close to its point of generation (Kärkönen and Kuchitsu 2015). The different generation sites for ROS production are discussed in this section, and a summary is shown in Fig. 10.1.

10.2.1 Mitochondria

Multiple stresses can generate ROS in mitochondria by hindering and modifying the electron transport chain (ETC) and ATP generation, resulting in a surplus reduction in electron carriers and the generation of ROS (Noctor et al. 2007). Electron leakage from complex I and III of the ETC generates $O_2{\bullet}^-$, which is catalyzed into H_2O_2 by Mn-SOD and CuZn-SOD. The flavoprotein region of complex I is responsible for the direct reduction of O_2 to $O_2{\bullet}^-$. Reverse electron transport from complex III to complex I as a result of an insufficiency of NAD^+-linked substrate enhances ROS generation at complex I, and this reverse electron movement is controlled by ATP hydrolysis (Turrens 2003). The ETC has a ubiquinone-cytochrome region (complex

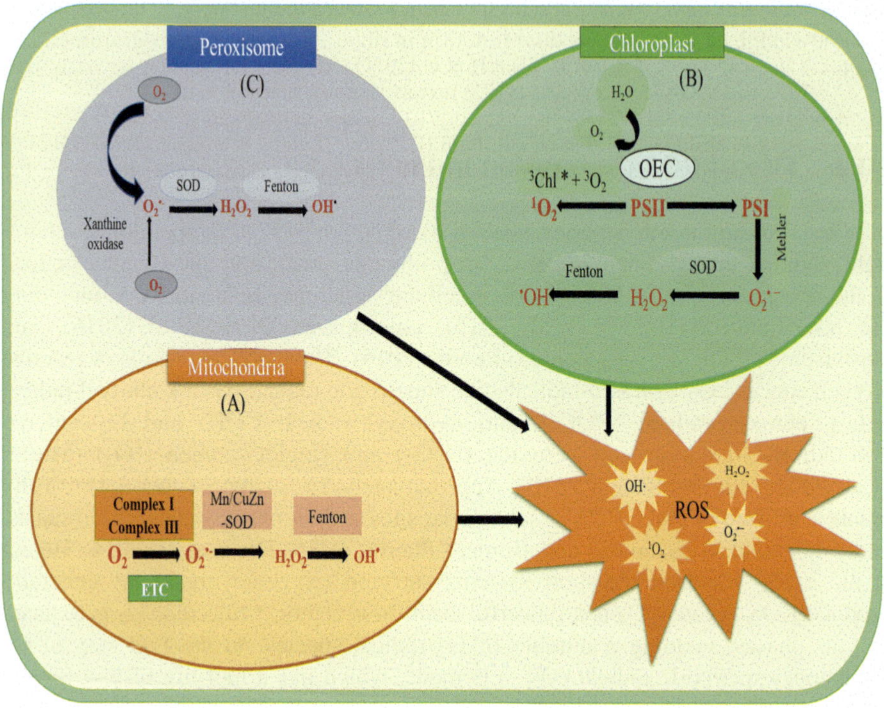

Fig. 10.1 Diagrammatic representation of the generation of reactive oxygen species (ROS) in mitochondria (a), chloroplast (b), and peroxisomes (c) in plants

III) which also generates $O_2\bullet^-$ from O_2. Completely reduced ubiquinone is thought to donate an electron to cytochrome c_1 and leave behind an unstable, extremely reducing ubisemiquinone radical that is conducive to electron leakage to O_2, subsequently leading to the generation of $O_2^{\bullet-}$ (Murphy 2009). Moreover, mitochondrial matrix enzymes that contribute to ROS production include aconitase, which produces ROS directly, and 1-galactono-lactone dehydrogenase, which contributes to ROS production indirectly by providing electrons to the ETC (Rasmusson et al. 2008). $O_2\bullet^-$ is the predominant ROS in mitochondria. However, Mn-SOD and ascorbate peroxidase (APX) quickly convert it into stable and membrane-permeable H_2O_2, which undergoes further conversion to the extremely reactive $^{\bullet}OH$ through the Fenton reaction (Fig. 10.1a).

10.2.2 Chloroplast

In chloroplast, triplet chlorophyll (^{3}Chl), ETC of photosystem I (PSI), and photosystem II (PSII) are the primary sources of ROS generation (Singh et al. 2019). Under the influence of light, the chlorophyll in the PSII light-harvesting complex reaches a high-energy singlet state after becoming excited. By photochemical quenching, a portion of this energy is transferred to P680 to power the photosynthetic ETC. However, when the amount of energy absorbed surpasses the capacity of the photochemical quenching, in that case, the extra energy is released as heat, fluorescence, or through the intersystem crossing, forming triplet excited chlorophyll (^{3}Chl*) (Müller et al. 2001). If ineffective quenching of this triplet excited chlorophyll (^{3}Chl*) occurs, it will undergo a redox reaction with $^3O_2$ liberated during the splitting of water in the oxygen-evolving complex (OEC), producing $^1O_2$ (Li et al. 2009). Furthermore, P680 is excited to the singlet state (^{1}P680*) after absorbing light in the PSII reaction center (RC), after which it couples with pheophytin (Pheo) to form ^{1}(P680$^+$ Pheo$^-$) and finally transfers an electron to the quinone (Q_A) to form P680$^+$ $Q_A{}^-$. In an adverse circumstance, if Q_A has been reduced to the point where it cannot accept any additional electrons, the excited state ^{3}P680* is formed when ^{3}(P680$^+$ Pheo$^-$) recombines with P680 (Krieger-Liszkay 2005). Two molecules of β-carotene are present in the PSII RC and can quench this high-energy ^{3}P680*; since they are too far apart, quenching is unable to take place, which results in the production of $^1O_2$ (Krieger-Liszkay et al. 2008). Additionally, in PSII, stomatal closure caused due to the abiotic stresses minimizes the chloroplastic CO_2 concentration, leading to the overproduction of the ETC and increasing the chance of electronic conductivity between ^{1}P680* and Q_A, resulting in an increase in $^1O_2$ generation (Flors et al. 2006). On the other hand, $O_2{}^{\bullet -}$ can be formed through the Mehler reaction and then transformed into H_2O_2 by SOD, as opposed to $^1O_2$ being made at PSII (Bose et al. 2014). Subsequently, metal catalysts like Fe^{2+} transform $O_2{}^{\bullet -}$ and H_2O_2 into the much more unstable and reactive $^{\bullet}$OH (Singh et al. 2019, Fig. 10.1b).

10.2.3 Peroxisomes

Similarly to mitochondria and chloroplasts, peroxisomes generate ROS as a byproduct of regular metabolism. The enzyme glycolate oxidase, often known as GOX, is the principal source of ROS in peroxisomes (Kerchev et al. 2016). Xanthine oxidase in the peroxisomal matrix transforms xanthine and hypoxanthine to uric acid and generates $O_2{}^{\bullet -}$ as a byproduct (Fig. 10.1c). The peroxisomal membrane-located NADPH-dependent small ETC constituted of NADH, and Cytb uses O_2 as the electron acceptor and releases $O_2{}^{\bullet -}$ into the cytosol. Furthermore, the three integral membrane proteins responsible for $O_2{}^{\bullet -}$ generation are peroxisomal membrane polypeptides (PMPs) and have molecular masses of 18, 29, and 32 kDa, respectively. The 18 and 32 kDa PMPs reduce cytochrome c by utilizing NADH as an electron donor, while the 29 kDa PMP uses NADPH as an electron donor to accomplish the same task (López-Huertas et al. 1999).

10.3 ROS Signaling in Plants

In the past few years, significant development has been made in identifying ROS as a key signal in many biological processes. Several genes and signal transduction pathways can be affected by ROS because ROS are small molecules and they can diffuse over limited distances; thus, they are ideal for use as signaling molecules (Mittler et al. 2022). They are also important for stress sensing, which is the process of putting together multiple stress-response signaling networks and turning on and off the plant's defense systems and acclimatization (Hipsch et al. 2021). A wide range of biomolecules can interact with ROS, which may result in permanent harm, necrosis, and death (Girotti 2001). It is crucial to remember that aerobic life existed during the occurrence of ROS when researching ROS signaling in plant cells, indicating that ROS have a profound role in stress sensing and signaling, and that is why the majority of cells can protect themselves from ROS damage (Inupakutika et al. 2016). ROS, which plants can produce in response to environmental stimuli, can regulate an array of physiological processes, like activation of programmed cell death (PCD), germination, stomatal regulation, the timing of sexual reproduction, and responses to biotic and abiotic stresses (Simon et al. 2000; Huang et al. 2019). These are a few illustrations of the several processes that ROS metabolism regulates. These processes are regulated by ROS-mediated signal transduction pathways. The detection of ROS and the activation of multiple signal transduction pathways can occur in separate compartments in plant cells due to the complex nature of their subcellular environments. ROS signaling can be classified as extrinsic (cell wall and apoplast), intrinsic (nucleus and cytosol), and organellar (peroxisomes, mitochondria, chloroplast, and other compartments) ROS signaling (Mittler 2017; Mittler et al. 2022).

Extrinsic ROS Signaling The cell wall and apoplast contain several enzymes that either neutralize or actively generate ROS and also have numerous nonenzymatic antioxidants. The most significant contribution to ROS signaling at the apoplast is made by respiratory burst oxidase homologues (RBOHs), aquaporins (AQPs), and cell wall-bound peroxidases (Torres et al. 2002; Rodrigues et al. 2017; Mittler et al. 2022). The carefully regulated transmembrane proteins known as RBOHs use NADPH from the cytosol to create ROS in the apoplast, which then gets converted into H_2O_2 by SOD (Smirnoff and Arnaud 2019). RBOHs have been called "the engines of ROS signaling" due to the fact that they are switched "on" or "off" in response to various stressors and other stimuli, resulting in the production of ROS signatures at the apoplast (Torres et al. 2002; Miller et al. 2009). Phosphorylation, acetylation, and guanidinylation control the opening and closing of AQPs and link many signaling pathways with ROS transport (Bienert et al. 2007). Apoplastic ROS synthesis and cytosolic entrance proteins are controlled by posttranslational improvements to the RBOHs' and AQPs' cytoplasmic side, and ROS buildup at the apoplast can result in cell death (Golani et al. 2013). Events of cytosolic phosphorylation is mediated by receptors and modifying Ca^{2+} fluxes via plasma membrane channels. During stress, the interface between apoplast and cytosol becomes an increasingly critical hub for several ROS-related signal transduction events (Mittler et al. 2022).

Intrinsic ROS Signaling There are various ROS-scavenging mechanisms and a few ROS-generating enzymes in the cytosol. These are thought to keep an eye on ROS signals that come from the cytosol and ROS signals that come from the apoplast or other organelles and are sent to the nucleus through the cytosol (Davletova et al. 2005). There are many signaling hubs in the cytosol, such as calcium-dependent protein kinases, calcineurin B-like-interacting protein kinases, MAPK cascades, different phosphatases (PP2A, PP2C, and PTPs), ROP/RAC small GTPases, and multiple redox-sensing networks (such as PrxRs, GRXs, and TRXs), which use ROS signals with other signaling molecules such as Ca^{2+} and various hormones (Zhu 2016). Between organelles and the nucleus, the cytosol mediates the transmission of both retrograde and anterograde signals in both directions (Nomura et al. 2012). The cytosol is essential for analyzing and incorporating the different ROS signatures made in different parts of the cell. This is done by sending the information in these signatures to the nucleus (Shapiguzov et al. 2019). Numerous redox-responsive transcriptional regulators which control plant stress are activated in response to ROS in the cytoplasm before accessing the nucleus, suggesting that ROS levels in the nucleus are regulated to avoid unnecessary alternations that could result in DNA damage and mutations (Exposito-Rodriguez et al. 2017).

Organelle ROS Signaling Network The various organelles found in plant cells each have a variety of ROS-scavenging and ROS-producing processes that regulate ROS signaling inside each organelle and take part in communication between organelles and with the nucleus (Smirnoff and Arnaud 2019). The reactive oxygen species concentrations in one organelle might affect those in another organelle or the nucleus via a series of intermediate metabolites, hormones, and/or protein mobilization (Giesguth et al. 2015). Mitochondria, chloroplasts, and peroxisomes have received the most attention for their roles in ROS signaling and metabolism. However, research on ROS signaling and metabolism in the endoplasmic reticulum, plasmodesmata, and vacuole is limited. ROS signatures may be generated in the cell in a variety of different "maps" or "landscapes," depending on the kind of pressure that is applied. It is possible that many ROS sensors scattered across the various compartments can decode these (Mittler et al. 2022).

10.4 ROS Signaling-Mediated Tolerance in Plants

It is well-known that various stressors, including heavy metals, drought, temperature, salinity, and others, may throw off the balance normally maintained between ROS formation and its removal. The magnitude and duration of stressors, changes in growth behavior, and the capacity of plants to adapt to changing environmental circumstances are the primary contributors to plant tolerance in situations associated with those stresses (Miller et al. 2004; Yamasaki et al. 2019). Several genes are

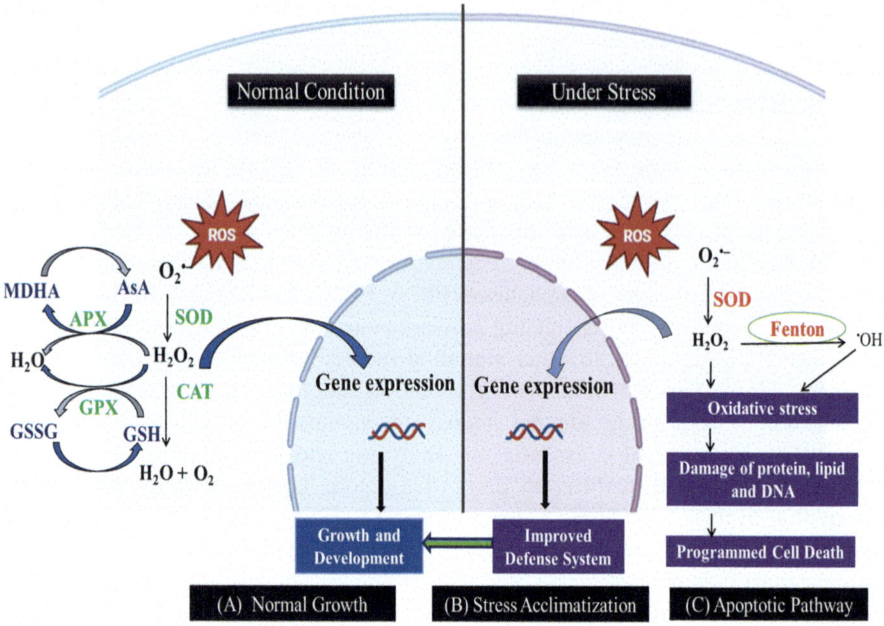

Fig. 10.2 Schematic representation to explain the ROS-mediated response under normal and stress conditions. Under normal conditions (**a**), ROS are generated as a byproduct of normal metabolism and are proficiently removed by both enzymatic and nonenzymatic antioxidant defense systems (normal growth). On the other hand, under stress (**b**), increased generation of ROS triggers a signal for defense genes that result in an improved defense system and acclimatized plant under stress (stress acclimatization). However, when the ROS level is very high, that leads to oxidative stress (**c**), a potentially harmful process that damages biomolecules and cellular structures and contributes to programmed cell death (PCD), which involves defense withdrawal

encoded in the various proteins that regulate ROS metabolism and its signaling. ROS are either produced passively or actively. Actively produced ROS are primarily supposed to be produced for the signaling purpose and may also be employed to the plant's benefit (Halliwell and Gutteridge 2015; Mittler 2017; Smirnoff and Arnaud 2019; Sies and Jones 2020). At the same time, ROS are constantly cleaned up by enzymes like superoxide dismutase (SOD), catalase (CAT), ascorbate peroxidase (APX), guaiacol peroxidase (GPX), and glutathione-S-transferase (GST), as well as nonenzymatic low molecular compounds like reduced glutathione (GSH) and ascorbic acid (AsA), which are mostly found in all the cellular compartments (Gill and Tuteja 2010; Halliwell and Gutteridge 2015; Sies and Jones 2020) (Fig. 10.2a). Due to their ability to regulate the levels of $O_2^{\bullet-}$ and H_2O_2 in cells, SOD is considered an essential component of the antioxidant defense system. Catalases get rid of most of the H_2O_2, but ascorbate peroxidases (APX) could get rid of H_2O_2 that catalase cannot get to because they are more attracted to it and are in different places inside the cell (Gill and Tuteja 2010). Other parts of the plant have a water-water cycle to scavenge ROS that involves many enzymes and low-molecular-weight antioxidant molecules like AsA and GSH (Halliwell and Gutteridge 2015). Under low-stress levels, this antioxidant defense mechanism successfully eliminates free radicals (Fig. 10.2b). However, the antioxidant defense system may become overwhelmed during acute stress by the increasing radical production frequency. Excessive concentrations of ROS cause damage to the photosynthetic system, macromolecules like lipids, proteins, and DNA, resulting in severe cellular damage or death in plants (Gill and Tuteja 2010) (Fig. 10.2c). The antioxidant defense system's significance is illustrated by the overproduction of many ROS scavengers in various transgenic plants, which substantially protect against oxidative stress (Martins et al. 2020; Liu et al. 2021).

Additionally, ROS are transported to various cells, tissues, or organs for its elimination, accumulation, and signaling purposes (Rodrigues et al. 2017). So, ROS function locally as well as distantly. ROS generation and accumulation during the various environmental stress conditions, resulting in the imbalance in the redox status of various proteins, enzymes, and macro and micromolecules, cause the alteration or integration of various stress responses and their associated signal transduction pathways. These pathways modulate the expression of several genes and prepare the plants to deal with the stress condition (Zou et al. 2015; Yin et al. 2018; Martins et al. 2020; Liu et al. 2021). This means that cells learned how to avoid ROS toxicity from that time, and ROS started being used mainly as a signaling and stress-sensing molecule (Mittler 2017). In contrast to other traditional signaling molecules like phytohormones, changes in the level of ROS can change the structure and function of several proteins and enzymes, which alter the different signaling pathways. These different signaling pathways of ROS are mainly mediated by oxidative posttranslational modification (Chan et al. 2016; De Smet et al. 2019; Huang et al. 2019). In this process, ROS interacts with the thiol group of cysteine and methionine residue of several proteins and finally converts into irreversible sulfonic acid, which results in the degradation of proteins (Zaffagnini et al. 2016; Akter et al. 2018). Thus, ROS changes the properties of cysteine- and methionine-containing proteins which can further accelerate or suppress the stress-responsive signal transduction pathways. For example, some protein phosphatase activities like class 2 serine/threonine protein phosphatase, tyrosine phosphatase, and catabolic phosphatase SAL1 are inhibited by ROS-induced oxidative posttranslational modification, and these enzymes are involved in several signaling processes (Gupta and Luan 2003; Silver et al. 2013; Chan et al. 2016).

In plant cells, the extracellular region contains multiple enzymes and nonenzymatic antioxidants that actively regulate the level of ROS. Respiratory burst oxidase homologues (RBOHs), cell wall-bound peroxidases, and aquaporins (AQPs) have a significant role in extracellular ROS signaling. RBOHs, also called "the engine of ROS signaling," become active or inactive in response to various stresses (Torres et al. 2002; Miller et al. 2009; Kaya et al. 2014; Mittler 2017; Mittler et al. 2022). Moreover, ROS also requires AQPs to enter the intracellular region, where it regulates the intracellular signaling pathways, and the opening and closing of AQPs are also mediated by ROS (Rodrigues et al. 2017; Maurel et al. 2015). Apoplastic ROS can accelerate the phosphorylation of cytosolic residue via receptors and alter the flow of Ca^{2+}. ROS also plays a significant role as a secondary (2°) messenger. Hydrogen peroxide (H_2O_2), produced as a byproduct in several oxidative processes, performs the signaling role because of its high stability compared to other ROS (Martins et al. 2020). H_2O_2, secreted by polyamine oxidase in the apoplast, takes part in the balancing of stress tolerance and cell death (Chae et al. 2013). ROS can also modulate the signal transduction pathway through proline and Ca^{2+} (Wang et al. 2020). Moreover, it also modulates the abscisic acid (ABA) signaling by regulating the phosphatases enzymes, level of microRNA, and mRNA splicing (Tran et al. 2013; Iyer et al. 2012; He et al. 2021). An increased level of ROS downregulates the expression of some housekeeping genes required for microRNA splicing and other complexes (Ohama et al. 2017).

10.5 ROS Signaling-Mediated Programmed Cell Death in Plants

Cell death is an important part of how a plant grows and lives. In plants, two main ways have been identified: programmed cell death (PCD) and necrosis. PCD is controlled by genes and has some of the same characteristics as apoptotic cell death in animal cells, such as shrinking cells, clumping of the cytoplasm, clumping of the chromatin, and breaking up of the DNA. Necrosis is caused by severe and long-lasting trauma and is not thought to be caused by genes. Through the selective involvement of particular proteases and nucleases, some cells are removed in a highly sophisticated and multistep manner during PCD, which is a highly active and genetically managed process. As a result, the cells that are going to die are eliminated, and the surrounding cells are not harmed (Gadjev et al. 2008). Developmental processes in plants that are linked to programmed cell death (dPCD) include embryo formation, the disintegration of the aleurone layer in monocot seed germination, degeneration of the anther tapetum, pollen self-incompatibility, differentiation of tracheary components in xylem tissues, production of aerenchyma and epidermal trichomes of the root, abscission of floral organs, remodeling of certain kinds of leaf morphology, leaf senescence, and response to environment (Gechev et al. 2006; Thomas and Franklin-Tong 2004). Many biotic and abiotic conditions, such as high temperatures, high salinity, and pollution, may potentially trigger programmed cell death (ePCD) that is not desirable (Koukalova et al. 1997; Overmyer et al. 2000; Swidzinski et al. 2002). ROS like hydrogen peroxide (H_2O_2) is well-known to play a pivotal role in regulating PCD and a wide variety of other biological processes, including growth, development, and stress adaption (Gechev et al. 2006). The first experimental evidence that showed ROS could possibly work as signals in plant PCD, in addition to only being an activator of oxidative activities, was found when it was shown that inhibitors of protein synthesis might reduce H_2O_2-induced cell death (Levine et al. 1994). Cell death caused by exogenous H_2O_2 is dose- and time-dependent (Tochigi et al. 2013). Protein kinases, protein phosphatases, and transcription factors are downstream components of H_2O_2 and ROS signal transduction networks that govern plant PCD (Van Hautegem et al. 2015). The presence of H_2O_2 has been linked to the occurrence of PCD during HR in both plants and suspension cultures (Pennell and Lamb 1997). Harpin causes PCD in *Arabidopsis* suspension cultures through H_2O_2 (Desikan et al. 2001). Harpin was shown to rapidly activate a 44 KDa protein, as was the previously reported kinase protein that shares properties with mitogen-activated protein kinases, and H_2O_2 can also activate the same or a related kinase. However, it has not been determined whether or not protein kinase activation is crucial for both H_2O_2 PCD and triggered gene expression (Desikan et al. 2001). The need for transcription and translation and exposure to H_2O_2 suggests that this type of cell death is predetermined (Desikan et al. 1998). Numerous abiotic and biotic stresses activate mitogen-activated protein kinase (MAPK) (Mittler and Zilinskas 1992). In the absence of a pathogen, cell death in *Arabidopsis* lsd1 mutants is caused by an aberrant buildup of $O_2^{\bullet-}$ and a lack of receptivity to signals arising from it. In these mutants, $O_2^{\bullet-}$ accumulates prior to the initiation of cell death and subsequently in cells next to spreading lsd1 lesions, which is both essential and sufficient to start lesion formation (Jabs et al. 1996). There are reports of H_2O_2 involvement in numerous PCD subtypes brought on by abiotic stress. The significance of time and severity of oxidative stress required for PCD induction complicates our understanding of the ROS-dependent PCD network. In tobacco BY-2 cells, H_2O_2 contributes to PCD caused by heat shock (HS) (Vacca et al. 2007; Locato et al. 2008). It has been shown that the level of oxidative stress and the timing of ROS formation determine the metabolic responses that are triggered in tobacco BY-2 cells (de Pinto et al. 2006). Cells detect the rise in ROS levels caused by stress to initiate intercellular signaling and subsequent changes in gene expression. Apart from receptors, plants also use redox-sensitive compounds like transcription factors and phosphatase inhibition to detect ROS (Liang and Zhou 2018). Changes in gene expression result from the detection of ROS at the apoplast, which then triggers intracellular and intercellular signaling. Various parts of a cell produce the four types of reactive oxygen species (Mittler 2017; Mhamdi and Van Breusegem 2018). Every type of ROS used in signaling has a specific cellular target with which it communicates. Leucine-rich repeat and cysteine-rich receptor-like kinases (RLKs) have been recognized as ROS sensors in *Arabidopsis*; both of these possess an extracellular domain that alters after being exposed to ROS through redox reactions (Gauthier et al. 2011). In organized pathways known as MAPK cascades, phosphorylation of the upstream-activated MAPK kinase kinase (MAPKKK) results in the activation of downstream MAPK kinases (MAPKKs), which in turn activate particular MAPKs. Through the posttranslational modification of target proteins, MAPK cascades mediate the transduction of environmental signals, which ultimately reorganizes gene expression and promotes stress adaptation (Saxena et al. 2016). More than 60 MAPKKKs, 10 MAPKKs, and 20 MAPKs are all encoded in the *Arabidopsis* genome, which controls how plants react to biotic and abiotic stresses (MAPK Group 2002). ROS response starts when MAPK cascade components are changed (Pitzschke and Hirt 2009). H_2O_2-dependent sulfenylation of three members of the *Arabidopsis* MAPK family (MPK2, MPK4, and MPK7) has been found, which may affect their function in response to changes in cellular redox state (Waszczak et al. 2014). H_2O_2 has been demonstrated to activate ANP1, oxidative signal-inducible 1 (OXI1) kinase, and one of the

MAPKKKs in *Arabidopsis*, all of which start a phosphorylation cascade involving MPK3 and MPK6 (Rentel et al. 2004). *Arabidopsis* plants regulate the expression of many APETALA2/ethylene response factor family transcription factors, including ERF6 and ERF104, by activating downstream retrograde communication between the chloroplast and nucleus during the transition from low- to high-light levels (Vogel et al. 2014). The MEKK1-MKK1/MKK2-MPK4 cascade also controls the homeostasis of ROS and PCD (Pitzschke et al. 2009). Both abiotic and biotic stresses can activate *Arabidopsis* MPK4 (Zhang et al. 2012). According to a theory, MPK4 is a ROS/hormonal rheostat that positively controls photosynthesis, ROS metabolism, and growth, while negatively regulating immune defenses in an SA-dependent manner (Gawroński et al. 2014). The previously mentioned OXI1 kinase plays a crucial role in both "^1O and H_2O_2 signaling," which has an impact on the JA-dependent pathway and programmed cell death (Shumbe et al. 2016). All these studies imply a dual role of ROS in the PCD: first, as a developmental PCD (dPCD) for the development of the plant, and second, PCD as a common consequence of environmental stress (ePCD; Lam 2004).

10.6 Conclusion

In this chapter, we have attempted to summarize the ROS-mediated signal response in plants, focusing on tolerance and PCD. The signaling role of ROS is more versatile since it may stimulate various genes in a variety of combinations or alone. ROS triggers defense genes and adaptive responses when present in low quantities. Sublethal ROS levels may acclimatize plants to different stress conditions and limit plant development, most likely as part of an adaptive response. At higher concentrations, ROS causes a genetically controlled programmed cell death. In Fig. 10.2, we have tried to summarize the ROS-mediated signal's role in stress acclimatization and programmed cell death. ROS also creates a network with other signal molecules and pathways to coordinate subsequent responses. Because of their newly proven function in growth and morphogenesis, ROS are likely not just stress signal molecules but also a fundamental signal in plant development and growth. There is still a lot of work to be done to figure out how the complex regulatory network links ROS signals with parts of different signaling pathways for the onset of growth and developmental process in plants.

References

Akter S et al (2018) Chemical proteomics reveals new targets of cysteine sulfinic acid reductase. Nat Chem Biol 14:995–1004

Apel K, Hirt H (2004) Reactive oxygen species: metabolism, oxidative stress, and signaling transduction. Annu Rev Plant Biol 55:373

Bendall DS, Howe CJ, Nisbet EG, Nisbet RE (2008) Photosynthetic and atmospheric evolution. Introduction. Philos Trans R Soc Lond Ser B Biol Sci 363(1504):2625–2628

Bhattacharjee S (2005) Reactive oxygen species and oxidative burst: roles in stress, senescence and signal. Curr Sci 89:1113–1121

Bienert GP et al (2007) Specific aquaporins facilitate the diffusion of hydrogen peroxide across membranes. J Biol Chem 282:1183–1192

Bose J, Rodrigo-Moreno A, Shabala S (2014) ROS homeostasis in halophytes in the context of salinity stress tolerance. J Exp Bot 65:1241–1257

Breygina M, Klimenko E (2020) ROS and ions in cell Signaling during sexual plant reproduction. Int J Mol Sci 21:9476

Chae HB et al (2013) Thioredoxin reductase type C (NTRC) orchestrates enhanced thermotolerance to *Arabidopsis* by its redox-dependent holdase chaperone function. Mol Plant 6:323–336

Chan KX et al (2016) Sensing and signaling of oxidative stress in chloroplasts by inactivation of the SAL1 phosphoadenosine phosphatase. Proc Natl Acad Sci U S A 113:E4567–E4576

Das K, Roychoudhury A (2014) Reactive oxygen species (ROS) and response of antioxidants as ROS-scavengers during environmental stress in plants. Front Environ Sci 2:53

Davletova S et al (2005) Cytosolic ascorbate peroxidase 1 is a central component of the reactive oxygen gene network of *Arabidopsis*. Plant Cell 17:268–281

De Smet B, Willems P, Fernandez-Fernandez AD, Alseekh S, Fernie AR, Messens J, Van Breusegem F (2019) In vivo detection of protein cysteine sulfenylation in plastids. Plant J 97(4):765–778

Desikan R, Hancock JT, Ichimura K, Shinozaki K, Neill SJ (2001) Harpin induces activation of the *Arabidopsis* mitogen-activated protein kinases AtMPK4 and AtMPK6. Plant Physiol 126(4): 1579–1587

Desikan R, Reynolds A, Hancock TJ, Neill JS (1998) Harpin and hydrogen peroxide both initiate programmed cell death but have differential effects on defence gene expression in *Arabidopsis* suspension cultures. Biochem J 330(1):115–120

Dietz KJ, Turkan I, Krieger-Liszkay A (2016) Redox-and reactive oxygen species-dependent signaling into and out of the photosynthesizing chloroplast. Plant Physiol 171(3):1541–1550

Exposito-Rodriguez M, Laissue PP, Yvon-Durocher G, Smirnoff N, Mullineaux PM (2017) Photosynthesis-dependent H2O2 transfer from chloroplasts to nuclei provides a high-light signalling mechanism. Nat Commun 8:49

Flors C, Fryer MJ, Waring J, Reeder B, Bechtold U, Mullineaux PM, Nonell S, Wilson MT, Baker NR (2006) Imaging the production of singlet oxygen in vivo using a new fluorescent sensor, singlet oxygen sensor green®. J Exp Bot 57(8):1725–1734

Foyer CH, Noctor G (2005) Oxidant and antioxidant signalling in plants: a re-evaluation of the concept of oxidative stress in a physiological context. Plant Cell Environ 28(8):1056–1071

Gadjev I, Stone JM, Gechev TS (2008) Programmed cell death in plants: new insights into redox regulation and the role of hydrogen peroxide. Int Rev Cell Mol Biol 270:87–144. https://doi.org/10.1016/S1937-6448(08)01403-2

Gauthier NC, Fardin MA, Roca-Cusachs P, Sheetz MP (2011) Temporary increase in plasma membrane tension coordinates the activation of exocytosis and contraction during cell spreading. Proc Natl Acad Sci 108(35):14467–14472

Gawroński P, Witoń D, Vashutina K, Bederska M, Betliński B, Rusaczonek A, Karpiński S (2014) Mitogen-activated protein kinase 4 is a salicylic acid-independent regulator of growth but not of photosynthesis in *Arabidopsis*. Mol Plant 7(7):1151–1166

Gechev TS, Van Breusegem F, Stone JM, Denev I, Laloi C (2006) Reactive oxygen species as signals that modulate plant stress responses and programmed cell death. BioEssays 28(11): 1091–1101

Giesguth M, Sahm A, Simon S, Dietz KJ (2015) Redox-dependent translocation of the heat shock transcription factor AtHSFA8 from the cytosol to the nucleus in *Arabidopsis thaliana*. FEBS Lett 589:718–725

Gill SS, Tuteja N (2010) Reactive oxygen species and antioxidant machinery in abiotic stress tolerance in crop plants. Plant Physiol Biochem 48(12):909–930

Girotti AW (2001) Photosensitized oxidation of membrane lipids: reaction pathways, cytotoxic effects and cytoprotective mechanisms. J Photochem Photobiol 63:103–113

Golani Y, Kaye Y, Gilhar O, Ercetin M, Gillaspy G, Levine A (2013) Inositol polyphosphate phosphatidylinositol 5-phosphatase9 (At5ptase9) controls plant salt tolerance by regulating endocytosis. Mol Plant 6(6):1781–1794

Gull A, Lone AA, Wani NUI (2019) Biotic and abiotic stresses in plants. In (Ed.), Abiotic and Biotic stress in plants. IntechOpen. https://doi.org/10.5772/intechopen.85832

Gupta R, Luan S (2003) Redox control of protein tyrosine phosphatases and mitogen-activated protein kinases in plants. Plant Physiol 132:1149–1152

Halliwell B (2006) Reactive species and antioxidants. Redox biology is a fundamental theme of aerobic life. Plant Physiol 141(2):312–322

Halliwell B, Gutteridge JM (2015) Free radicals in biology and medicine. Oxford university press, USA

He H, Denecker J, Van Der Kelen K, Willems P, Pottie R, Phua SY, Hannah MA, Vertommen D, Van Breusegem F, Mhamdi A (2021) The *Arabidopsis* mediator complex subunit 8 regulates oxidative stress responses. Plant Cell 33(6):2032–2057

Hipsch M, Lampl N, Zelinger E, Barda O, Waiger D, Rosenwasser S (2021) Sensing stress responses in potato with whole-plant redox imaging. Plant Physiol 187(2):618–631

Huang H, Ullah F, Zhou DX, Yi M, Zhao Y (2019) Mechanisms of ROS regulation of plant development and stress responses. Front Plant Sci 10:800

Huang S, Van Aken O, Schwarzländer M, Belt K, Millar AH (2016) The roles of mitochondrial reactive oxygen species in cellular signaling and stress response in plants. Plant Physiol 171(3): 1551–1559

Inupakutika MA, Sengupta S, Devireddy AR, Azad RK, Mittler R (2016) The evolution of reactive oxygen species metabolism. J Exp Bot 67:5933–5943

Iyer NJ, Jia X, Sunkar R, Tang G, Mahalingam R (2012) microRNAs responsive to ozone-induced oxidative stress in *Arabidopsis thaliana*. Plant Signal Behav 7:484–491

Jabs T, Dietrich RA, Dangl JL (1996) Initiation of runaway cell death in an *Arabidopsis* mutant by extracellular superoxide. Science 273:1853–1856

Kärkönen A, Kuchitsu K (2015) Reactive oxygen species in cell wall metabolism and development in plants. Phytochemistry 112:22–32

Kaya H, Nakajima R, Iwano M, Kanaoka MM, Kimura S, Takeda S, Kawarazaki T, Senzaki E, Hamamura Y, Higashiyama T, Takayama S (2014) Ca_2^+-activated reactive oxygen species production by *Arabidopsis* RbohH and RbohJ is essential for proper pollen tube tip growth. Plant Cell 26(3):1069–1080

Kerchev P, Waszczak C, Lewandowska A, Willems P, Shapiguzov A, Li Z, Alseekh S, Mühlenbock P, Hoeberichts FA, Huang J, Van Der Kelen K (2016) Lack of GLYCOLATE OXIDASE1, but not GLYCOLATE OXIDASE2, attenuates the photorespiratory phenotype of CATALASE2-deficient *Arabidopsis*. Plant Physiol 171(3):1704–1719

Koukalova B, Kovarik A, Fajkus J, Siroky J (1997) Chromatin fragmentation associated with apoptotic changes in tobacco cells exposed to cold stress. FEBS Lett 414:289–292

Krieger-Liszkay A (2005) Singlet oxygen production in photosynthesis. J Exp Bot 56:337–346

Krieger-Liszkay A, Fufezan C, Trebst A (2008) Singlet oxygen production in photosystem II and related protection mechanism. Photosynth Res 98:551–564

Lam E (2004) Controlled cell death, plant survival and development. Nat Rev Mol Cell Biol 5:305–315

Levine A, Tenhaken R, Dixon R, Lamb C (1994) H_2O_2 from the oxidative burst orchestrates the plant hypersensitive disease resistance response. Cell 79:583–593

Li Z, Wakao S, Fischer BB, Niyogi KK (2009) Sensing and responding to excess light. Annu Rev Plant Biol 60:239–260

Liang X, Zhou JM (2018) Receptor-like cytoplasmic kinases: central players in plant receptor kinase–mediated signaling. Annu Rev Plant Biol 69:267–299

Liu D et al (2021) Tobacco transcription factor bHLH123 improves salt tolerance by activating NADPH oxidase NtRbohE expression. Plant Physiol 186:1706–1720

Locato V, Gadaleta C, De Gara L, De Pinto MC (2008) Production of reactive species and modulation of antioxidant network in response to heat shock: a critical balance for cell fate. Plant Cell Environ 31(11):1606–1619

López-Huertas E, Corpas FJ, Sandalio LM, Del Río LA (1999) Characterization of membrane polypeptides from pea leaf peroxisomes involved in superoxide radical generation. Biochem J 337(3):531–536

MAPK Group (2002) Mitogen-activated protein kinase cascades in plants: a new nomenclature. Trends Plant Sci 7:301–308

Martins L et al (2020) Redox modification of the iron-sulfur glutaredoxin GRXS17 activates holdase activity and protects plants from heat stress. Plant Physiol 184:676–692

Maurel C et al (2015) Aquaporins in plants. Physiol Rev 95:1321–1358

Mhamdi A, Van Breusegem F (2018) Reactive oxygen species in plant development. Development 145(15):164376

Miller G, Schlauch K, Tam R, Cortes D, Torres MA, Shulaev V, Dangl JL, Mittler R (2009) The plant NADPH oxidase RBOHD mediates rapid systemic signaling in response to diverse stimuli. Sci Signal 2(84):45

Miller GA, Suzuki N, Ciftci-Yilmaz SU, Mittler RO (2004) Reactive oxygen species homeostasis and signalling during drought and salinity stresses. Plant Cell Environ 33:453–467

Mittler R (2017) ROS are good. Trends Plant Sci 22:11–19

Mittler R, Vanderauwera S, Gollery M, Van Breusegem F (2004) Reactive oxygen gene network of plants. Trends Plant Sci 9:490–498

Mittler R, Zandalinas SI, Fichman Y, Van Breusegem F (2022) Reactive oxygen species signalling in plant stress responses. Nat Rev Mol Cell Biol 23:663–679

Mittler R, Zilinskas BA (1992) Molecular cloning and characterization of a gene encoding pea cytosolic ascorbate peroxidase. J Biol Chem 267:21802–21807

Müller P, Li XP, Niyogi KK (2001) Non-photochemical quenching. A response to excess light energy. Plant Physiol 125:1558–1566

Murphy MP (2009) How mitochondria produce reactive oxygen species. Biochem J 417:1–3

Noctor G, De Paepe R, Foyer CH (2007) Mitochondrial redox biology and homeostasis in plants. Trends Plant Sci 12:125–134

Nomura H, Komori T, Uemura S, Kanda Y, Shimotani K, Nakai K, Furuichi T, Takebayashi K, Sugimoto T, Sano S, Suwastika IN (2012) Chloroplast-mediated activation of plant immune signalling in *Arabidopsis*. Nat Commun 3:926

Ohama N, Sato H, Shinozaki K, Yamaguchi-Shinozaki K (2017) Transcriptional regulatory network of plant heat stress response. Trends Plant Sci 22:53–65

Overmyer K, Tuominen H, Kettunen R, Betz C, Langebartels C, Sandermann H Jr, Kangasjärvi J (2000) Ozone-sensitive *Arabidopsis* rcd1 mutant reveals opposite roles for ethylene and jasmonate signaling pathways in regulating superoxide-dependent cell death. Plant Cell 12: 1849–1862

Pennell RI, Lamb C (1997) Programmed cell death in plants. Plant Cell 9:1157–1168

de Pinto MC, Paradiso A, Leonetti P, De Gara L (2006) Hydrogen peroxide, nitric oxide and cytosolic ascorbate peroxidase at the crossroad between defence and cell death. Plants J 48:784–795

Pitzschke A, Djamei A, Bitton F, Hirt H (2009) A major role of the MEKK1-MKK1/ 2-MPK4 pathway in ROS signalling. Mol Plant 2:120–137

Pitzschke A, Hirt H (2009) Disentangling the complexity of mitogen-activated protein kinases and reactive oxygen species signalling. Plant Physiol 149:606–615

Podolyan A, Maksimov N, Breygina M (2019) Redox-regulation of ion homeostasis in growing lily pollen tubes. J Plant Physiol 243

Rasmusson AG, Geisler DA, Møller IM (2008) The multiplicity of dehydrogenases in the electron transport chain of plant mitochondria. Mitochondrion 8:47–60

Raza A, Razzaq A, Mehmood SS, Zou X, Zhang X, Lv Y, Xu J (2019) Impact of climate change on crops adaptation and strategies to tackle its outcome: a review. Plants (Basel) 8(2):34.s

Rentel MC, Lecourieux D, Ouaked F, Usher SL, Petersen L, Okamoto H, Knight H, Peck SC, Grierson CS, Hirt H, Knight MR (2004) OXI1 kinase is necessary for oxidative burst-mediated signalling in *Arabidopsis*. Nature 427(6977):858–861

Rodrigues O et al (2017) Aquaporins facilitate hydrogen peroxide entry into guard cells to mediate ABA and pathogen-triggered stomatal closure. Proc Natl Acad Sci U S A 114:9200–9205

Sandalio LM, Romero-Puertas MC (2015) Peroxisomes sense and respond to environmental cues by regulating ROS and RNS signalling networks. Ann Bot 116:475–485

Saxena I, Srikanth S, Chen Z (2016) Cross talk between H_2O_2 and interacting signal molecules under plant stress response. Front Plant Sci 28:570

Shapiguzov A et al (2019) *Arabidopsis* RCD1 coordinates chloroplast and mitochondrial functions through interaction with ANAC transcription factors. Elife 8:e43284

Shumbe L, Chevalier A, Legeret B, Taconnat L, Monnet F, Havaux M (2016) Singlet oxygen-induced cell death in *Arabidopsis* under high-light stress is controlled by OXI1 kinase. Plant Physiol 170:1757–1771

Sies H, Jones DP (2020) Reactive oxygen species (ROS) as pleiotropic physiological signalling agents. Nat Rev Mol Cell Biol 21:363–383

Silver DM et al (2013) Insight into the redox regulation of the phosphoglucan phosphatase SEX4 involved in starch degradation. FEBS J 280:538–548

Simon HU, Haj-Yehia A, Levi Schaffer F (2000) Role of reactive oxygen species in apoptosis induction. Apoptosis 5:415–418

Singh A, Kumar A, Yadav S, Singh IK (2019) Reactive oxygen species-mediated signaling during abiotic stress. Plant Gene 18:100173

Smirnoff N, Arnaud D (2019) Hydrogen peroxide metabolism and functions in plants. New Phytol 221:1197–1214

Swidzinski JA, Sweetlove LJ, Leaver CJ (2002) A custom microarray analysis of gene expression during programmed cell death in *Arabidopsis* thaliana. Plant J 30(4):431–446

Thomas SG, Franklin-Tong VE (2004) Self-incompatibility triggers programmed cell death in Papaver pollen. Nature 429(6989):305–309

Tochigi M, Inoue T, Suzuki-Karasaki M, Ochiai T, Ra C, Suzuki-Karasaki Y (2013) Hydrogen peroxide induces cell death in human TRAIL-resistant melanoma through intracellular superoxide generation. Int J Oncol 42(3):863–872

Torres MA, Dangl JL, Jones JDGG (2002) *Arabidopsis* gp91 phox homologues AtrbohD and AtrbohF are required for accumulation of reactive oxygen intermediates in the plant defense response. Proc Natl Acad Sci U S A 99:517–522

Tran D et al (2013) Post-transcriptional regulation of GORK channels by superoxide anion contributes to increases in outward-rectifying K^+ currents. New Phytol 198:1039–1048

Turrens JF (2003) Mitochondrial formation of reactive oxygen species. J Physiol 552:335–344. https://doi.org/10.1113/jphysiol.2003.049478

Vacca RA, Valenti D, Bobba A, de Pinto MC, Merafina RS, De Gara L, Passarella S, Marra E (2007) Proteasome function is required for activation of programmed cell death in heat shocked tobacco bright-yellow 2 cells. FEBS Lett 581:917–922

Van Hautegem T, Waters AJ, Goodrich J, Nowack MK (2015) Only in dying, life: programmed cell death during plant development. Trends Plant Sci 20(2):102–113

Vogel MO, Moore M, König K, Pecher P, Alsharafa K, Lee J, Dietz KJ (2014) Fast retrograde signaling in response to high light involves metabolite export, MITOGEN-ACTIVATED PROTEIN KINASE6, and AP2/ERF transcription factors in *Arabidopsis*. Plant Cell 26:1151–1165

Wang L et al (2020) The *Arabidopsis* SAFEGUARD1 suppresses singlet oxygen-induced stress responses by protecting grana margins. Proc Natl Acad Sci U S A 117:6918–6927

Waszczak C, Akter S, Eeckhout D, Persiau G, Wahni K, Bodra N, Van Molle I, De Smet B, Vertommen D, Gevaert K, De Jaeger G (2014) Sulfenome mining in *Arabidopsis thaliana*. Proc Natl Acad Sci U S A 111:11545–11550

Xie HT, Wan ZY, Li S, Zhang Y (2014) Spatiotemporal production of reactive oxygen species by NADPH oxidase is critical for tapetal programmed cell death and pollen development in *Arabidopsis*. Plant Cell 26(5):2007–2023

Yamasaki H, Ogura MP, Kingjoe KA, Cohen MF (2019) D-cysteine-induced rapid root abscission in the water fern azolla pinnata: implications for the linkage between d-amino acid and reactive sulfur species (RSS) in plant environmental responses. Antioxidants 8:411

Yin Y et al (2018) BZR1 transcription factor regulates heat stress tolerance through FERONIA receptor-like kinase-mediated reactive oxygen species signaling in tomato. Plant Cell Physiol 59:2239–2254

Zaffagnini M, De Mia M, Morisse S, Di Giacinto N, Marchand CH, Maes A, Lemaire SD, Trost P (2016) Protein S-nitrosylation in photosynthetic organisms: a comprehensive overview with future perspectives. Biochim Biophys Acta (BBA)-Proteins and Proteomics 1864(8):952–966

Zhang Z, Wu Y, Gao M, Zhang J, Kong Q, Liu Y, Ba H, Zhou J, Zhang Y (2012) Disruption of PAMP-induced MAP kinase cascade by a *Pseudomonas syringae* effector activates plant immunity mediated by the NB-LRR protein SUMM2. Cell Host Microbe 11:253–263

Zhu JK (2016) Abiotic stress signaling and responses in plants. Cell 167:313–324

Zou JJ et al (2015) *Arabidopsis* CALCIUM-DEPENDENT PROTEIN KINASE8 and CATALASE3 function in abscisic acid-mediated signaling and H_2O_2 homeostasis in stomatal guard cells under drought stress. Plant Cell 27:1445–1460

11 Insight into the Interaction of Strigolactones, Abscisic Acid, and Reactive Oxygen Species Signals

Hanan A. Hashem and Radwan Khalil

Abstract

Stress disrupts many physiological processes in plants by causing an increase in reactive oxygen species (ROS), which leads to serious damage to DNA, RNA, lipids, and proteins. Meanwhile, the generated ROS play an important role as signaling molecules, inducing a metabolic cascade that allows plants to tolerate stress. Plant development and stress adaptation are known to be monitored by phytohormones by regulating ROS levels to modulate signaling and prevent oxidative stress. Strigolactones (SLs) and abscisic acid (ABA) are carotenoid-derived hormones that play an active role in stress responses. A strong relationship was discovered between ABA and SLs biosynthesis. ABA was revealed to be involved in the regulation of SL production, and *max2* mutants (MAX2 is an F-box protein required for SL signaling) were found to be more sensitive to osmotic stress, with an impaired ABA response. The cross talk between SL and ABA signaling during stress is of major importance. SL signaling has been linked to ROS responses during osmotic stress and nutrient deprivation. Similarly, in water-stressed plants, ABA signals promote stomatal closure via ROS generated by respiratory burst oxidases (RBOH). This chapter summarizes the current understanding of how ROS production, detoxification, and signaling interact with SLs and ABA action to regulate plant growth and metabolism under acute stress conditions.

H. A. Hashem
Botany Department, Faculty of Science, Ain Shams University, Cairo, Egypt
e-mail: hananhashem@Sci.asu.edu.eg

R. Khalil (✉)
Botany Department, Faculty of Science, Benha University, Benha, Egypt
e-mail: Radwan.aboelabbas@fsc.bu.edu.eg

M. Faizan et al. (eds.), *Reactive Oxygen Species*,
https://doi.org/10.1007/978-981-19-9794-5_11

Keywords

Abscisic acid · ROS · Strigolactones · Stresses

11.1 Introduction

A variety of abiotic and biotic stressors activate signal transduction networks in plants, altering a variety of metabolic, molecular, and physiological responses and regulating acclimatization, growth, and development. Drought, heat, cold, and salinity are examples of abiotic stressors that have a negative impact on plant growth, reproduction, and survival, affecting agricultural crop output and yield (Choudhury et al. 2017; Ding et al. 2019; Gupta et al. 2020; Lamers et al. 2020). In plants, stress is sensed by receptors such as histidine kinases, receptor-like kinases, and G-protein-coupled receptors, as well as receptors for reactive oxygen species (ROS) and other stress-related metabolites and signaling molecules. These receptors detect stress and activate multiple complex communication networks in plants, including the formation of reactive oxygen species (ROS), changes in phytohormone levels, gene regulation, and various kinase/phosphatase signaling cascades that regulate stress responses (Xia et al. 2015; Noctor et al. 2018).

ROS, partial reductions, and derivatives of free radicals are highly reactive and toxic and can induce oxidative cell death. ROS, in addition to being hazardous byproducts of aerobic metabolism, play a role in the control and regulation of biological processes such as growth, the cell cycle, programmed cell death, hormone signaling, biotic and abiotic stress reactions, and development. ROS are produced in plants as a result of the inevitable escape of electrons to oxygen from the electron transport activities of chloroplasts, mitochondria, and plasma membranes. Chloroplasts, mitochondria, plasma membranes, peroxisomes, apoplasts, endoplasmic reticulum, and cell walls all produce reactive species. For efficient scavenging of ROS generated under diverse environmental stresses, the action of several nonenzymatic and enzymatic antioxidants present in tissues is necessary (Mansoor et al. 2022).

Plant resistance to environmental challenges is boosted by phytohormones, which have a negative impact on plant output and pose a threat to future food security. Strigolactones (SLs), a carotenoid-derived phytohormone, were first identified as an "ecological signal" promoting parasite seed germination and the creation of a symbiotic interaction between plants and beneficial bacteria. Because of their critical involvement in the control of many physiological and molecular processes during plant adaptation to abiotic stressors, SLs have recently received a lot of attention. Cross talk between SLs and other phytohormones, such as abscisic acid, in response to abiotic stimuli implies that SLs are active participants in phytohormone-controlled regulatory networks of plant stress adaptation. Furthermore, it has been postulated that SLs could have a function in managing plant growth and development under severe environmental conditions (Mostofa et al. 2018).

This chapter delves into the creation of reactive oxygen species (ROS) as a result of stresses on plants and their biology, as well as the signaling mechanisms that plants use to effectively cope with abiotic stress. Cross talk between reactive oxygen species (ROS) and phytohormone signaling molecules such as strigolactones and abscisic acid is also a concern.

11.2 Reactive Oxygen Species

11.2.1 Oxidative Stress and Detoxification

The production of reactive oxygen species (ROS) is a natural byproduct of plant cellular metabolism. Excessive formation of reactive oxygen species (ROS) is induced by a variety of environmental stressors, resulting in increasing oxidative damage and, eventually, cell death. They are well-known second messengers in a range of cellular functions, including conferring tolerance to diverse environmental stimuli, despite their destructive activity. The delicate balance between ROS generation and scavenging determines whether ROS will act as signaling molecules or cause oxidative damage to cells. The action of many nonenzymatic as well as enzymatic antioxidants present in the tissues is required for the efficient scavenging of ROS produced during diverse environmental stressors (Sharma et al. 2012).

Free radicals such as superoxide anion (O2•) and hydroxyl radical (•OH) and non-radical molecules such as hydrogen peroxide (H2O2) and singlet oxygen are examples of ROS (1O2). ROS are always generated in plants as a byproduct of multiple metabolic processes located in different cellular compartments or as a result of the unavoidable leaking of electrons onto O2 from the electron transport activities of chloroplasts, mitochondria, and plasma membranes (Foyer and Harbinson 1994; Blokhina and Fagerstedt 2010). Drought, salt, cold, metal toxicity, and UV-B radiation, as well as pathogen infection, cause an increase in ROS production in plants due to cellular homeostasis disruption (Shah et al. 2001; Sharma and Dubey 2007).

In large quantities, all ROS are extremely hazardous to organisms. A cell is said to be in a state of "oxidative stress" when the level of ROS exceeds the defense systems. Increased ROS production amid environmental conditions can endanger cells by triggering lipid peroxidation, protein oxidation, nucleic acid damage, enzyme inhibition, activation of the programmed cell death (PCD) pathway, and finally cell death (Sharma and Dubey 2005; Mishra et al. 2011; Srivastava and Dubey 2011). The precise balance between ROS generation and scavenging determines whether ROS will serve as a harmful or signaling molecule. Because of the multifunctional roles of ROS, it is vital for cells to precisely manage ROS levels in order to minimize oxidative harm rather than entirely eliminating them.

11.2.2 Sites of Production of ROS

ROS are formed in chloroplasts, mitochondria, plasma membranes, peroxisomes, apoplast, endoplasmic reticulum, and cell walls in both stressed and unstressed cells. ROS are always produced as a result of electron leakage onto O_2 from electron transport processes in chloroplasts, mitochondria, and plasma membranes or as a byproduct of numerous metabolic pathways located in different cellular compartments.

11.2.3 Signal Transduction

In plants, the role of reactive oxygen species (ROS) as a second messenger in hormone-mediated cellular responses is well established. Plants use redox-sensitive proteins, calcium mobilization, protein phosphorylation, and gene expression to sense, transduce, and transform ROS signals into suitable physiological responses. ROS have been implicated as second messengers in intracellular signaling cascades that mediate a variety of plant responses in plant cells, including stomatal closure (Yan et al. 2007), programmed cell death (Mittler 2002), gravitropism (Jung et al. 2001), and the acquisition of tolerance to both biotic and abiotic stresses (Fig. 11.1) (Miller et al. 2008).

Plants' adaptation to environmental stimuli is aided by reactive oxygen species, which behave as main molecules. They primarily function as signal transduction molecules, controlling a variety of pathways during the plant's adaptation to stress conditions (Choudhury et al. 2017; Antoniou et al. 2016). According to several studies (Mittler 2017), ROS is required for the success of a variety of fundamental natural processes, including cellular proliferation and differentiation. Furthermore, it has been proven that, in addition to ROS, reactive nitrogen species (RNS), reactive sulfur species (RSS), and reactive carbonyl species (RCS) all play a role in plant abiotic stress tolerance and are all involved in cross talk (Yamasaki et al. 2019). As a result, ROS plays a dual role in plant biology, making it an exciting field of study for plant biologists.

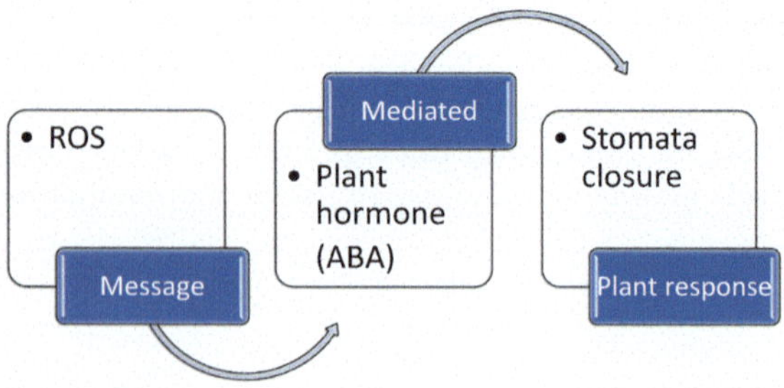

Fig. 11.1 ROS as second messenger to plant hormone (ABA) responses for stomatal closure

11.3 Abscisic Acid

11.3.1 Metabolism and Transport

Because abiotic stressors have such a profound effect on practically every aspect of plant growth, development, and reproduction, they cause changes in the levels and distribution of almost every plant hormone examined to date, including abscisic acid (ABA) (Nolan et al. 2020). Furthermore, to coordinate plant development and stress responses, vast crosstalk networks exist between ROS and hormone signaling pathways (Xia et al. 2015; Mignolet-Spruyt et al. 2016). Hormones can cause a further alteration in ROS levels, primarily by activating RBOHs (Kollist et al. 2019; Lamers et al. 2020) (Fig. 11.2). In addition to the altered production of ROS triggered directly by abiotic stresses, for example, due to altered metabolic reactions (Choudhury et al. 2017), once induced, hormones can cause a further alteration in ROS levels, primarily by activating.

11.3.2 ABA Signaling and Function in Plants

ABA is one of the key hormones implicated in the induction of stress responses. Plants use ABA to coordinate a complex regulatory network that allows them to cope with abiotic challenges like drought, salt, and temperature variations (Cutler et al. 2010; Kim et al. 2010a, b; Miyakawa et al. 2013). ABA is also important for plant growth and development in nonstress situations, such as embryo, seed, and seedling development (Finkelstein et al. 2002) and seed dormancy (Finkelstein et al. 2002). (Finkelstein et al. 2008). In recent investigations (Fujii et al. 2009; Umezawa et al. 2009), type 2C protein phosphatases (PP2Cs), protein kinases [SNF1 (sucrose non-fermenting 1)-related kinases subfamily 2, SnRK2s], and downstream targets have all been discovered in recent investigations (Fujii et al. 2009).

Abscisic acid functions in plants by recognizing intracellular receptors called PYLs (Ma et al. 2009; Park et al. 2009). PYLs bound to ABA form complexes with clade A PP2Cs, allowing the inhibition of SnRK2 protein kinases by PP2Cs to be released (Fujii et al. 2009; Ma et al. 2009; Park et al. 2009; Rubio et al. 2009). SnRK2s are then activated either by autophosphorylation or by other protein kinases, such as Raf-like MAPKKKs (Lee et al. 2015; Nguyen et al. 2019). SnRK2s control a

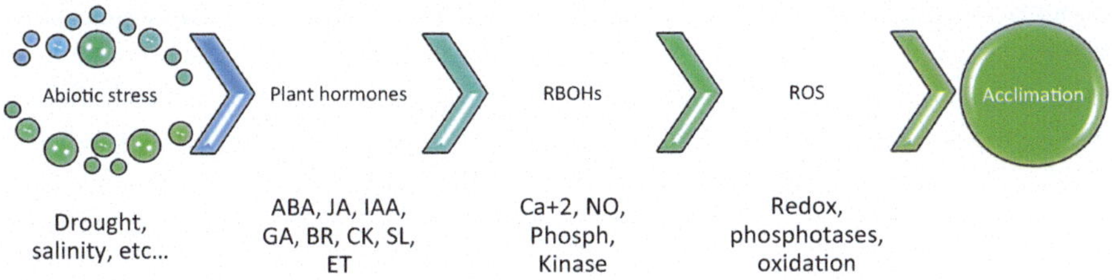

Fig. 11.2 Hormone biosynthesis, degradation, and transport

variety of physiological responses by phosphorylating a variety of target substrates, including ion channels, transcription factors, and transporters (Umezawa et al. 2013; Wang et al. 2013). PP2Cs interact with and inhibit SnRK2s to prevent ABA signaling in the absence of ABA.

PYL8 regulates root ABA sensitivity in a nonredundant manner (Antoni et al. 2013) and increases lateral root growth by boosting MYB77-dependent auxin-responsive gene transcription (Zhao et al. 2014). Other intracellular and extracellular ABA receptors, in addition to PYLs, may need to be found (Cutler et al. 2010; Klingler et al. 2010). When plants are exposed to a variety of abiotic and biotic challenges, endogenous ABA levels rise, which then exquisitely activates a cellular signaling network to turn on adaptive responses and govern a variety of developmental processes. ABA binds to PYLs, and the binary complex subsequently interacts physically with PP2Cs. Because the PYL-PP2C heterodimer prevents substrate SnRK2s from attaching to PP2Cs, SnRK2 kinase activity is stimulated, which was previously inhibited by PP2Cs (Park et al. 2009; Santiago et al. 2009; Umezawa et al. 2009).

11.3.3 ABA-Induced Stomatal Closure

Plants use guard cell turgor pressure to control stomatal aperture in response to environmental and hormonal inputs (Nilson and Assmann 2007). In guard cells, ROS are considered second messengers in the abscisic acid (ABA) transduction pathway (Neill et al. 2002; Yan et al. 2007). Through the activation of calcium-permeable channels in the plasma membrane, ABA-induced H2O2 is an important signal in inducing stomatal closure to minimize water loss (Pel et al. 2000). Jannat et al. (2011) discovered that ABA-inducible cytosolic H2O2 elevation is required for ABA-induced stomatal closure, whereas H2O2 elevation in the absence of ABA causes stomatal closure. To stimulate stomatal closure, ABA signaling increases the buildup of reactive oxygen species (ROS) in guard cells.

Plants have been studied for ROS bursts caused by the activation of respiratory burst oxidase homolog (RBOH) enzymes in response to a variety of developmental and environmental signals (Chapman et al. 2019). These enzymes transfer electrons from NADPH or FADH2 to molecular oxygen, resulting in the generation of extracellular superoxide (Suzuki et al. 2011). We were able to overlay the site of ROS accumulation with organelle-specific markers to reveal ABA-elevated ROS in the mitochondria after adding genetically encoded biosensors to our previous toolset of ROS-responsive chemical probes. We found that RBOH enzymes are essential for ABA-induced ROS buildup in the cytoplasm and mitochondria and that H2O2 is enough to close stomata. A mutant with higher mitochondrial ROS production had a faster rate of ABA-induced stomatal closure, showing that ROS generation in this organelle plays a role in ABA signaling. These findings suggest that ABA-induced H2O2 buildup is tightly regulated spatially, affecting stomatal closure and guard cell physiology (Postliglione and Gloria 2022).

11.4 Strigolactones

11.4.1 Discovery, Biosynthesis, and Physiological Functions

Strigolactones (SLs) are carotenoid derivatives found in a variety of plants. SLs were first discovered as root exudates in cotton plants by Cook et al. (1966). The late discovery of SLs can be attributed to their nature as chemicals that were released in very small quantities and disappeared quickly due to environmental instability. "Strigol" was the first naturally occurring SL to be identified (Cook et al. 1966). It is established that SLs are synthesized in roots and stems and transported in the xylem (Kohlen et al. 2011). Several monocots and dicots have been identified as SL producers, including sorghum, maize, cotton, cowpea, and red clover. To date, approximately 25 SLs have been extracted from various plants, including strigol, orobanchol, sorgolactone, 20-epi-orobanchol, solanacol, and sorgomol (Faizan et al. 2020). In recognition of their various roles in controlling aboveground plant architecture (e.g., by inhibiting bud outgrowth) and underground communication with adjacent organisms, SLs were classified as a new class of plant hormone in 2008 (Gomez-Roldan et al. 2008; Umehara et al. 2008). SLs have two functions: they are both endogenous and exogenous signaling molecules.

11.4.1.1 Discovery of SLs

Plants produce and release a wide range of chemicals, as well as primary and secondary metabolites, into their surroundings. Because some of these chemicals are only released in trace amounts and disappear quickly due to environmental instability, their broader contributions to plant-to-plant communication have only recently been revealed. Many of the key chemical players in plant-plant, plant-microbe, and plant-insect chemical communication are likely to be unknown. To identify these hidden signaling molecules and understand their modes of action, new highly sensitive analytical methods must be developed. SLs were initially isolated as germination stimulants for seeds of parasitic weeds in the Orobanchaceae family. Cook et al. reported in 1966 that a crystalline germination stimulant (trivial name strigol) for the root parasite, witchweed (*Striga lutea* Lour.), had been isolated and characterized as a $C_{19}H_{22}O_6$ compound from cotton root exudates. Although the stimulant appears to be distinct from known plant hormones, it is active at hormonal levels, causing germination at concentrations as low as 10^{-5} parts per million. As a result, these compounds were labeled as harmful secondary metabolites because they were harmful to the producing plant. They were later shown to be essential chemical signals for root colonization by symbiotic arbuscular mycorrhizal (AM) fungi and thus were recognized as beneficial plant metabolites (Akiyama et al. 2005).

However, nonhosts of AM fungi, such as *Arabidopsis* (which hosts *Orobanche* and *Phelipanche* spp.) and white lupin, have been shown to produce SLs, implying that SLs have other unknown functions in plants, possibly in normal growth and development. Recent research has shown that SLs or their metabolites are important plant hormones involved in the control of shoot branching (Gomez-Roldan et al. 2008). Furthermore, there are strong evidence that SLs play important biological roles in rhizosphere communications as well as plant growth and development under normal and stress conditions (Koltai et al. 2009; Soto et al. 2010).

11.4.1.2 Structure and Biosynthesis of SLs

SLs are divided into two types: canonical SLs and noncanonical SLs (Yoneyama et al. 2007). Canonical SLs have a tricyclic lactone (three-ring ABC structure) that is linked by enol-ether bond to a methyl butanolide (D ring) (Fig. 11.3). Currently, 23 distinct canonical SLs have been identified (Yoneyama et al. 2018). However, as more species are studied, the number of canonical SLs in the plant kingdom may increase. Canonical SLs are classified into strigol and orobanchol types based on the stereochemical orientation of the C ring. Noncanonical SLs, on the other hand, lack the ABC ring system but contain the enol-ether D ring which is essential for biological activities of SLs. Although both canonical and noncanonical SLs are chemically unstable, they may last longer in the mildly acidic rhizosphere than in bulk soil (Bertin et al. 2003). Canonical SLs have seemed to be slightly more stable than noncanonical SLs in general.

The SL biosynthetic pathway is becoming better understood, with several key genetic determinants identified. Mutations in DWARF27 (D27), a β-carotene isomerase (Alder et al. 2012; Lin et al. 2009), more auxiliary branching (MAX3)/carotenoid cleavage dioxygenase 7 (CCD7) and MAX4/CCD8 enzymes (Beveridge and Kyozuka 2009; Booker et al. 2004), and MAX1, a cytochrome P450 (Booker et al. 2005; Cardoso et al. 2014), have been linked to a hyper-branching phenotype as well as decreased SL biosynthesis and/or exudation in several species.

In vitro evidence of the activity of D27, CCD7, and CCD8 (plastid localized enzymes) was also provided, demonstrating that these enzymes can produce carlactone (CL) – the precursor of SLs – from all-trans β-carotene (Alder et al. 2012; Seto et al. 2014). CCD7 degrades 9-cis-carotene to yield 9-cis-apo-10′--carotenal, which is then degraded by CCD8 to yield CL, a biosynthetic precursor for SLs. SL biosynthesis occurs in plastids, and when CL, a mobile product, is formed, it is transferred to the cytoplasm. The CL is then oxidized further by cytochrome P450 monooxygenase MAX1 or other homologous genes into various forms of SLs via a few other unidentified steps catalyzed by novel unidentified

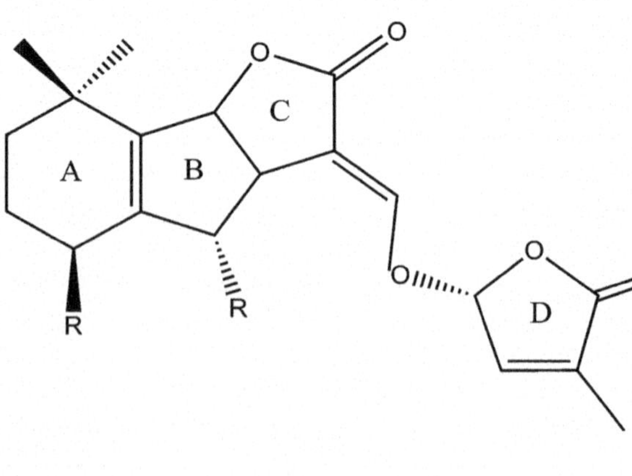

Fig. 11.3 General structure of canonical strigolactones

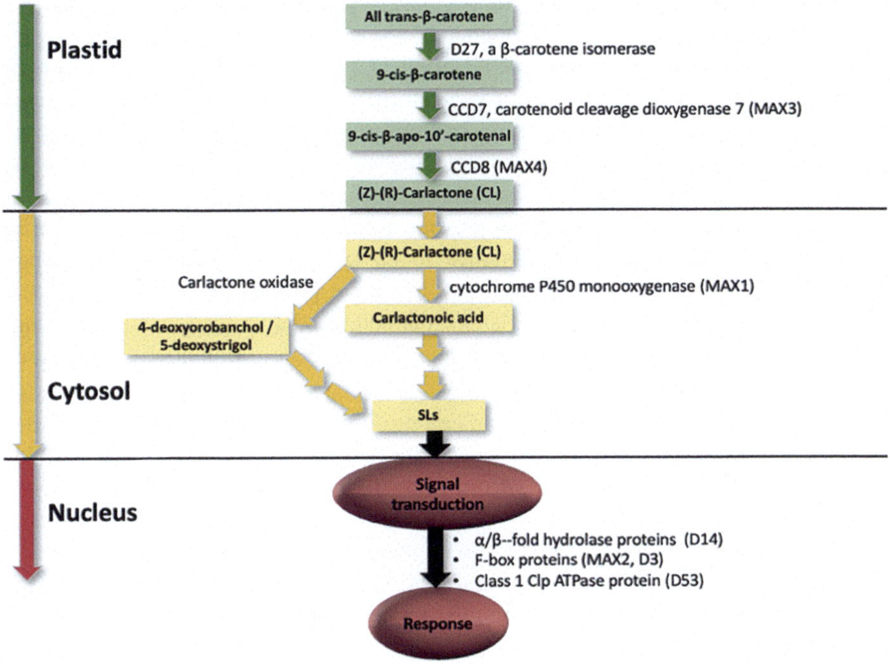

Fig. 11.4 SL biosynthetic pathway

enzymes (Zhang et al. 2014). MAX1 can also catalyze the conversion of carlactone to carlactonoic acid and its methyl ester (MeCLA), which has been shown to have SL-like activity in *Arabidopsis* (Abe et al. 2014). Further action of oxidase leads to the formation of the B and C rings in the structure of 4-deoxy-orobanchol or 5-deoxystrigol (Brewer et al. 2016). The enzymes responsible for additional SL structural diversification are still undisclosed.

The pathway of SL biosynthesis is shown in Fig. 11.4.

11.4.1.3 Physiological Functions of SLs

Arbuscular Mycorrhizal Fungi (AMF) and the Rhizosphere

The rhizosphere is a region of soil surrounding the roots that is critical for AMF (Bais et al. 2006). Nearly 80% of terrestrial plants have symbiotic relationships with AM fungi (soilborne microorganisms) (Parniske 2008), in which these fungi enter and colonize plant roots, where they form highly branched structures known as arbuscules, which serve as nutrient exchange sites. The fungus provides water and nutrients to their hosts, especially phosphate and nitrogen, which are received through the hyphae that live outside in the soil, while the AM fungi get photosynthates from their host plants. AM symbiosis is a collaborative association of plants with fungi that is regarded as one of the most significant plant-microorganism associations, which improve plant nutrient uptake (Bouwmeester et al. 2007). In this symbiosis, SLs play a critical role (Bonfante and Genre 2010). AMF and plants communicate via SLs which act as signaling molecules released by plant roots in the rhizosphere (Bais et al. 2006). SLs promote hyphal branching in the vicinity of the host roots, increasing the possibility of contact between the roots and fungi. The use of a synthetic analog of SL (GR24) improved mycorrhiza colonization in *Petunia hybrida* (Breuillin et al. 2010). SLs have been demonstrated to not only cause hyphal branching but also to activate mitochondrial activity and spore germination (Besserer et al. 2006, 2009), generating fast changes in mitochondrial shape, density, and motility 1 hour after treatment. GR24 treatment increased NADH concentrations, NADH dehydrogenase activity, and ATP content in *Gaertnera rosea* cells within minutes, demonstrating a substantial and quick acceleration of energy consumption in the fungal mitochondria. This early mitochondrial response occurred in the absence of novel gene expression or elevation of mitochondrial metabolism-related genes, and hyphal branching and cell proliferation were detected 5 days after treatment. As a result, SLs appear to initiate a series of molecular and cellular activities required for hyphae to become infective, with the initial fast phase of these events governed by posttranscriptional regulation. In accordance with this role, SL synthesis and release into the soil are increased in response to phosphate deficit (Yoneyama et al. 2007; Umehara et al. 2008; Kohlen et al. 2011). These findings clearly show that SLs are critical signaling molecules in the establishment of the AM fungi-plant symbiosis (Foo and Davies 2011).

Seed Germination

SLs were first recognized as germination stimulants for the root parasitic plants witchweeds (*Striga* spp.) and broomrapes (*Orobanche* and *Phelipanche* spp.). Extensive research on the structure-activity relationship (SAR) of SLs on the stimulation of germination of seed of root parasitic weeds has been undertaken, and the C-D ring moiety has been identified as the key structure for germination stimulation activity (Zwanenburg et al. 2009). Indeed, natural and synthesized SLs containing this moiety exhibit moderate to strong germination activity. Germination activity is also influenced by other components of the SL structure. The inclusion of the 4-hydroxyl group appears to boost activity (Xie et al. 2007). In general, hydroxy-SLs are 10- to 100-fold more active than acetates (and likely other conjugates) (Xie et al. 2008). In this respect, canonical and noncanonical SLs stimulate seed germination in *Striga* and *Orobanche* species at doses ranging from picomolar to micromolar. The seeds of several *Striga* and *Orobanche* species are sensitive to diverse SLs (Kim et al. 2010a, b; Kisugi et al. 2013), and plants do not release a single SL but rather a combination of at least two SLs. For example, 11 distinct canonical SLs were found in tobacco root exudate (Xie et al. 2013; Xie et al. 2016). As a result, it is probable that a variety of SLs in the root exudate lead to host-specific germination of root parasitic plant seeds rather than a single SL. So far, no studies have been conducted to investigate the effects of SL mixes on seed germination stimulation. On their germination-stimulating actions, combinations of SLs may have additive, synergistic, and antagonistic effects. For example, root exudates of legumes that generate a lot of SL did not stimulate *O. cumana* seed germination, demonstrating that SLs produced by leguminous plants have an antagonistic impact on *O. cumana* seed germination (Fernández-Aparicio et al. 2011). This shows that modifying a crop's SL profile, as in *Striga*-resistant sorghum cultivars, might limit or hinder seed germination of a specific root parasite (Gobena et al. 2017). In contrast, fractions from reverse-phase HPLC separation of English ivy (*Hedera helix*) root exudates that were active for *Orobanche hederae* germination were not active for *O. minor* germination and vice versa, indicating the presence of host-specific germination stimulants (Yoneyama et al. 2018).

Branching of Shoots

Branching is an important developmental process in which axillary buds develop into flowers; it is influenced by several external and internal factors that govern how energy is distributed within the plant (Leyser 2009). Hormones, among other internal factors, play critical roles in shoot branch management (Umehara et al. 2008). Auxin (Aux) and cytokinin (CK), two well-known phytohormones, are known to regulate bud growth; cytokinin promotes growth (Sachs and Thimann 1967), while auxin acts as a repressor (Thimann and Skoog 1934). SLs were identified by two research groups independently as the most portable phytohormones involved in shoot branching by suppressing bud growth (Gomez-Roldan et al. 2008; Umehara et al. 2008). As a result, cytokinins have an antagonistic effect on SLs (Dun et al. 2012). Most studies on plant architecture control are conducted with increased branching mutants, most notably ramosus (rms) in garden pea (*Pisum sativum*), more axillary growth (max) in *Arabidopsis* (*Arabidopsis thaliana*), decreased apical dominance (dad) in *Petunia hybrida*, dwarf (d), and high tillering dwarf (htd) in rice (*Oryza sativa*) (Domagalska and Leyser 2011). In all these cases, an inhibitor of branch formation is involved (Simons et al. 2007). The application of specific concentrations of an exogenous synthetic SL (GR24) to the rms1 mutant plant slowed lateral bud growth. Similarly, exogenous SL application inhibited shoot branching (Dun et al. 2013; Barbier et al. 2019), stimulated internode growth (De Saint Germain et al. 2013), inhibited the outgrowth of axillary buds (Minakuchi et al. 2010), increased stem thickness, and induced secondary growth (Agusti et al. 2011) and other morphological changes. Multiple levels of auxin-SL interactions were discovered to be critical for branching control (Koltai et al. 2010). Canalization is one of the hypothesized models. Bud outgrowth in this model is dependent on the bud's ability to transfer auxin from a high concentration (the source) to a lower concentration (the main stem) (the sink). Coordinated export needs the PIN1 transporter to be mobilized subcellularly and supracellularly. SLs from the root suppress PIN1 activity in both the bud and main stem, impacting both source and sink strength and decreasing the bud's auxin export capability (Waters et al. 2017). However, it is still unknown how these inhibitory processes work at the molecular level.

Root Development

According to research, SLs promote the elongation of primary roots and root hairs (Sun et al. 2014) but inhibit the formation of lateral roots (Ruyter-Spira et al. 2011). MAX2-dependent enhancement in primary root development was observed at all

GR24 concentrations tested. Lateral roots were shown to be more abundant in mutants lacking in SL biosynthesis (i.e., max3 and max4) and signaling (i.e., max2) than in the wild type (WT) (Kapulnik et al. 2011; Ruyter-Spira et al. 2011). Under favorable development circumstances (sufficient Pi and sugar), application of GR24 resulted in a MAX2-dependent increase in primary root length. In a MAX2-independent manner, the opposite effect – suppression of primary root – was seen for treatments with substantially greater dosages of GR24. However, under carbohydrate restriction circumstances (Jain et al. 2007), GR24 treatments at all doses showed a favorable, MAX2-dependent impact on primary root elongation. Taken together, these findings indicate that SLs influence primary root and lateral root growth, as well as root hair elongation, hence impacting both root and shoot development in plants. The variations in SL activity under various sugar and Pi circumstances suggest that SLs may govern root growth in response to external environmental factors. However, the usage of the synthetic SL GR24 may sometimes be deceptive due to its greater stability in aqueous solution than natural SLs (Akiyama et al. 2010). Additionally, SL control of root growth is thought to occur through cross talk with other phytohormones that influence primary root, lateral roots, and root hair development, including auxin and ethylene. Concerning the interaction between SLs and auxin, it has been demonstrated that auxin induces SL synthesis in the root via induction of MAX3 and MAX4 expression (Beveridge and Kyozuka 2010); on the other hand, auxin and SLs have been suggested to interact in the regulation of root hairs, primary root, and lateral root development (Kapulnik et al. 2011; Ruyter-Spira et al. 2011). Additionally, SLs may influence auxin efflux during lateral root formation by controlling the PIN proteins, which regulate the positioning, initiation, and length of lateral roots (Koltai 2014). Concerning the interaction between SLs and ethylene, the significantly decreased SL response in the ethylene-signaling mutants etr and ein (Stepanova and Alonso 2009) revealed that ethylene signaling is involved in the SL response (Kapulnik et al. 2011; Koltai 2011). SLs have been found to stimulate ethylene biosynthesis in the seeds of the parasitic plant *Striga*, resulting in seed germination (Sugimoto et al. 2003). Additionally, this result supports the notion that the influence of SLs on root hair elongation is mediated through ethylene production. Taken together, the findings show a widespread effect of SLs on ethylene production, which may have an effect on plant growth.

Leaf Senescence

Leaf senescence is an important process that happens at the termination of leaf development (Nooden 1988). This includes leaf yellowing caused by chlorophyll degradation, degeneration of the chloroplast structure and concomitant lipid degradation, degradation of photosynthetic proteins, and reduction in photosynthetic activity (Noodén, 2004).

It is a complicated process regulated by a number of elements, including age and flowering, dark treatment, nutrient deficiencies, and a range of stresses (Lim et al. 2007). During the process of leaf senescence, nutrients are reallocated from older to younger tissue (Brundrett and Tedersoo 2018). Several phytohormones are involved in the control of leaf senescence (Kusaba et al. 2013). ABA, jasmonic acid, salicylic acid, and ethylene may accelerate leaf senescence, but cytokinins can suppress it (Jibran et al. 2013). SLs also govern the process of leaf senescence since both SL-deficient and SL-insensitive mutants display senescence retardation (Hamiaux et al. 2012).

Ueda and Kusaba (2015) proved that strigolactone biosynthesis mutant strains of *Arabidopsis* (*Arabidopsis thaliana*) showed a delayed senescence phenotype during dark incubation. The strigolactone biosynthesis genes MORE AXIALLY GROWTH3 (MAX3) and MAX4 were drastically induced during dark incubation and treatment with the senescence-promoting phytohormone ethylene, suggesting that strigolactone is synthesized in the leaf during leaf senescence. SLs accelerate the aging of leaves. GR24 was sprayed on rice leaves to evaluate these effects on plants (Tsuchiya et al. 2010). Exogenous GR24 treatment accelerated leaf senescence in rice (d27, d17, and d10) SL-deficient mutants (Yamada and Umehara 2015). According to Ueda and Kusaba (2015), dark treatment promotes ethylene production and activation of the ethylene signaling pathway, hence initiating senescence signaling. Senescence signaling activation results in the transcription of MAX3 and MAX4, which results in strigolactone production in the leaf. Strigolactone further increases senescence signaling by increasing senescence-promoting pathways that are ethylene-dependent and ethylene-independent. Finally, active senescence signaling results in the leaf's senescence syndrome. Ha et al. (2014) found that abscisic acid, sodium chloride, and dehydration stressors stimulate MAX3 and MAX4 expression in the leaf. According to Ha et al. (2014) and Ueda and Kusaba (2015) findings, strigolactone is produced in the leaf under particular stress circumstances.

As more plant scientists become interested in SL research, further roles and consequences of SLs are expected to be discovered. Figure 11.5 shows the physiological functions of SLs in plants that have been revealed so far.

11.4.2 Positive Effect of SLs on Stress Tolerance

Climate change exacerbates drought and salt in the soil, lowering crop yields in impacted regions. Typically, crops react to stress by perceiving it and then transducing signals, which are then regulated at the molecular level. While there were several promising prospects, phytohormones stood out as a critical tool for dealing with stress management. Plant hormones are important factors in how plants behave, and they work with other signals like reactive oxygen species (ROS), Ca^{2+}, and other signaling molecules to help plants adapt to hostile environments. With their dual roles as endogenous hormones and signaling molecules in the rhizosphere, SLs are well positioned as regulators of changes in plant development as affected by external cues. To do that, SL metabolism or signaling must be environmentally sensitive. Several lines of evidence point to a link between SLs, or genes that encode their metabolism, and light—in the form of photoperiod, intensity, and wavelength (Liu et al. 2013). A link with nutrient availability has also been established: phosphate (P) and/or nitrogen (N) deficiency promotes SL production and exudation in a

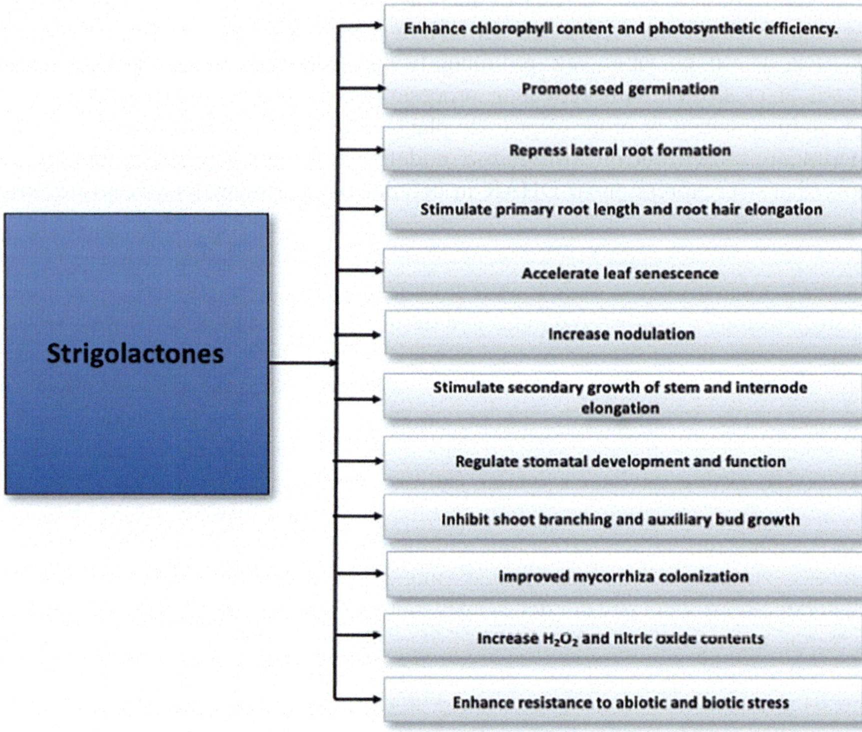

Fig. 11.5 The physiological functions of SLs in plants that have been revealed so far

variety of species. The responses to P and N availability exhibit species-specific diversity, and it has thus been proposed to regulate nutrient acquisition strategies utilizing AMF or rhizobia symbionts (Xie and Yoneyama 2010). Drought, freezing, or salt can all cause osmotic stress, which is a major limitation to plant growth and yield around the world. Plants under water stress rapidly accumulate ABA, which is required for stomatal closure (Zhu 2002). The increased activity of carotenoid cleavage enzymes of the 9-cis-epoxycarotenoid dioxygenase family (NCED), which catalyze the rate-limiting step in ABA biosynthesis, is largely responsible for ABA accumulation (Tan et al. 1997). Catabolism also helps to fine-tune ABA levels under stress conditions (Nambara and Marion-Poll 2005). The involvement of SLs in response to osmotic stress has been proposed based on the interactions between ABA and SL metabolism due to the shared carotenoid precursors and the involvement of carotenoid-cleaving enzymes (Tsuchiya and McCourt 2009).

11.4.2.1 Salinity and Drought

Kong et al. (2017) discovered that SL may provide salt stress resistance in plants through the production of H_2O_2 as a signaling molecule. Exogenously administered SL greatly ameliorated the detrimental effects of salt stress on common sage (*Salvia nemorosa*), as seen by increases in photosynthesis, the *Fv/Fm* ratio, and gas-exchange indices (Sharifi and Bidabadi 2020). Ma et al. (2017), working on rapeseed, found that SL increased chlorophyll content and the *Fv/Fm* ratio in comparison to control plants. The protective effect of SL against abiotic stress may be attributed to the reduction of ROS in plants caused by SL. Permeability of the plasma membrane is a critical criterion for salt tolerance in plants. Sharifi and Bidabadi (2020) found that supplementing with SL increased the plasma membrane permeability of salinity-stressed *Salvia* plants by decreasing the electrolyte leakage index. Koltai (2014), on the other hand, argued that SL signaling modulates auxin flow, hence modulating the permeability of plant cells' plasma membranes during salinity stress, resulting in salt tolerance. Exogenous SL dramatically reduced ROS production and mitigated the oxidative stress produced by KCl stress (Zheng et al. 2021). To avoid oxidative damage caused by salt stress, plants evolve an antioxidant mechanism that regulates the dynamic equilibrium of ROS. SOD, POD, and CAT are the three primary antioxidant enzymes that scavenge reactive oxygen species (Ismail et al. 2014). Under salt and drought stress, overexpression of *Sapium sebiferum* SsMAX2 (a SL signaling gene) in *Arabidopsis* increased the activities of SOD, POD, and CAT in the cleavage of ROS, while the max2 mutant displayed significantly decreased enzyme activity (Wang et al. 2019).

Additionally, SLs control the drought stress response in part through ABA signaling, as shown by the reduced sensitivity of all max mutants to ABA during germination under drought stress conditions (Ha et al. 2014). Additional evidence for an ABA-mediated SL response in *Arabidopsis* comes from higher transpiration rates and stomata density, as well as a change in ABA-mediated stomata closure. Microarray study of max2 and wild-type plants showed an SL network involved in abiotic stress tolerance. This network included previously identified abiotic stress-sensitive genes and phytohormones (ABA and CK). Plants with the max2 mutation exhibit downregulation of ABA import genes (ABCG22 and ABCG40) and CK catabolism genes (CKX1, CKX2, CKX3, and CKX5) and upregulation of genes that respond to ABA and osmotic stress (CIPK1) and abiotic stress-sensitive genes (AtNAC2; Ha et al. 2014). Bu et al. (2014) demonstrated that under drought stress, max2 mutant plants exhibit a thinner cuticle and wider stomata opening.

MAX2 expression is stimulated during seed germination and seedling stage by ABI3 and ABI5, two TFs involved in ABA signaling, but ABA modestly downregulates MAX2 expression during the mature stage, suggesting that MAX2 functions downstream of ABA signaling (Bu et al. 2014).

Zheng et al. (2021) hypothesized that when exposed to KCl-induced salt stress, SL may reduce the K content by expelling K^+ from the cytoplasm and compartmentalizing K+ into the vacuole, therefore maintaining Na^+ and K^+ homeostasis in the cytoplasm. Apart from the ion transporters, exogenous SL altered the expression of the enzyme MdMAPKKKa and the transcription factors MdbHLH162 and MdERF109. According to reports, the MAPK pathway is involved in the signaling pathway associated with salt stress in *Arabidopsis*, peppermint, soybean, cucumber, and other plants (Mehlmer et al. 2010; Xu et al. 2011; Im et al. 2012; Li et al. 2016).

11.4.2.2 Extreme Temperature

Strigolactones have been shown to favorably influence dark chilling resistance in pea and *Arabidopsis* (Cooper et al. 2018), as well as cold tolerance in rice (Liu et al. 2020), and to mitigate the negative effects of heat on tall fescue leaf development (Hu et al. 2019).

Study by Chi et al. (2021) provided numerous lines of evidence showing that strigolactones favorably influence tomato responses to heat and cold stress. To begin, heat or cold stress enhanced the accumulation of transcripts for both strigolactone biosynthesis genes (CCD7, CCD8, MAX1) and the signaling gene MAX2, as well as the accumulation of solanacol in tomato roots. By easing oxidative stress, strigolactones enhanced *Fv/Fm* and PSII, suggesting that strigolactones may prevent stress-induced photoinhibition (*Fv/Fm*) and decreases in photosynthetic electron transport at PSII (PSII) through oxidative stress relief (Chi et al. 2021).

There is evidence that strigolactones have an effect on the transcript levels of heat shock proteins (HSPs) and heat shock transcription factor A6B (HSFA6B) under stress (Ha et al. 2014) or normal conditions (Wang et al. 2020). HSPs have the ability to protect cellular proteins from severe harm in hot circumstances, which is critical for plant survival under situations of heat stress. Under heat and cold stresses, both the activity of antioxidant enzymes (SOD, superoxide dismutase; APX, ascorbate peroxidase; GR, glutathione reductase; MDAR, monodehydroascorbate reductase; and DHAR, dehydroascorbate reductase) and the transcript levels of the corresponding genes were decreased in ccd7 mutants compared to WT. Regulation of enzyme activity occurs at three levels: gene transcription, protein turnover, and stability. In keeping with this, we discovered that changes in the growth temperature and strigolactone level influenced the activity of antioxidant enzymes and/or the transcription of antioxidant genes (Chi et al. 2021).

Exogenous strigolactone was found to improve the cold resistance of rape seedlings in several ways: it increases cell viability, inhibits ROS production, decreases MDA content and electrolyte leakage, increases soluble protein and proline content, increases antioxidant enzyme activity, promotes photosynthesis, and induces gene expression associated with low-temperature stress. Additionally, treatment with two H_2O_2 scavengers, DPI (diphenyleneiodonium chloride) and DMTU (N,N′-dimethylthiourea), significantly inhibits strigolactone's ability to protect rape seedlings from low-temperature damage, indicating that H_2O_2, as a critical signaling molecule, plays a critical role in strigolactone's ability to improve rape seedlings' cold resistance (Zhang et al. 2020a).

11.4.2.3 Nutrient Deficiency

Under nutrient-deficient environments, the plant accumulates an abundance of SLs, which suppresses shoot branching and promotes symbiosis (Umehara et al. 2008). In this respect, P and N deprivation in *Medicago truncatula* results in increased expression of the SL biosynthesis genes CCD7, CCD8, D27, and MAX1 (Bonneau et al. 2013). SLs play a critical function in nitrogen and phosphorus deficit due to their activity through root and shoot architecture alteration and promotion of rhizobial bacteria and AM fungus symbiosis (Marzec 2016). Through their hyphal extensions, fungi assure the provision of water and nutrients (especially phosphate and nitrogen) in AM symbiotic relationships. Under low-phosphate conditions, natural plants treated with GR24 displayed a significant number of lateral roots; however, the SL mutants exhibited a low number of lateral roots. GR24 recovers the Pi-deficient symptoms in all genotypes except those with SL signaling defects, indicating that SL biosynthesis is essential under this condition. In contrast, when GR24 is applied in a high-phosphate environment, it results in a reduction in lateral root density by reducing both outgrowth and lateral root-forming potential (Ruyter-Spira et al. 2011). Auxins are proved to regulate root architecture through SL by influencing the localization of PIN proteins (Shinohara et al. 2013). Additionally, in the presence of Pi deficiency, max2 and max4 exhibit altered expression of Pi-deficient signature genes such as acid phosphatase type 5 (ACP5), phosphate transporter 1;5 (PHT1;5), and PHT1;4. TIR1 was shown to be implicated in the SL-mediated response to low Pi, suggesting that SLs and auxins coordinate the response to low Pi (Mayzlish-Gati et al. 2012).

Apart from their function in maintaining phosphate homeostasis, SLs have been identified as a potential mechanism for controlling plant growth in response to nitrogen availability. The altered response to nitrogen deprivation was detected in *Arabidopsis* mutants of SL biosynthesis (max1-1) and SL insensitivity (atd14-1). Additionally, it was shown that when nitrogen was scarce, the expression levels of SL biosynthesis genes (MAX3 and MAX4) were altered (Ito et al. 2016).

11.4.2.4 Biotic Stress

SLs have been identified as a factor in conferring resistance to certain infections (Marzec 2016). The first piece of evidence came with the discovery of pathogen-associated transcription factor domains in the promoter regions of genes involved in SL production (Torres-Vera et al. 2014). The work using strigolactone-deficient *A. thaliana* mutants established the significance of strigolactone in plant resistance to bacteria such as *Rhodococcus fascians*, *Pectobacterium carotovorum*, and *Pseudomonas syringae* (Piisilä et al. 2015; Stes et al. 2015). This might be explained by the fact that SL/ABA closed the stomata, thereby inhibiting *Pectobacterium* and *Pseudomonas* entry (Piisilä et al. 2015). Infection with *R. fascians* resulted in a more severe leafy gall syndrome in mutants lacking SL biosynthesis (max1, max3, and max4) and signaling (max2). Additionally, infection induces upregulation of genes involved in the synthesis of SL (max1, max3, and max4; Stes et al. 2015). By using the proteinase inhibitor II (PIN II) gene, SLs have the potential to escape the transcriptional stress induced by pathogens (Torres-Vera et al. 2014). However, the protective activity of AM fungus against infections (Harrier and Watson 2004) may possibly augment the beneficial effects of SLs.

The treatment of GR24 prevented the radial development of a variety of phytopathogenic fungus (Dor et al. 2011). Increased vulnerability to *Alternaria alternata* and *Botrytis cinerea* was also reported in the *Lycopersicon esculentum* SL biosynthesis mutant slccd8, which was associated with a decrease in the concentration of plant defense hormones (Torres-Vera et al. 2014). SLs, on the other hand, were not shown to have a role in resistance to other diseases, including *Pythium irregulare* and *Fusarium oxysporum* (Blake et al. 2016; Foo et al. 2016). Thus, the SL may participate in plant immune responses against just particular bacterial and fungal diseases.

11.4.3 Transportation and Signaling Pathways

Plant roots are the primary site of SL biosynthesis, but synthesis in the lower part of the shoot has also been proposed (Dun et al. 2009). The xylem appeared to be the most likely route of hormone transport, and canonical SL orobanchol was found in tomato and *Arabidopsis* xylem sap (Kohlen et al. 2011; Kohlen et al. 2012). However, no other researchers were able to confirm this discovery. Yoneyama et al. (2018) also collected large amounts of xylem sap from several plant species but found no known SLs in them. Furthermore, SLs fed to rice plant roots were detected in shoots harvested 20 hours after treatment but not in xylem sap (Xie et al. 2015). Nonetheless, it is proposed that SLs, their metabolites, or other unknown secondary messengers move in the root-to-shoot direction to confer a significant reduction in shoot branching (Dun et al. 2009).

The PLEIOTROPIC DRUG RESISTANCE 1 (PDR1) protein is required for SL root exudates into the soil in *Petunia* (Kretzschmar et al. 2012). PDR1 is the only protein identified to date with SL transport function, and mutant and overexpression lines with altered SL exudation confirm this claim. PDR1 is a member of the ATP-binding cassette (ABC) transporter family, which has been implicated in the transfer of other hormones such as ABA and auxin (Petrášek and Friml 2009; Kuromori et al. 2010). PDR1 is significantly expressed in the root tip, which is consistent with a function in SL exudation into the soil.

Two protein classes are essential for SL perception and signaling: the DWARF14 (D14) family of α/β-fold hydrolase proteins (Hamiaux et al. 2012) and the family of F-box proteins (MORE AXILLARY GROWTH2 (MAX2) in *Arabidopsis*; D3 in rice) (Arite et al. 2012) (Fig. 11.2). The D14 protein has been postulated as an SL receptor, while MAX2 is a component of a Skp-Cullin-F-box (SCF) E3 ligase complex (Hamiaux et al. 2012). The D14 protein has a conserved catalytic triad (Ser-His-Asp) that functions as a hydrolase capable of recognizing and deactivating SL (Yao et al. 2016). SL molecules bind to D14, altering its conformation and enhancing D14's interaction with F-box protein MAX2 (Zhao et al. 2015). This complex catalyzes the ubiquitination of transcriptional repressor D53; the *D53* gene product shares predicted features with the class I Clp ATPase proteins (in *Arabidopsis*, this repressor is designated SMXL6, SMXL7, or SMXL8), leading in its 26S proteasomal degradation and hence the transcription of SL-sensitive genes (Yao et al. 2018; Wu et al. 2022). D53 is a critical target for axillary bud outgrowth regulation in rice (Fang et al. 2020). In rice, the SQUAMOSA promoter binding protein-like (SPL) family transcription factor Ideal Plant Architecture1 (IPA1) and D53 work in concert to activate genes involved in the SL regulation pathway (Song et al. 2017). Both D53 and D53-like SMXL proteins interact with TPL/TPR (TOPLESS/TOPLESS RELATED) proteins in the absence of SLs and suppress downstream target genes by inhibiting the activity of unknown transcription factors (TFs) (Soundappan et al. 2015).

11.4.4 Signaling Cross Talk Between ROS, Strigolactones, and Abscisic Acid in Response to Developmental and Environmental Cues

Previous studies reveal that the connection between stress signaling pathways and developmental inputs is not linear but rather comprises a complex network of interactions between several hormone signaling pathways, with considerable metabolic cross talk and sites of reciprocal regulation. Specifically, in this section, the signaling interactions that occur between ROS, strigolactones, and abscisic acid in response to developmental and environmental stimuli will be discussed (Fig. 11.6).

Exogenous GR24 acts as a positive regulator in response to stress, enhancing *Arabidopsis*' drought and salt tolerance (Kapulnik and Koltai 2014; Ha et al. 2014). Additionally, it may change the opening of the plant's stomata in response to

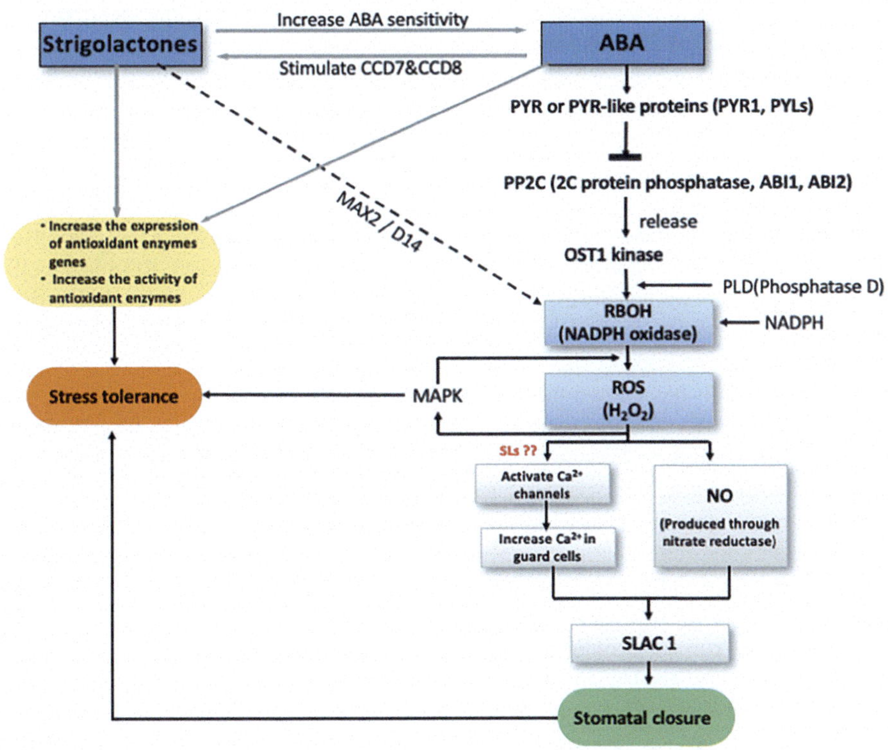

Fig. 11.6 Crosstalk mechanisms between SLs, ABA, and ROS in plants. Dotted arrow refers to pathway that has not been fully investigated. ?, unknown if activation of Ca^{2+} channels is involved in SL-mediated stomatal closure

environmental stress by perceiving and reacting to external stimuli (Lv et al. 2018). Interestingly, SL and abscisic acid (ABA) share their biosynthetic precursor, both being carotenoid-derived terpenoid lactones. In response to water stress, SLs have been shown to interact with ABA. Additionally, the ABA mutants had decreased LeCCD7 and LeCCD8 transcript levels (López-Ráez et al. 2010). Interestingly, it was shown that either localized improvement of SL synthesis in response to drought or exogenous SL administration to shoots resulted in increased stomatal sensitivity to ABA (Visentin et al., 2016).

Consistent data reveals that MAX2 (MAX2 is an F-box protein essential for SL signal transduction), independently of ABA, has a significant effect on stomatal response to external stress, leading us to believe that MAX2 and SLs may influence stomatal movement intrinsically (Lv et al. 2018).

The connection between SLs and ROS was proved when it was discovered that FAR-RED ELONGATED HYPOCOTYL3 (FHY3) functions as a negative regulator of RBOH genes. FHY3, a transposase-related transcription factor (TF), is a critical component of phytochrome A signaling and the circadian clock, since it regulates the far-red (FR) light response (Lin et al. 2007). FHY3 inhibits root and shoot branching in *Arabidopsis* fhy3max2 double mutant plants, indicating that FHY3 functions as a MAX2 suppressor (Ouyang et al. 2011). It has been shown that inactivating FHY3 results in increased expression of respiratory burst oxidase homolog (RBOH) genes, which may be responsible for branching suppression. Additionally, it has been shown that RBOH controls shoot branching in the tomato, *Lycopersicon esculentum*, where antisense RBOH expression results in enhanced shoot branching (Koltai et al. 2011). Another connection between SLs and ROS derives from the fact that SLs are involved in drought and salt stress. *Max2* mutant plants exhibit greater susceptibility to various stimuli and decreased ABA response, which affects stomata closure and stress-responsive gene expression. ROS is a well-known second messenger during ABA signaling, and it is extremely probable that RBOH is involved in the control of shoot and root branching and other stress responses in the presence of SL (Xia et al. 2015).

Recent research has shown that exogenous GR24 stimulates growth and photosynthesis, alleviates oxidative stress, and enhances rape's resistance to salt stress through gene regulation (Ma et al. 2017). By spraying GR24 on tomato seedlings exposed to low light, it is possible to boost photosynthesis, antioxidant enzyme activity, gene expression, and low-light tolerance (Lu et al. 2019). Additionally, strigolactone may neutralize excess ROS, including H_2O_2, engage in the signaling cascade, and modulate plants' stress tolerance. NADPH oxidases (RBOHs) can oxidize NADPH to form hydrogen peroxide (H_2O_2) in the cytoplasm and are involved in the response of plants to biotic or abiotic stress (Suzuki et al. 2011; Marino et al. 2012). To determine whether H_2O_2 plays a role in SL-induced stomatal closure, stomata were examined after SL application in the presence or absence of ascorbic acid (ASA; a substrate for H_2O_2 removal), catalase (CAT; an H_2O_2 scavenger enzyme), and diphenylene iodonium (DPI; a NADPH oxidase inhibitor), all of which reduce H_2O_2 concentrations (Khokon et al. 2010). In the presence of ASA, CAT, and DPI, SL-induced stomatal closure was totally prevented. These findings indicated that H_2O_2 is necessary for SL-induced stomatal closure and bolstered the argument that NADPH oxidase-dependent H_2O_2 is required for guard cell SL signaling (Shi et al. 2015).

According to Lv et al. (2018), SLs operate as important regulators of the stomatal aperture independently of ABA when combined with H_2O_2 and NO generation and SLOW ANION CHANNEL-ASSOCIATED 1 (SLAC1) activation.

The MAPK (mitogen-activated protein kinase) cascade pathway, as a ubiquitous signal transduction system in plants, also plays a critical role in plants' response to biotic or abiotic stress (Kumar et al. 2008). Strigolactone regulates salt tolerance in *Sesbania cannabis* seedlings through NADPH oxidase-induced H_2O_2 (Kong et al. 2017). Strigolactone increased the accumulation of photosynthetic pigments, stomata opening, and photosynthetic efficiency in rape seedlings subjected to low-temperature stress and reduced PSII damage. Studies have shown that strigolactone may induce *Arabidopsis* stomata to shut and that this mechanism is mediated by NO and H_2O_2 (Lv et al. 2018).

GR24 considerably increases the SOD and POD activities of rape seedlings exposed to salt stress and aids in the removal of ROS (Ma et al. 2017). The administration of GR24 improves the activity of antioxidant enzymes and the production of antioxidant enzyme genes, therefore reducing the damage caused by ROS to the photosystem (Lu et al. 2019). Strigolactone may activate NADPH oxidase, generating reactive oxygen species (ROS), which can be used to control plant growth and increase plant tolerance (Sagi and Fluhr 2006). RNA-seq analysis revealed that pretreatment of rape seedlings with GR24 increases the expression of genes implicated in the ROS and MAPK cascade signaling pathways, including NTRC, PXG3, ISU1, ISCA, VTC2, NFXL1, and PS2, whereas PAB2, SF3B1, and ESM1 are downregulated. The foregoing data suggest that complicated connections exist between strigolactone, NADPH oxidase genes, and MAP kinase cascade pathways and that H_2O_2 plays a critical part in the mechanism by which GR24 alleviates cold damage to rape seedlings (Zhang et al. 2020b). In roots, nutrient deficiency activates NADPH oxidases and regulates the expression of genes involved in SL biosynthesis (Bonneau et al. 2013). However, the link between SLs and ROS generation, as well as the root response to nutrient deficiency stressors, still needs further investigations.

Similarly, in guard cells, NADPH oxidase-dependent ROS generation is critical for ABA-mediated stomatal closure (Kwak et al. 2003). ABA sensing and signaling have proven to be elusive owing to the network's exceptional complexity, which incorporates a huge number of components. The pyrabactin resistance protein (PYR) or PYR-like proteins (PYR1/PYLs) detect ABA (Park et al. 2009). ABA binding to the PYR1/PYL receptor complex inhibits protein phosphatase 2Cs (PP2Cs), which subsequently enhances downstream ABA signaling by inhibiting the positive regulator of the ABA response sucrose non-fermenting 1 (SNF1)-related protein kinase 2 family (SnRK2s), Open Stomata 1 (OST1) (Mustilli et al. 2002). In *Arabidopsis*, OST1 is thought to control ROS production through a direct contact with and phosphorylation of the RBOHF NADPH oxidase component (Sirichandra et al. 2009). The second messenger phospholipase D (PLD) may also be involved in the modulation of NADPH oxidase activity during ABA-mediated responses in guard cells (Zhang et al. 2009). PLD1 loss resulted in reduced ROS accumulation and ABA-induced stomatal closure. ROS produced by ABA may activate Ca^{2+} channels and raise cytosolic Ca^{2+} levels in guard cells, hence mediating ABA-induced stomatal closure (Pei et al. 2000; Kwak et al. 2003). NO generation through nitrate reductase operates downstream of H_2O_2 in ABA-induced stomatal closure (Bright et al. 2006). Experiments using ABA, ROS, NO, and Ca^{2+} indicate that SLAC1 is a critical protein involved in mediating stomatal closure downstream of these mediators. Ca^{2+} channels stimulated by H_2O_2 have been shown to be critical for ABA-induced stomatal closure (Pei et al. 2000). Thus, it would be fascinating to determine if H_2O_2-activated Ca^{2+} channels are also essential for SL-induced stomatal closure.

Apart from their involvement in ABA-induced stomatal closure, ROS generation is required for seedlings to tolerate ABA-mediated stress. In maize, water stress or ABA treatment results in the formation of ROS, as well as an increase in the expression of genes encoding antioxidant enzymes and antioxidant enzyme activity (Jiang and Zhang 2002). Results suggest that ABA-induced ROS generation and the stress response are mediated via control of photosynthetic electron transport and/or carbon metabolism in chloroplasts. MAPK cascades may function both upstream and downstream of ROS generation during the ABA-mediated stress response (Xing et al. 2008).

ROS generation and related redox processing are essential components of hormone regulation and play a role in plant growth and stress tolerance. We have only lately started to comprehend the intricate network of contacts and processes that permit the combination of ROS and hormone signaling that underpins this regulation. Much remains to be discovered in this area, most notably the spatial-temporal control of ROS generation and the discovery of the proteins that detect variations in ROS and utilize this information to allow cross talk across the various hormone signaling pathways.

11.5 Perspectives and Future Directions

Although significant progress has been made in recent years, there are still several uncertainties and gaps in our knowledge of how ROS impact plants' stress responses.

In general, ROS have been postulated to have a dual effect on stress responses. ROS react with a wide range of biomolecules, causing irreversible damage that may result in tissue necrosis and eventually plant death. On the other hand, ROS influence the expression of a variety of genes and signaling pathways, which imply that cells have developed ways to use ROS as environmental indicators and biological marker that activate and regulate a variety of genetic stress response programs.

Plants include a large number of redox-sensitive proteins, each of which has the potential to play a role in redox sensing and signal transduction. Redox control of proteins that make up important parts of plant development and defense systems is a quick and simple way to control these processes. Thiol enzymes modulating carbon absorption, metabolism, and transport are examples of these proteins. Heat shock proteins are another possible target for redox control.

ROS generation and related redox processing are required for hormone regulation and play a role in plant growth and stress tolerance. Researchers have only lately started to comprehend the intricate network of elements and processes that regulate the combination of ROS and hormone signaling. Much remains unknown, most notably the spatial-temporal control of ROS generation and the discovery of the proteins that detect variations in ROS and utilize this information to allow cross talk across the various hormone signaling pathways. The mechanisms by which variations in ROS generation and their relative concentrations in distinct cellular compartments regulate hormone-dependent activities remain unknown.

Abscisic acid and SLs are both carotenoid derivative hormones that proved to have a number of regulatory functions in the growth, development, and response of plants to a variety of environmental stimuli. These functions have been intensively investigated, with a particular emphasis on stomatal regulation. It is vital to decode the function of ABA and SLs in a variety of physiological processes in plants and to provide a theoretical foundation for addressing major agricultural production, quality, and resistance challenges.

Exogenously, SLs stimulate hyphal branching during AMF symbiosis, leguminous plant nodulation, and parasitic weed seed development. SLs govern shoot and root architecture, secondary growth, senescence, and fruit ripening on an endogenous level. Endogenous and exogenous signaling pathways are both triggered in response to a variety of environmental cues, including light, temperature, nutrition availability, and abiotic and biotic stressors. Due to their involvement in stress tolerance development, SLs may be used to make genetically engineered agricultural plants, which may aid in resolving the world food grain crisis.

Numerous significant puzzles of SL signaling remain unsolved, for instance, their significance in multiple stress management to address the specific field problems, the possibility of cross talk with other stress-related hormones, interaction with plant transcription factors and related signaling pathways, and their modulation of plasma membrane NADPH oxides. Additionally, future research should focus on identifying the downstream signaling component(s) that transduce the SL signal in guard cells. In this regard, it has been proposed that guard cell-expressed calcium-dependent protein kinases (CDPKs) contribute to guard cell signal transduction and ion channel regulation, making them excellent candidates as downstream components of the SL-induced stomatal closure signaling pathway.

References

Abe S, Sado A, Tanaka K, Kisugi T, Asami K, Ota S, Kim HI, Yoneyama K, Xie X, Ohnishi T, Seto Y, Yamaguchi S, Akiyama K, Yoneyama K, Nomura T (2014) Carlactone is converted to carlactonoic acid by MAX1 in Arabidopsis and its methyl ester can directly interact with AtD14 in vitro. Proc Natl Acad Sci U S A 111(50):18084–18089

Agusti J, Herold S, Schwarz M, Sanchez P, Ljung K, Dun EA, Brewer PB, Beveridge CA, Sieberer T, Sehr EM, Greb T (2011) Strigolactone signaling is required for auxin-dependent stimulation of secondary growth in plants. Proc Natl Acad Sci U S A 108:20242–20247. https://doi.org/10.1073/pnas.1111902108

Akiyama K, Matsuzaki K, Hayashi H (2005) Plant sesquiterpenes induce hyphal branching in arbuscular mycorrhizal fungi. Nature 435:824–827

Akiyama K, Ogasawara S, Ito S, Hayashi H (2010) Structural requirements of strigolactones for hyphal branching in AM fungi. Plant Cell Physiol 51:1104–1117

Alder A, Jamil M, Marzorati M, Bruno M, Vermathen M, Bigler P, Ghisla S, Bouwmeester H, Beyer P, Al-Babili S (2012) The path from beta-carotene to carlactone, a strigolactone-like plant hormone. Science 335(6074):1348–1351

Antoni R, Gonzalez-Guzman M, Rodriguez L, Peirats-Llobet M, Pizzio GA, Fernandez MA et al (2013) PYRABACTIN RESISTANCE1-LIKE8 plays an important role for the regulation of abscisic acid signaling in root. Plant Physiol 161:931–941. https://doi.org/10.1104/pp.112.208678

Antoniou C, Savvides A, Christou A, Fotopoulos V (2016) Unravelling chemical priming machinery in plants: the role of reactive oxygen–nitrogen–sulfur species in abiotic stress tolerance enhancement. Curr Opin Plant Biol 33:101–107

Arite T, Kameoka H, Kyozuka J (2012) Strigolactone positively controls crown root elongation in rice. J Plant Growth Regul 31:165–172. https://doi.org/10.1007/s00344-011-9228-6

Bais HP, Weir TL, Perry LG, Gilroy S, Vivanco JM (2006) The role of root exudates in rhizosphere interactions with plants and other organisms. Ann Rev Plant Biol 57:233–266

Barbier FF, Dun EA, Kerr SC, Chabikwa TG, Beveridge CA (2019) An update on the signals controlling shoot branching. Trends Plant Sci 24(3):220–236

Bertin C, Yang X, Weston LA (2003) The role of root exudates and allelochemicals in the rhizosphere. Plant Soil 67:67–83

Besserer A, Becard G, Roux C, Sejalon-Delmas N (2009) Role of mitochondria in the response of arbuscular mycorrhizal fungi to strigolactones. Plant Signal Behav 4:75–77

Besserer A, Puech-Pages V, Kiefer P, Gomez-Roldan V, Jauneau A et al (2006) Strigolactones stimulate arbuscular mycorrhizal fungi by activating mitochondria. PLoS Biol 4:1239–1247

Beveridge CA, Kyozuka J (2009) New genes in the strigolactone-related shoot branching pathway. Curr Opin Plant Biol 13(1):34–39

Beveridge CA, Kyozuka J (2010) New genes in the strigolactone-related shoot branching pathway. Curr Opin Plant Biol 13:34–39

Blake SN, Barry KM, Gill WM, Reid JB, Foo E (2016) The role of strigolactones and ethylene in disease caused by *Pythium irregulare*. Mol Plant Pathol 17:680–690. https://doi.org/10.1111/mpp.12320

Blokhina O, Fagerstedt KV (2010) Reactive oxygen species and nitric oxide in plant mitochondria: origin and redundant regulatory systems. Physiol Plant 138(4):447–462

Bonfante P, Genre A (2010) Mechanisms underlying beneficial plant-fungus interactions in mycorrhizal symbiosis. Nat Commun 1:1–11

Bonneau L, Huguet S, Wipf D, Pauly N, Truong HN (2013) Combined phosphate and nitrogen limitation generates a nutrient stress transcriptome favorable for arbuscular mycorrhizal symbiosis in *Medicago truncatula*. New Phytol 199:188–202. https://doi.org/10.1111/nph.12234

Booker J, Auldridge M, Wills S, McCarty D, Klee H, Leyser O (2004) MAX3/CCD7 is a carotenoid cleavage dioxygenase required for the synthesis of a novel plant signaling molecule. Curr Biol 14(14):1232–1238

Booker J, Sieberer T, Wright W, Williamson L, Willett B, Stirnberg P, Turnbull CGN, Srinivasan M, Goddard P, Leyser O (2005) *MAX1* encodes a cytochrome P450 family member that acts downstream of *MAX3/4* to produce a carotenoid-derived branch-inhibiting hormone. Dev Cell 8:7

Bouwmeester HJ, Roux C, Lopez-Raez JA, Becard G (2007) Rhizosphere communication of plants, parasitic plants and AM fungi. Trends Plant Sci 12:224–230

Breuillin F, Schramm J, Hajirezaei M, Ahkami A, Favre P, Druege U et al (2010) Phosphate systemically inhibits development of arbuscular mycorrhiza in *Petunia hybrida* and represses genes involved in mycorrhizal functioning. Plant J 64:1002–1017

Brewer PB, Yoneyama K, Filardo F, Meyers E, Scaffidi A, Frickey T, Akiyama K et al (2016) LATERAL BRANCHING OXIDOREDUCTASE acts in the final stages of strigolactone biosynthesis in Arabidopsis. Proc Natl Acad Sci U S A 113:6301–6306

Bright J, Desikan R, Hancock JT, Weir IS, Neill SJ (2006) ABA- induced NO generation and stomatal closure in *Arabidopsis* are dependent on H_2O_2 synthesis. Plant J 45:113–122

Brundrett MC, Tedersoo L (2018) Evolutionary history of mycorrhizal symbioses and global host plant diversity. New Phytol. https://doi.org/10.1111/nph.14976

Bu Q, Lv T, Shen H, Luong P, Wang J, Wang Z et al (2014) Regulation of drought tolerance by the F-box protein MAX2 in *Arabidopsis*. Plant Physiol 164:424–439. https://doi.org/10.1104/pp.113.226837

Cardoso C, Zhang Y, Jamil M, Hepworth J, Charnikhova T, Dimkpa SO, Meharg C, Wright MH, Liu J, Meng X, Wang Y, Li J, McCouch SR, Leyser O, Price AH, Bouwmeester HJ, Ruyter-Spira C (2014) Natural variation of rice strigolactone biosynthesis is associated with the deletion of two *MAX1* orthologs. Proc Natl Acad Sci U S A 111(6):2379–2384

Chapman JM, Muhlemann JK, Gayomba SR, Muday GK (2019) RBOHDependent ROS synthesis and ROS scavenging by plant specialized metabolites to modulate plant development and stress responses. Chem Res Toxicol 32:370–396

Chi C, Xu X, Wang M et al (2021) Strigolactones positively regulate abscisic acid-dependent heat and cold tolerance in tomato. Hortic Res 8:237. https://doi.org/10.1038/s41438-021-00668-y

Choudhury FK, Rivero RM, Blumwald E, Mittler R (2017) Reactive oxygen species, abiotic stress and stress combination. Plant J 90:856–867

Cook CE, Whichard LP, Turner B, Wall ME, Egley GH (1966) Germination of witchweed (*Striga lutea* Lour.): isolation and properties of a potent stimulant. Science 154:1189–1190

Cooper JW et al (2018) Strigolactones positively regulate chilling tolerance in pea and in Arabidopsis. Plant Cell Environ 41:1298–1310

Cutler SR, Rodriguez PL, Finkelstein RR, Abrams SR (2010) Abscisic acid: emergence of a core signaling network. Annu Rev Plant Biol 61:651–679. https://doi.org/10.1146/annurev-arplant-042809-112122

De Saint Germain A, Ligerot Y, Dun EA, Pillot JP, Ross JJ, Beveridge CA et al (2013) Strigolactones stimulate internode elongation independently of gibberellins. Plant Physiol 163:1012–1025. https://doi.org/10.1104/pp.113.220541

Ding Y, Shi Y, Yang S (2019) Advances and challenges in uncovering cold tolerance regulatory mechanisms in plants. New Phytol 222:1690–1704

Domagalska MA, Leyser O (2011) Signal integration in the control of shoot branching. Nat Rev Mol Cell Biol 12:211–221

Dor E, Joel DM, Kapulnik Y, Koltai H, Hershenhorn J (2011) The synthetic strigolactone GR24 influences the growth pattern of phytopathogenic fungi. Planta 234:419–427

Dun EA, Brewer PB, Beveridge CA (2009) Strigolactones: discovery of the elusive shoot branching hormone. Trends Plant Sci 14:364–372

Dun EA, de-Saint GA, Rameau C, Beveridge CA (2012) Antagonistic action of strigolactone and cytokinin in bud outgrowth control. Plant Physiol 158:487–498

Dun EA, de-Saint GA, Rameau C, Beveridge CA (2013) Dynamics of strigolactone function and shoot branching responses in *Pisum sativum*. Mol Plant 6:128–140

Faizan M, Faraz A, Sami F, Siddiqui H, Yusuf M, Gruszka D, Hayat S (2020) Role of strigolactones: signalling and crosstalk with other phytohormones. Open Life Sci 15:217–228

Fang Z, Ji Y, Hu J, Guo R, Sun S, Wang X (2020) Strigolactones and brassinosteroids antagonistically regulate the stability of the D53–OsBZR1 complex to determine FC1 expression in rice tillering. Mol Plant 13:586–597. https://doi.org/10.1016/j.molp.2019.12.005

Fernández-Aparicio M, Yoneyama K, Rubiales D (2011) The role of strigolactones in host specificity of *Orobanche* and *Phelipanche* seed germination. Seed Sci Res 21:55–61

Finkelstein R, Reeves W, Ariizumi T, Steber C (2008) Molecular aspects of seed dormancy. Annu Rev Plant Biol 59:387–415. https://doi.org/10.1146/annurev.arplant.59.032607.092740

Finkelstein RR, Gampala SS, Rock CD (2002) Abscisic acid signaling in seeds and seedlings. Plant Cell 14(Suppl):S15–S45

Foo E, Blake SN, Fisher BJ, Smith JA, Reid JB (2016) The role of strigolactones during plant interactions with the pathogenic fungus *Fusarium oxysporum*. Planta 243:1387–1396. https://doi.org/10.1007/s00425-015-2449-3

Foo E, Davies NW (2011) Strigolactones promote nodulation in pea. Planta 234:1073–1081

Foyer CH, Harbinson J (1994) Oxygen metabolism and the regulation of photosynthetic electron transport. In: Foyer CH, Mullineaux P (eds) Causes of photooxidative stresses and amelioration of defense systems in plants. CRC Press, Boca Raton, Fla, pp 1–42

Fujii H, Chinnusamy V, Rodrigues A, Rubio S, Antoni R, Park SY, Cutler SR, Sheen J, Rodriguez PL, Zhu JK (2009) In vitro reconstitution of an abscisic acid signalling pathway. Nature 462: 660–666

Gobena D, Shimels M, Rich PJ, Ruyter-Spira C, Bouwmeester H, Kanuganti S, Mengiste T, Ejeta G (2017) Mutation in sorghum *LOW GERMINATION STIMULANT 1* alters strigolactones and causes *striga* resistance. Proc Natl Acad Sci USA 114:4471–4476

Gomez-Roldan V, Fermas S, Brewer PB, Puech-Pages V, Dun EA, Pillot JP et al (2008) Strigolactone inhibition of shoot branching. Nature 455:189–194

Gupta A, Rico-Medina A, Cano-Delgado AI (2020) The physiology of plant responses to drought. Science 368:266–269

Ha CV, Leyva-González MA, Osakabe Y, Tran UT, Nishiyama R, Watanabe Y, Tanaka M, Seki M, Yamaguchi S, Dong NV, Yamaguchi-Shinozaki K, Shinozaki K, Herrera-Estrella L, Tran LS (2014) Positive regulatory role of strigolactone in plant responses to drought and salt stress. Proc Natl Acad Sci 111(2):851–856. https://doi.org/10.1073/pnas.1322135111

Hamiaux C, Drummond RS, Janssen BJ, Ledger SE, Cooney JM, Newcomb RD et al (2012) DAD2 is an α/β hydrolase likely to be involved in the perception of the plant branching hormone, strigolactone. Curr Biol 22(21):2032–2036

Harrier LA, Watson CA (2004) The potential role of arbuscular mycorrhizal (AM) fungi in the bioprotection of plants against soil-borne pathogens in organic and/ or other sustainable farming systems. Pest Manag Sci 60:149–157

Hu QN, Zhang SX, Huang BR (2019) Strigolactones promote leaf elongation in tall fescue through upregulation of cell cycle genes and downregulation of auxin transport genes in tall fescue under different temperature regimes. Int J Mol Sci 20:1836

Im JH, Lee H, Kim J, Kim HB, An CS (2012) Soybean MAPK, GMK1 is dually regulated by phosphatidic acid and hydrogen peroxide and translocated to nucleus during salt stress. Mol Cells 34:271–278. https://doi.org/10.1007/s10059-012-0092-4

Ismail A, Riaz M, Akhtar S, Ismail T, Amir M, Zafar-ul-Hye M (2014) Heavy metals in vegetables and respective soils irrigated by canal, municipal waste and tube well waters. Food Addit Contam B 7:213–219. https://doi.org/10.1080/19393210.2014.888783

Ito S, Ito K, Abeta N, Takahashi R, Sasaki Y, Yajima S (2016) Effects of strigolactone signaling on Arabidopsis growth under nitrogen deficient stress condition. Plant Signal Behav 11:e1126031. https://doi.org/10.1080/15592324.2015.1126031

Jain JA, Poling MD, Karthikeyan AS, Blakeslee JJ, Peer WA, Titapiwatanakun B, Murphy AS, Raghothama KG (2007) Differential effects of sucrose and auxin on localized pi-deficiency induced modulation of different traits of root system architecture in arabidopsis. Plant Physiol 144:232–247

Jannat R, Uraji MMM et al (2011) Roles of intracellular hydrogen peroxide accumulation in abscisic acid signaling in Arabidopsis guard cells. J Plant Physiol 168(16):1919–1926

Jiang MY, Zhang JH (2002) Water stress-induced abscisic acid accumulation triggers the increased generation of reactive oxygen species and up-regulates the activities of antioxidant enzymes in maize leaves. J Exp Bot 53:2401–2410

Jibran R, Hunter DA, Dijkwel PP (2013) Hormonal regulation of leaf senescence through integration of developmental and stress signals. Plant Mol Biol 82:547–561

Jung HJ, Bae YS, Lee JS (2001) Role of auxin-induced reactive oxygen species in root gravitropism. Plant Physiol 126(3):1055–1060

Kapulnik Y, Delaux PM, Resnick N, Mayzlish-Gati E, Wininger S, Bhattacharya C et al (2011) Strigolactones affect lateral root formation and root-hair elongation in *Arabidopsis*. Planta 233: 209–216

Kapulnik Y, Koltai H (2014) Strigolactone involvement in root development, response to abiotic stress, and interactions with the biotic soil environment. Plant Physiol 166(2):560–569. https://doi.org/10.1104/pp.114.244939

Khokon MA, Hossain MA, Munemasa S, Uraji M, Nakamura Y, Mori IC, Murata Y (2010) Yeast elicitor-induced stomatal closure and peroxidase- mediated ROS production in Arabidopsis. Plant Cell Physiol 51:1915–1921

Kim HI, Xie X, Kim HS, Chun JC, Yoneyama K, Nomura T, Takeuchi Y, Yoneyama K (2010a) Structure-activity relationship of naturally occurring strigolactones in *Orobanche minor* seed germination stimulation. J Pestic Sci 35:344–347

Kim TH, Bohmer M, Hu H, Nishimura N, Schroeder JI (2010b) Guard cell signal transduction network: advances in understanding abscisic acid, CO2, and Ca2+ signaling. Annu Rev Plant Biol 61:561–591. https://doi.org/10.1146/annurevarplant-042809-112226

Kisugi T, Xie X, Kim HI et al (2013) Strigone, isolation and identification as a natural strigolactone from *Houttuynia cordata*. Phytochemistry 87:60–64

Klingler JP, Batelli G, Zhu JK (2010) ABA receptors: the START of a new paradigm in phytohormone signalling. J Exp Bot 61:3199–3210. https://doi.org/10.1093/jxb/erq151

Kohlen W, Charnikhova T, Lammers M et al (2012) The tomato CAROT- ENOID CLEAVAGE DIOXYGENASE8 (SlCCD8) regulates rhizosphere signaling, plant architecture and affects reproductive development through strigolactone biosynthesis. New Phytol 196:535–547

Kohlen W, Charnikhova T, Liu Q, Bours R, Domagalska MA, Beguerie S, Verstappen F, Leyser O, Bouwmeester H, Ruyter-Spira C (2011) Strigolactones are transported through the xylem and play a key role in shoot architectural response to phosphate deficiency in nonarbuscular mycorrhizal host Arabidopsis. Plant Physiol 155:974–987

Kollist H, Zandalinas SI, Sengupta S, Nuhkat M, Kangasjarvi J, Mittler R (2019) Rapid responses to abiotic stress: priming the landscape for the signal transduction network. Trends Plant Sci 24: 25–37

Koltai H (2011) Strigolactones are regulators of root development. New Phytol 190:545–549

Koltai H (2014) Receptors, repressors, PINs: a playground for strigolactone signaling. Trends Plant Sci 19:727–733. https://doi.org/10.1016/j.tplants.2014.06.008

Koltai H, Cohen M, Chesin O, Mayzlish-Gati E, Bécard G, Puech V et al (2011) Light is a positive regulator of strigolactone levels in tomato roots. J Plant Physiol 168:1993–1996. https://doi.org/10.1016/j.jplph.2011.05.022

Koltai H, Dor E, Hershenhorn J, Joel D, Weininger S et al (2009) Strigolactones' effect on root growth and root-hair elongation may be mediated by auxin-efflux carriers. J Plant Growth Regul 29:129–136

Koltai H, Dor E, Hershenhorn J, Joel DM, Weininger S, Lekalla S, Shealtiel H, Bhattacharya C, Eliahu E, Resnick N, Barg R, Kapulnik Y (2010) Strigolactones' effect on root growth and root-hair elongation may be mediated by auxin-efflux carriers. J Plant Growth Regul 29:129–136. https://doi.org/10.1007/s00344-009-9122-7

Kong CC, Ren CG, Li RZ, Xie ZH, Wang JP (2017) Hydrogen peroxide and strigolactones signaling are involved in alleviation of salt stress induced by arbuscular mycorrhizal fungus in sesbania cannabina seedlings. J Plant Growth Regul 36(3):734–742. https://doi.org/10.1007/s00344-017-9675-9

Kretzschmar T, Kohlen W, Sasse J, Borghi L, Schlegel M, Bachelier JB, Reinhardt D, Bours R, Bouwmeester HJ, Martinoia E (2012) A petunia ABC protein controls strigolactone-dependent symbiotic signalling and branching. Nature 483:341–U135

Kumar K, Rao KP, Sharma P, Sinha AK (2008) Differential regulation of rice mitogen activated protein kinase kinase (MKK) by abiotic stress. Plant Physiol Biochem 46(10):891–897. https://doi.org/10.1016/j.plaphy.2008.05.014

Kuromori T, Miyaji T, Yabuuchi H, Shimizu H, Sugimoto E, Kamiya A, Moriyama Y, Shinozaki K (2010) ABC transporter AtABCG25 is involved in abscisic acid transport and responses. Proc Natl Acad Sci U S A 107:2361–2366

Kusaba M, Tanaka A, Tanaka R (2013) Stay-green plants: what do they tell us about the molecular mechanism of leaf senescence. Photosynth Res 117:221–234

Kwak JM, Mori IC, Pei ZM, Leonhardt N, Torres MA, Dangl JL, Bloom RE, Bodde S, Jones JD, Schroeder JI (2003) NADPH oxidase *AtrbohD* and *AtrbohF* genes function in ROS-dependent ABA signaling in *Arabidopsis*. EMBO J 22:2623–2633

Lamers J, van der Meer T, Testerink C (2020) How plants sense and respond to stressful environments. Plant Physiol 182:1624–1635

Lee SJ, Lee MH, Kim JI, Kim SY (2015) Arabidopsis putative MAP kinase kinase kinases Raf10 and Raf11 are positive regulators of seed dormancy and ABA response. Plant Cell Physiol 56: 84–97

Leyser O (2009) The control of shoot branching: an example of plant information processing. Plant Cell Environ 32:694–703

Li Z, Wang W, Li G, Guo K, Harvey P, Chen Q, Zhao Z, Wei Y, Li J, Yang H (2016) MAPK-mediated regulation of growth and essential oil composition in a salt- tolerant peppermint (*Mentha piperita* L.) under NaCl stress. Protoplasma 253:1541–1556. https://doi.org/10.1007/s00709-015-0915-1

Lim PO, Kim HJ, Nam HG (2007) Leaf senescence. Annu Rev Plant Biol 5:115–136

Lin H, Wang R, Qian Q, Yan M, Meng X, Fu Z, Yan C, Jiang B, Su Z, Li J, Wang Y (2009) DWARF27, an iron-containing protein required for the biosynthesis of strigolactones, regulates rice tiller bud outgrowth. Plant Cell 21(5):1512–1525

Lin R, Ding L, Casola C, Ripoll DR, Feschotte C, Wang H (2007) Transposase-derived transcription factors regulate light signaling in *Arabidopsis*. Science 318:1302–1305. https://doi.org/10.1126/science.1146281

Liu J, Lovisolo C, Schubert A, Cardinale F (2013) Signaling role of strigolactones at the interface between plants, (micro)organisms, and a changing environment. J Plant Interact 8(1):17–33. https://doi.org/10.1080/17429145.2012.750692

Liu X et al (2020) Carotene isomerase suppresses tillering in rice through the coordinated biosynthesis of strigolactone and abscisic acid. Mol Plant 13:1784–1801

López-Ráez JA, Kohlen W, Charnikhova T, Mulder P, Undas AK, Sergeant MJ et al (2010) Does abscisic acid affect strigolactone biosynthesis? New Phytol 187:343–354. https://doi.org/10.1111/j.1469-8137.2010.03291.x

Lu T, Yu H, Li Q, Chai L, Jiang WJ (2019) Improving plant growth and alleviating photosynthetic inhibition and oxidative stress from low-light stress with exogenous GR24 in tomato (*Solanum lycopersicum* L.) seedlings. Front Plant Sci 10:490

Lv S, Zhang YH, Li C, Liu ZJ, Yang N, Pan LX, Wu JB, Wang JJ, Yang JW, Lv YT, Zhang YT, Jiang WQ, She XP, Wang GD (2018) Strigolactone-triggered stomatal closure requires hydrogen peroxide synthesis and nitric oxide production in an abscisic acid- independent manner. New Phytol 217(1):290–304. https://doi.org/10.1111/nph.14813

Ma N, Hu C, Wan L, Hu Q, Xiong J, Zhang C (2017) Strigolactones improve plant growth, photosynthesis, and alleviate oxidative stress under salinity in rapeseed (*Brassica napus* L.) by regulating gene. Front Plant Sci 8:1671. https://doi.org/10.3389/fpls.2017.01671

Ma Y, Szostkiewicz I, Korte A, Moes D, Yang Y, Christmann A, Grill E (2009) Regulators of PP2C phosphatase activity function as abscisic acid sensors. Science 324:1064–1068

Mansoor S, Ali Wani O, Lone JK, Manhas S, Kour N, Alam P et al (2022) Reactive oxygen species in plants: from source to sink. Antioxidants 11(2):225

Marino D, Dunand C, Puppo A, Pauly N (2012) A burst of plant NADPH oxidases. Trends Plant Sci 17(1):9–15

Marzec M (2016) Strigolactones as part of the plant defence system. Trends Plant Sci 16:30121–30122. https://doi.org/10.1016/j.tplants.2016.08.010

Mayzlish-Gati E, De-Cuyper C, Goormachtig S, Beeckman T, Vuylsteke M, Brewer PB et al (2012) Strigolactones are involved in root response to low phosphate conditions in *Arabidopsis*. Plant Physiol 160:1329–1341. https://doi.org/10.1104/pp.112.202358

Mehlmer N, Wurzinger B, Stael S, Hofmann-Rodrigues D, Csaszar E, Pfister B, Bayer R, Teige M (2010) The Ca^{2+}-dependent protein kinase CPK3 is required for MAPK-independent salt-stress acclimation in *Arabidopsis*. Plant J 63:484–498. https://doi.org/10.1111/j.1365-313X.2010.04257.x

Mignolet-Spruyt L, Xu E, Idanheimo N, Hoeberichts FA, Muhlenbock P, Brosche M, Van Breusegem F, Kangasjarvi J (2016) Spreading the news: subcellular and organellar reactive oxygen species production and signalling. J Exp Bot 67:3831–3844

Miller G, Shulaev V, Mittler R (2008) Reactive oxygen signaling and abiotic stress. Physiol Plant 133(3):481–489

Minakuchi K, Kameoka H, Yasuno N, Umehara M, Luo L, Kobayashi K, Hanada A, Ueno K, Asami T, Yamaguchi S, Kyozuka J (2010) FINE CULM1 (FC1) works downstream of strigolactones to inhibit the outgrowth of axillary buds in rice. Plant Cell Physiol 51:1127–1135. https://doi.org/10.1093/pcp/pcq083

Mishra S, Jha AB, Dubey RS (2011) Arsenite treatment induces oxidative stress, upregulates antioxidant system, and causes phytochelatin synthesis in rice seedlings. Protoplasma 248(3): 565–577

Mittler R (2002) Oxidative stress, antioxidants and stress tolerance. Trends Plant Sci 7(9):405–410

Mittler R (2017) ROS are good. Trends Plant Sci 22:11–19

Miyakawa T, Fujita Y, Yamaguchi-Shinozaki K, Tanokura M (2013) Structure and function of abscisic acid receptors. Trends Plant Sci 18:259–266. https://doi.org/10.1016/j.tplants.2012.11.002

Mostofa MG, Li W, Nguyen KH, Fujita M, Tran LSP (2018) Strigolactones in plant adaptation to abiotic stresses: An emerging avenue of plant research. Plant Cell Environ 41(10):2227–2243

Mustilli AC, Merlot S, Vavasseur A, Fenzi F, Giraudat J (2002) *Arabidopsis* OST1 protein kinase mediates the regulation of stomatal aperture by abscisic acid and acts upstream of reactive oxygen species production. Plant Cell 14:3089–3099

Nambara E, Marion-Poll A (2005) Abscisic acid biosynthesis and catabolism. Annu Rev Plant Biol 56:165–185

Neill S, Desikan R, Hancock J (2002) Hydrogen peroxide signalling. Curr Opin Plant Biol 5(5): 388–395

Nguyen QTC, Lee SJ, Choi SW, Na YJ, Song MR, Hoang QTN, Sim SY, Kim MS, Kim JI, Soh MS, Kim SY (2019) Arabidopsis Raf-like kinase Raf10 is a regulatory component of core ABA signaling. Mol Cells 42:646–660

Nilson SE, Assmann SM (2007) The control of transpiration. Insights from arabidopsis. Plant Physiol 143:19–27

Noctor G, Reichheld JP, Foyer CH (2018) ROS-related redox regulation and signaling in plants. Semin Cell Dev Biol 80:3–12

Nolan TM, Vukasinovic N, Liu D, Russinova E, Yin Y (2020) Brassinosteroids: multidimensional regulators of plant growth, development, and stress responses. Plant Cell 32:298–318

Nooden LD (1988) The phenomena of senescence and aging. In: Leopold AC, Nooden LD (eds) Senescence and aging in plants. Academic Press, San Diego, CA, pp 1–50

Noodén LD (2004) Introduction. In: Noodén LD (ed) Plant cell death processes. Academic Press, London, pp 1–18

Ouyang X, Li J, Li G, Li B, Chen B, Shen H et al (2011) Genome-wide binding site analysis of FAR-RED ELONGATED HYPOCOTYL3 reveals its novel function in *Arabidopsis* development. Plant Cell 23:2514–2535. https://doi.org/10.1105/tpc.111.085126

Park SY, Fung P, Nishimura N, Jensen DR, Fujii H, Zhao Y, Lumba S, Santiago J, Rodrigues A, Chow TF, Alfred SE, Bonetta D, Finkelstein R, Provart NJ, Desveaux D, Rodriguez PL,

McCourt P, Zhu JK, Schroeder JI, Volkman BF, Cutler SR (2009) Abscisic acid inhibits type 2C protein phosphatases via the PYR/PYL family of START proteins. Science 324:1068–1071
Parniske M (2008) Arbuscular mycorrhiza: the mother of plant root endo-symbioses. Nat Rev Microbiol 6:763–775
Pei ZM, Murata Y, Benning G, Thomine S, Klusener B, Allen GJ, Grill E, Schroeder JI (2000) Calcium channels activated by hydrogen peroxide mediate abscisic acid signaling in guard cells. Nature 406:731–734
Pel ZM, Murata Y, Benning G et al (2000) Calcium channels activated by hydrogen peroxide mediate abscisic acid signalling in guard cells. Nature 406(6797):731–734
Petrášek J, Friml J (2009) Auxin transport routes in plant development. Development 136:2675–2688
Piisilä M, Keceli MA, Brader G, Jakobson L, Jõesaar I, Sipari N et al (2015) F-box protein MAX2 contributes to resistance to bacterial phytopathogens in *Arabidopsis thaliana*. BMC Plant Biol 15:53. https://doi.org/10.1186/s12870-015-0434-4
Postliglione AE, and Gloria KM (2022) Abscisic Acid Drives Stomatal Closure through Increases in Hydrogen Peroxide in Distinct Subcellular Compartments Including Mitochondria. ***bioRxiv***
Rubio S, Rodrigues A, Saez A, Dizon MB, Galle A, Kim TH, Santiago J, Flexas J, Schroeder JI, Rodriguez PL (2009) Triple loss of function of protein phosphatases type 2C leads to partial constitutive response to endogenous abscisic acid. Plant Physiol 150:1345–1355
Ruyter-Spira C, Kohlen W, Charnikhova T, van Zeijl A, van Bezouwen L, de-Ruijter N et al (2011) Physiological effects of the synthetic strigolactone analog GR24 on root system architecture in Arabidopsis: another belowground role for strigolactones? Plant Physiol 155:721–734
Sachs T, Thimann KV (1967) The role of auxins and cytokinins in the release of buds from apical dominance. Am J Bot 45:136–144
Sagi M, Fluhr R (2006) Production of reactive oxygen species by plant NADPH oxidases. Plant
Santiago J, Rodrigues A, Saez A, Rubio S, Antoni R, Dupeux F et al (2009) Modulation of drought resistance by the abscisic acid receptor PYL5 through inhibition of clade A PP2Cs. Plant J 60: 575–588. https://doi.org/10.1111/j.1365-313X.2009.03981.x
Seto Y, Sado A, Asami K, Hanada A, Umehara M, Akiyama K, Yamaguchi S (2014) Carlactone is an endogenous biosynthetic precursor for strigolactones. Proc Natl
Shah K, Kumar RG, Verma S, Dubey RS (2001) Effect of cadmium on lipid peroxidation, superoxide anion generation and activities of antioxidant enzymes in growing rice seedlings. Plant Sci 161(6):1135–1144
Sharifi P, Bidabadi SS (2020) Strigolactone could enhances gas-exchange through augmented antioxidant defense system in *Salvia nemorosa* L. plants subjected to saline conditions stress. Ind Crops Prod 151:112460
Sharma P, Dubey RS (2005) Drought induces oxidative stress and enhances the activities of antioxidant enzymes in growing rice seedlings. Plant Growth Regul 46(3):209–221
Sharma P, Dubey RS (2007) Involvement of oxidative stress and role of antioxidative defense system in growing rice seedlings exposed to toxic concentrations of aluminum. Plant Cell Rep 26(11):2027–2038
Sharma P, Jha AB, Dubey RS, Pessarakli M (2012) Reactive oxygen species, oxidative damage, and antioxidative defense mechanism in plants under stressful conditions. J Bot 2012
Shi CY, Qi C, Ren HY, Huang AX, Hei SM, She XP (2015) Ethylene mediates brassinosteroid-induced stomatal closure via ga protein-activated hydrogen peroxide and nitric oxide production in Arabidopsis. Plant J 82:280–301
Shinohara N, Taylor C, Leyser O (2013) Strigolactone can promote or inhibit shoot branching by triggering rapid depletion of the auxin efflux protein PIN1 from the plasma membrane. PLoS Biol 1:e1001474. https://doi.org/10.1371/journal.pbio.1001474
Simons JL, Napoli CA, Janssen BJ, Plummer KM, Snowden KC (2007) Analysis of the *DECREASED APICAL DOMINANCE* genes of petunia in the control of axillary branching. Plant Physiol 143:697–706

Sirichandra C, Gu D, Hu HC, Davanture M, Lee S, Djaoui M, Valot B, Zivy M, Leung J, Merlot S (2009) Phosphorylation of the *Arabidopsis* AtrbohF NADPH oxidase by OST1 protein kinase. FEBS Lett 583:2982–2986

Song X, Lu Z, Yu H, Shao G, Xiong J, Meng X (2017) IPA1 functions as a downstream transcription factor repressed by D53 in strigolactone signaling in rice. Cell Res 27:1128–1141. https://doi.org/10.1038/cr.2017.102

Soto MJ, Fernandez-Aparicio M, Castellanos-Morales V, Carcıa-Garrido JM, Ocampo JA et al (2010) First indications for the involvement of strigolactones on nodule formation in alfalfa (*Medicago sativa*). Soil Biol Biochem 42:383–385

Soundappan I, Bennett T, Morffy N, Liang Y, Stanga JP, Abbas A et al (2015) SMAX1-LIKE/D53 family members enable distinct MAX2-dependent responses to strigolactones and karrikins in *Arabidopsis*. Plant Cell 27:3143–3159. https://doi.org/10.1105/tpc.15.00562. Search in Google Scholar

Srivastava S, Dubey RS (2011) Manganese-excess induces oxidative stress, lowers the pool of antioxidants and elevates activities of key antioxidative enzymes in rice seedlings. Plant Growth Regul:1–16

Stepanova AN, Alonso JM (2009) Ethylene signaling and response: where different regulatory modules meet. Curr Opin Plant Biol 12:548–555

Stes E, Depuydt S, De Keyser A, Matthys C, Audenaert K, Yoneyama K (2015) Strigolactones as an auxiliary hormonal defence mechanism against leafy gall syndrome in *Arabidopsis thaliana*. J Exp Bot 66:5123–5134. https://doi.org/10.1093/jxb/erv309

Sugimoto Y, Ali AM, Yabuta S, Kinoshita H, Inanaga S, Itai A (2003) Germination strategy of Striga hermonthica involves regulation of ethylene biosynthesis. Physiol Plant 119:137–145

Sun H, Tao J, Liu S, Huang S, Chen S, Xie X et al (2014) Strigolactones are involved in phosphate- and nitrate-deficiency-induced root development and auxin transport in rice. J Exp Bot 65: 6735–6746

Suzuki N, Miller G, Morales J, Shulaev V, Torres MA, Mittler R (2011) Respiratory burst oxidases: the engines of ROS signaling. Curr Opin Plant Biol 14:691–699

Tan BC, Schwartz SH, Zeevaart JA, McCarty DR (1997) Genetic control of abscisic acid biosynthesis in maize. Proc Natl Acad Sci U S A 94(22):12235–12240

Thimann KV, Skoog F (1934) On the inhibition of bud development and other functions of growth substances in *Vicia faba*. Proc R Soc London Ser B 114:317–339

Torres-Vera R, Garcia JM, Pozo MJ, Lopez-Raez JA (2014) Do strigolactones contribute to plant defense? Mol Plant Pathol 15:211–216. https://doi.org/10.1111/mpp.12074

Tsuchiya Y, McCourt P (2009) Strigolactones: a new hormone with a past. Curr Opin Plant Biol 12: 556–561

Tsuchiya Y, Vidaurre D, Toh S, Hanada A, Nambara E, Kamiya Y et al (2010) A small-molecule screen identifies new functions for the plant hormone strigolactone. Nat Chem Biol 6:741–749

Ueda H, Kusaba M (2015) Strigolactone regulates leaf senescence in concert with ethylene in Arabidopsis. Plant Physiol 169:138–147

Umehara M, Hanada A, Yoshida S, Akiyama K, Arite T, Takeda- Kamiya N, Magome H, Kamiya Y, Shirasu K, Yoneyama K, Kyozuka J, Yamaguchi S (2008) Inhibition of shoot branching by new terpenoid plant hormones. Nature 455:195–200. https://doi.org/10.1038/nature07272

Umezawa T, Sugiyama N, Mizoguchi M, Hayashi S, Myouga F, YamaguchiShinozaki K et al (2009) Type 2C protein phosphatases directly regulate abscisic acid-activated protein kinases in Arabidopsis. Proc Natl Acad Sci U S A 106:17588–17593. https://doi.org/10.1073/pnas.0907095106

Umezawa T, Sugiyama N, Takahashi F, Anderson JC, Ishihama Y, Peck SC, Shinozaki K (2013) Genetics and phosphoproteomics reveal a protein phosphorylation network in the abscisic acid signaling pathway in Arabidopsis thaliana. Sci Signal 6:rs8

Visentin I, Vitali M, Ferrero M, Zhang Y, Ruyter-Spira C, Novák O, Strnad M, Lovisolo C, Schubert A, Cardinale F (2016) Low levels of strigolactones in roots as a component of the systemic signal of drought stress in tomato. New Phytol 212(4):954–963. https://doi.org/10.1111/nph.14190. Epub 2016 Sep 26. PMID: 27716937

Wang L, Wang B, Yu H et al (2020) Transcriptional regulation of strigolactone signaling in *Arabidopsis*. Nature 583:272–281

Wang P, Xue L, Batelli G, Lee S, Hou YJ, Van Oosten MJ, Zhang H, Tao WA, Zhu JK (2013) Quantitative phosphoproteomics identifies SnRK2 protein kinase substrates and reveals the effectors of abscisic acid action. Proc Natl Acad Sci U S A 110:11205–11210

Wang Q, Ni J, Shah F, Liu W, Wang D, Yao Y, Hu H, Huang S, Hou J, Fu S, Wu L (2019) Overexpression of the stress inducible *SsMAX2* promotes drought and salt resistance via the regulation of redox homeostasis in *Arabidopsis*. Int J Mol Sci 20:837. https://doi.org/10.3390/ijms20040837

Waters MT, Gutjahr C, Bennett T, Nelson DC (2017) Strigolactone signaling and evolution. Annu Rev Plant Biol 68:8.1–8.31

Wu F, Gao Y, Yang W, Sui N, Zhu J (2022) Biological functions of strigolactones and their crosstalk with other phytohormones. Front Plant Sci 13:821563. https://doi.org/10.3389/fpls.2022.821563

Xia X, Zhou Y-H, Shi K, Zhou J, Foyer CH, Yu J-Q (2015) Interplay between reactive oxygen species and hormones in the control of plant development and stress tolerance. J Exp Bot 66(10):2839–2856

Xie X, Kusumoto D, Takeuchi Y, Yoneyama K, Yamada Y, Yoneyama K (2007) 2′-Epi-orobanchol and solanacol, two unique strigolactones, germination stimulants for root parasitic weeds, produced by tobacco. J Agric Food Chem 55:8067–8072

Xie X, Yoneyama K (2010) The strigolactone story. Annu Rev Phytopathol 48:93–117

Xie X, Yoneyama K, Kisugi T, Nomura T, Akiyama K, Asami T, Yoneyama K (2016) Structure- and stereospecific transport of strigolactones from roots to shoots. J Pestic Sci 41:55–58

Xie X, Yoneyama K, Kisugi T, Nomura T, Akiyama K et al (2015) Strigolactones are transported from roots to shoots, although not through the xylem. J Pestic Sci 40:214–216

Xie X, Yoneyama K, Kisugi T, Uchida K, Ito S, Akiyama K, Hayashi H, Yokota T, Nomura T, Yoneyama K (2013) Confirming stereochemical structures of strigolactones produced by rice and tobacco. Mol Plant 6:153–163

Xie X, Yoneyama K, Kusumoto D, Yamada Y, Yokota T et al (2008) Isolation and identification of alectrol as (+)-orobanchyl acetate, a novel germination stimulant for root parasitic plants. Phytochemistry 69:427–431

Xing Y, Jia W, Zhang J (2008) AtMKK1 mediates ABA-induced *CAT1* expression and H_2O_2 production via AtMPK6-coupled signaling in *Arabidopsis*. Plant J 54:440–451

Xu H, Sun X, Wang X, Shi Q, Yang X, Yang F (2011) Involvement of a cucumber MAPK gene (*CsNMAPK*) in positive regulation of ROS scavengence and osmotic adjustment under salt stress. Sci Hortic 127:488–493. https://doi.org/10.1016/j.scienta.2010.11.013

Yamada Y, Umehara M (2015) Possible roles of Strigolactones during leaf senescence. Plan Theory 4:664–677

Yamasaki H, Ogura MP, Kingjoe KA, Cohen MF (2019) D-cysteine-induced rapid root abscission in the water fern Azolla pinnata: implications for the linkage between d-amino acid and reactive sulfur species (RSS) in plant environmental responses. Antioxidants 8:411

Yan J, Tsuichihara N, Etoh T, Iwai S (2007) Reactive oxygen species and nitric oxide are involved in ABA inhibition of stomatal opening. Plant Cell Environ 30(10):1320–1325

Yao R, Ming Z, Yan L, Li S, Wang F, Ma S et al (2016) DWARF14 is a non-canonical hormone receptor for strigolactone. Nature 536:469–473. https://doi.org/10.1038/nature19073

Yao R, Wang L, Li Y, Chen L, Li S, Du X et al (2018) Rice DWARF14 acts as an unconventional hormone receptor for strigolactone. J Exp Bot 69:2355–2365. https://doi.org/10.1093/jxb/ery014

Yoneyama K, Xie X, Kusumoto D, Sekimoto H, Sugimoto Y et al (2007) Nitrogen deficiency as well as phosphorus deficiency in sorghum promotes the production and exudation of

5-deoxystrigol, the host recognition signal for arbuscular mycorrhizal fungi and root parasites. Planta 227:125–132

Yoneyama K, Xie X, Yoneyama K, Kisugi T, Nomura T, Nakatani Y, Akiyama K, Christopher SP, McErlean CSP (2018) Which are the major players, canonical or non-canonical strigolactones? J Exp Bot 69(9):2231–2239. https://doi.org/10.1093/jxb/ery090

Zhang TG, Shi ZF, Zhang XH, Zheng S, Wang J, Mo JN (2020b) Alleviating effects of exogenous melatonin on salt stress in cucumber. Sci Hortic 262:109070. https://doi.org/10.1016/j.scienta.2019.109070

Zhang X, Zhang L, Sun Y, Zheng S, Wang J, Zhang T (2020a) Hydrogen peroxide is involved in strigolactone induced low temperature stress tolerance in rape seedlings (*Brassica rapa* L.). Plant Physiol Biochem. https://doi.org/10.1016/j.plaphy.2020.11.006

Zhang Y, van Dijk ADJ, Scaffidi A, Flematti GR, Hofmann M, Charnikhova T, Verstappen F, Hepworth J, Van der Krol S, Leyser O, Smith SM, Zwanenburg B, Al-Babili S, Ruyter-Spira C, Bouwmeester HJ (2014) Rice cytochrome P450 MAX1 homologs catalyze distinct steps in strigolactone biosynthesis. Nat Chem Biol 10:1028–1033

Zhang YY, Zhu HY, Zhang Q, Li MY, Yan M, Wang R, Wang LL, Welti R, Zhang WH, Wang XM (2009) Phospholipase Dα1 and phosphatidic acid regulate NADPH oxidase activity and production of reactive oxygen species in ABA-mediated stomatal closure in *Arabidopsis*. Plant Cell 21:2357–2377

Zhao LH, Zhao LH, Zhou XE, Yi W, Wu Z, Liu Y et al (2015) Destabilization of strigolactone receptor DWARF14 by binding of ligand and E3-ligase signaling effector DWARF3. Cell Res 25:1219–1236. https://doi.org/10.1038/cr.2015.122

Zhao Y, Xing L, Wang X, Hou YJ, Gao J, Wang P et al (2014) The ABA receptor PYL8 promotes lateral root growth by enhancing MYB77-dependent transcription of auxin-responsive genes. Sci Signal 7:ra53. https://doi.org/10.1126/scisignal.2005051

Zheng X, Li Y, Xi X, Ma C, Sun Z, Yang X, Li X, Tian Y, Wang C (2021) Exogenous Strigolactones alleviate KCl stress by regulating photosynthesis, ROS migration and ion transport in *Malus hupehensis* Rehd. Plant Physiol Biochem 159:113–122

Zhu JK (2002) Salt and drought stress signal transduction in plants. Annu Rev Plant Biol 53:247–273

Zwanenburg B, Mwakaboko AS, Reizelman A, Anilkumar G, Sethumadhavan D (2009) Structure and function of natural and synthetic signaling molecules in parasitic weed germination. Pest Manag Sci 65:478–491

Hydrogen Peroxide: Regulator of Plant Development and Abiotic Stress Response 12

Ajmat Jahan, M. Masroor A. Khan, Bilal Ahmad, Khan Bilal Mukhtar Ahmed, Ram Prakash Pandey, and Mohd Gulfishan

Abstract

Due to their immobility both in vitro and in vivo, crop plants are constantly exposed to abiotic and biotic factors. As a result, they have more sophisticated immune defences than animals. They might experience the mixture of these stressors concurrently or successively. The study of hydrogen peroxide (H_2O_2) is becoming more popular in the realm of molecular biology. It is a significant redox (reduction-oxidation reaction) metabolite that causes oxidative injury to biomolecules at high quantities, which can lead to cell death. Conversely, at low concentrations, H_2O_2 functions as a signalling molecule and mimics plant hormones in several ways. The hazardous nature of hydrogen peroxide was first understood to result in cell viability losses due to injury at several levels of cell organisation. It is now well-known that H_2O_2 has a positive role as a major hub integrating signalling network in response to abiotic stress and during developmental processes. In this chapter, the production, scavenging and the dual role of hydrogen peroxide from the point of view of its role in plant growth and developmental process and in abiotic stress tolerance have been presented.

A. Jahan · M. M. A. Khan · B. Ahmad
Department of Botany, Aligarh Muslim University, Aligarh, India

K. B. M. Ahmed
Department of Biotechnology and Life Sciences, IBMER, Mangalayatan University, Aligarh, India

R. P. Pandey
Department of Biotechnology, Chandigarh University, Chandigarh, India

M. Gulfishan (✉)
School of Agricultural Sciences, Glocal University, Mirzapur, Saharanpur, India

M. Faizan et al. (eds.), *Reactive Oxygen Species*,
https://doi.org/10.1007/978-981-19-9794-5_12

Keywords

Hydrogen peroxide (H_2O_2) · Reactive oxygen species (ROS) · Abiotic stress · Plant development · Priming · Stress tolerance

12.1 Introduction

H_2O_2, a reactive chemical, aids in both the developmental and physiological processes of plants along with their resilience to stress. Initially, hydrogen peroxide was considered harmful to plants due to being a member of reactive oxygen species group which was linked with the toxicity, cell death, necrosis and peroxidation of lipids. Since last decade, its positive role in plant growth and development and a regulator of abiotic stress has also been confirmed. As a result, it might be difficult to discern between H_2O_2's useful (signalling) and detrimental (damaging) functions. The function of H_2O_2 and other reactive oxygen species (ROS), that include the singlet oxygen (O_2^-) and the hydroxyl radical (^-OH), which could also interact and be changed between each other via serendipitous and catalysed reactions, must also be distinguished. Recent research has demonstrated that one of plant's most significant molecules, H_2O_2, is essential for regulating plant development, photosynthetic function and antioxidant properties (Khan et al. 2018). As a result, it has received a great deal of attention. Under either normal or stressful circumstances, it is created both inside and outside of the cells (Dietz et al. 2016; Munne-Bosch and Pinto-Marijuan 2017; Khan et al. 2018). The non-enzymatic and enzymatic defence mechanisms in agricultural plants are in charge of controlling hydrogen peroxide, a resultant of biological and biochemical metabolisms. To avoid stress in crop plants under vital biological conditions, a fine balance involving production of H_2O_2, as reactive oxygen species, and its elimination should be preserved (Martinez et al. 2018). Cells generate reactive oxygen species (ROS) under challenging stressful situations which includes peroxy radicals, superoxide (O_2^{--}), hydroperoxyl, hydroxyl (OH^-), as well as nonradical molecules such singlet oxygen (1O) and hydrogen peroxide (H_2O_2) (Dikilitas et al. 2018; Noctor et al. 2018). A number of cell organelles, such as the mitochondria, plasma membrane, chloroplasts, endoplasmic reticulum, cell wall, apoplast and peroxisomes, create these radicals (Das and Roychoudhury 2014; You and Chan 2015). Elevated ROS levels including H_2O_2 result in the oxidation of carbohydrate, protein and lipids and disintegration of pigments which subsequently leads to DNA methylation with its single- and double-strand breaks and degraded enzyme activity (Bose et al. 2014; Rani et al. 2016; Kocyigit et al. 2017; Banerjee and Roychoudhury 2018). Research findings on plants have shown that pre-treatment with an optimum range of H_2O_2 can increase abiotic stress tolerance by attenuating a wide range of biochemical and physiological mechanisms, including photosynthesis, and a variety of stress-responsive pathways, including the methylglyoxal and reactive oxygen species detoxification pathways (Azevedo-Neto et al. 2005; Chao et al. 2009; Liu et al. 2010; Xu et al. 2010; Wang et al. 2010a). Even though, it is well documented that H_2O_2 functions as a signalling molecule, facilitating a variety of defence responses that strengthen plants' ability to withstand various adverse environmental conditions (Petrov and Van Breusegem 2012), remarkably little is understood about the mechanisms whereby the plants interpret H_2O_2 and how the mechanism of sensing is synchronised within a plant's developmental programme. After several well-documented studies, it is comprehended that H_2O_2 is not just a free radical produced as a by-product of oxidative damage but also useful in the regulation of cellular equilibrium in agricultural plants and its beneficial effects on growth and development of the plants exposed to a variety of abiotic conditions. It is also an essential molecule that regulates senescence of leaves, generation of pollens, stomata movement, flower and fruit development and photosynthesis as well. Keeping in view of the dual role H_2O_2 played in plant system, crop plants have the chance to develop stress resistance under a myriad of abiotic stress circumstances by adjusting the balance between the production and scavenging of H_2O_2.

12.2 Generation and Scavenging of H_2O_2 in Plants

12.2.1 Production of H_2O_2

In plants H_2O_2 is produced as a by-product of oxidative metabolism (Mittler 2002). Plants can either produce H_2O_2 through enzymatic or non-enzymatic means. In plant cells, H_2O_2 is produced via a variety of mechanisms, including redox reactions, electron transport chains (ETC) and photorespiration. Various enzymes such as oxalate (Hu et al. 2003), cell wall peroxidases (Francoz et al. 2015), amine oxidases and flavin-containing enzymes have been shown to produce H_2O_2 in plants (Cona et al. 2006; Fig. 1). Additionally, NADPH oxidases, which produce superoxide which can then be reduced to H_2O_2 by superoxide dismutases (SOD), may also raise the amount of H_2O_2 (Grivennikova and Vinogradov 2013; Brewer et al. 2015). According to Remans et al. (2010), the main cause of ROS build-up, particularly H_2O_2 production, in plants exposed to heavy metals is the activation of NADPH oxidase. Furthermore, during salt or mannitol stress, proline accumulation in *Arabidopsis thaliana* may be greatly increased by H_2O_2 generated by NADPH oxidases (Ben Rejeb et al. 2015). A few additional oxidases, including glucose oxidases, glycolate oxidases and sulphite oxidases (Chang and Tang 2014; Brychkova et al. 2012), may also oxidise their own substrates to create H_2O_2 (Fig. 12.1). There are also a number of non-enzymatic processes that can generate H_2O_2. For instance, numerous processes related to both respiration and photosynthesis are in charge of producing H_2O_2. It is continuously produced by electron transport processes that take place in both chloroplasts and mitochondria. In a plant cell, the peroxisome is thought to be the location of photorespiration because it requires light for O_2 uptake and releases CO_2 along with the production of H_2O_2. According to Foyer and Noctor (2003), the photosynthetic carbon oxidation cycle entails the oxidisation of glycolate, which is thought to contribute to H_2O_2 production (Fig. 12.1). Photosynthetic activity, in plants, takes place in the chloroplast. The

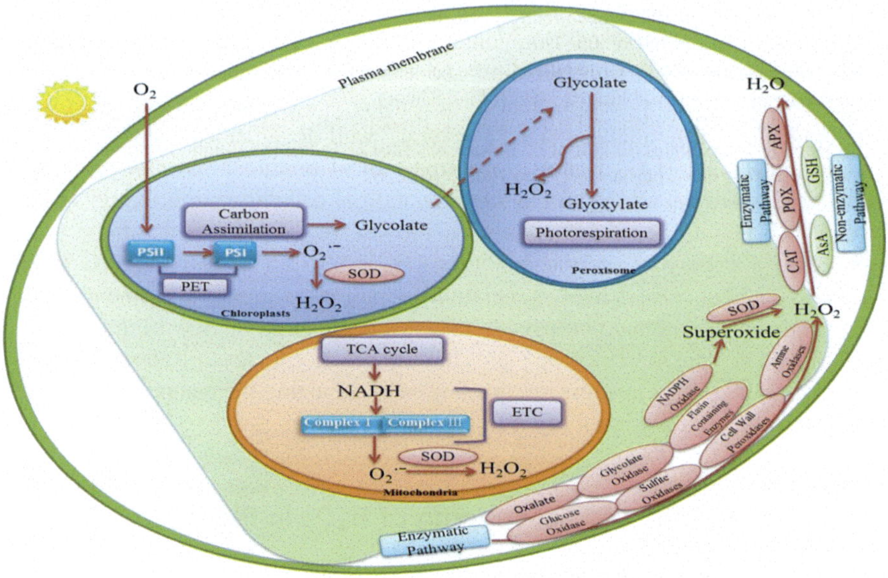

Fig. 12.1 Different pathways for H_2O_2 production and its scavenging in plants (adapted from Niu and Liao 2016)

essential locations for the generation of H_2O_2 throughout photosynthesis are chloroplasts. Oxygen reduction in chloroplasts is linked to H_2O_2 production (Fig. 12.1). Mehler (1951) found that when O_2 was reduced in chloroplasts with light present, H_2O_2 was produced. Additionally, photosynthetic electron transport (PET) chain elements like reduced thioredoxin (TRX), Fe-S centres, ferredoxin and reduced plastoquinone in the chloroplast can reduce O_2 to create H_2O_2 (Dat et al. 2000). The manganese-containing oxygen-generating complex that serves as the donor site of photosynthetic apparatus II can be used to identify non-enzymatic generation of H_2O_2 in chloroplast (Fig. 12.1). However, under physiological circumstances, this procedure is presumably overlooked in the majority of cases. Mitochondria is another important source for endogenous production of H_2O_2 (Dickinson and Chang 2011). When photosystem complexes I and III in the electron transport system produce oxygen during aerobic cellular respiration, the enzyme superoxide dismutase quickly transformed the oxygen to H_2O_2 (Fig. 12.1).

12.2.2 Scavenging of H_2O_2

Enzymatic and non-enzymatic H_2O_2 scavengers make up the antioxidant system that modulates H_2O_2 levels. Catalase (CAT; Willekens et al. 1997), peroxidase (POX; Fan and Huang 2012), ascorbate peroxidase (APX) and glutathione reductase (GR; Jahan and Anis 2014) are enzymes that scavenge H_2O_2. According to certain research, APX was present in the mitochondria, cytosol and chloroplasts (Begara-

Morales et al. 2013; Asada 2006; Navrot et al. 2007). In the meantime, CAT can break down H_2O_2 in peroxisomes (Nyathi and Baker 2006).

It is obvious that these enzymes are present in several organelles, where they may effectively reduce the H_2O_2 level and preserve membrane integrity. Non-enzymatic molecules including ascorbate (AsA) and glutathione (GSH) are continually involved in controlling the production of ROS (Kapoor et al. 2015). AsA, a crucial antioxidant for removing H_2O_2, can directly interact with the H_2O_2. A vital antioxidant called GSH quickly oxidises excessive H_2O_2 and could be linked to AsA regeneration. Consequently, GSH controls H_2O_2 levels and redox equilibrium in plant cells (Krifka et al. 2012). H_2O_2 homeostasis actually appears to have certain biological effects on plant cells that could serve as a signalling sign in the pathway of signal transduction.

12.3 H_2O_2 and Plant Growth and Development

H_2O_2 has an effect on plant growth according to its concentration (Xiong et al. 2015). Lower concentration (0–100 mM) of H_2O_2 increased cellular proliferation and root width when added to rice seedlings, while a higher quantity (100–500 mM) reduced root development (Xiong et al. 2015, Table 12.1). In addition, H_2O_2 promoted the growth of wheat roots (Hameed et al. 2004). Lower concentrations of H_2O_2 (0.1 or 0.5 mM) may be favourable in affecting plant growth when applied to the roots of *Solanum lycopersicum*. From these concentrations, 0.1 mM of H_2O_2 produced the greatest rise in the length of the root and shoot, dry mass and fresh mass (Nazir et al. 2019). Low levels of H_2O_2 in cucumber enhanced germination capacity, germination percentage, root size and hypocotyledonary axis, while high levels of H_2O_2 inhibited the same (Sun et al. 2009, Table 12.1). Improvements in plant height, shoot diameter and the dry masses of the shoot and roots were seen after treating pistachio seedlings with 1–10 mM of H_2O_2 (Bagheri et al. 2019, Table 12.1). Exogenous H_2O_2 treatment in cucumber dramatically accelerated root development and plant biomass (Sun et al. 2016; Li et al. 2016). Sweet potato seedlings' adventitious root growth and metabolic processes are significantly impacted by lower levels of H_2O_2 (Deng et al. 2012). In *Solanum lycopersicum*, root dipping treatment with H_2O_2 has been shown to increase photosynthetic pigments, rate of photosynthesis and its related properties (gs, Ci and E) (Nazir et al. 2019; Table 12.1). Likewise, elevated level of H_2O_2 causes chlorophyll breakdown in wheat genotypes (Nazir et al. 2019, Table 12.1). *Brassica juncea* plants benefited from the exogenous applications of H_2O_2 (25, 50 or 100 mM), with 50 mM being the optimal concentration, to improve photosynthesis in the ensuing plants (Khan et al. 2016, Table 12.1). H_2O_2 treatment of wheat seedlings enhances stomatal conductance, chlorophyll content and net photosynthetic rate (Ashfaque et al. 2014). The primary structures in plants that efficiently use CO_2 for photosynthesis and regulate water evaporation are the stomata surrounded by a few guard cells. This suggests that CO_2 comes from the environment, and water vapour is emitted into the environment from plants. As a result, maintaining photosynthesis and water-use efficiency (WUE) depends on

Table 12.1 Effects of exogenous application H_2O_2 on different parameters in plants (Nazir et al. 2020)

Physiological processes	H_2O_2-mediated effects (Dose-dependent manner)	Plant species	References
Seed germination	Lower concentration of H_2O_2 improved the germination capability and germination rate, where as its surplus levels suppressed the same	*Cucumis sativus*	Sun et al. (2009)
Root growth	Elevated level of 100–500 mM inhibited root elongation; however, 0–100 mM, a lower concentration enhanced cell expansion and root diameter	*Oryza sativa*	Xiong et al. (2015)
Photosynthesis	Root dipping treatment of 0.1, or 0.5 mM improves chlorophyll content, net photosynthesis rate and its related attributes (gs, Ci and E) Seed soaking treatment of H_2O_2 (0.01, 0.1 or 0.5 mM for 2, 4 or 8 h) improves the rate of photosynthesis, chlorophyll content and stomatal conductance	*Solanum lycopersicum* *Vigna radiata*	Nazir et al. (2019) Khan et al. (2018)
Stomatal movement	Root dipping treatment of 0.1 or 0.5 mM modulates opening of stomata. An increase in concentration of H_2O_2 decreased stomatal size and stomatal development	*Solanum lycopersicum* Wheat	Nazir et al. (2019) Kumari and Verma (2019)
Pollen development	H_2O_2 and other ROS play a significant function in pollen functionality and gametophyte integration H_2O_2 suppressed pollen germination and tube growth	Propagating plants *Paulownia tomentosa*	Cerny et al., (2018) Zhang and She (2012)
Flowering	Promotes reproductive development	*Litchi chinensis*	Zhou et al. (2014)
Fruit growth and development	Exogenous application of 5 mM triggered bigger fruit size, higher fruit set, fruit number and fruit biomass	Wax apple	Khandaker et al. (2012)

stomatal behaviour (Antunes et al. 2017). In *Solanum lycopersicum*, H_2O_2 increases stomatal conductance, which is thought to be influenced by the interaction of BR and H_2O_2. This interaction may have resulted in osmotic changes that subsequently decreased stomatal opening (Nazir et al. 2019, Table 12.1). Additionally, H_2O_2 administration significantly changed GSH content via regulating GSH output, which caused stomatal opening (Gondim et al. 2013). It has been established that the closure of stomata due to externally applied H_2O_2 occurred as a result of drop in K^+ concentration and an increase in K^+ discharge at the plasma membrane of guard cells via K^+ (An et al. 2016).

One crucial stage in the life cycle of seed plants that determines the efficacy of sexual reproduction is the pollen tube growth. H_2O_2 prevented germination of pollen and pollen tube formation in *Paulownia tomentosa* (Zhang and She 2012,

Table 12.1). Foliar application of H_2O_2 decreased pollen germination rate and pollen tube growth in *Paeonia suffruticosa* based on dose administered, whereas the inclusion of two H_2O_2 scavengers, i.e. ascorbic acid and catalase, encouraged pollen germination and pollen tube growth (He et al. 2006). Fruit growth could be enhanced by using low concentrations of H_2O_2. H_2O_2 inclusion in the fruit may aid in its maturity (Geros et al. 2012). Wax apple trees treated with 5 mM H_2O_2 exhibited enhanced fruit biomass, fruit size, greater fruit number, fruit set and yield as compared to the control plants (Khandaker et al. 2012, Table 12.1). In general, postharvest H_2O_2 treatments (0.5 or 15 mM) proved effective for preserving tomato fruits' quality under storage circumstances (Al-Saikhan and Shalaby 2019).

12.4 H_2O_2 Versus Abiotic Stress

Numerous abiotic stressors that are present all the time pose a threat to plant yield and growth. Even though extent of adaptability or sensitivity to particular stresses varies from one species to another, plants have evolved with sophisticated methods to sense external signals that enable them to respond adequately to climatic conditions. Under abiotic stressors, overproduction of reactive oxygen species is exacerbated which can damage plants or hinder their growth by causing oxidative damage to the macromolecules and structures of cell of plants. Of the various ROS, H_2O_2, freely diffusible and reasonably long-lived, plays a key role in the signalling pathways that deal with stress. Various approbatory responses that strengthen tolerance to different abiotic stresses can then be activated by these pathways. Recent research has shown that H_2O_2 priming can increase tolerance to abiotic stress by regulating various pathways and gene expression responsive to stress, altering ROS detoxification and increasing resistance to oxidative stress. When H_2O_2 is applied exogenously, it increases the plant's tolerance to environmental stresses such as drought, salinity, cold, extreme temperatures and heavy metals (Gong et al. 2001; Uchida et al. 2002; Azevedo-Neto et al. 2005; Chao et al. 2009; Liu et al. 2010; Wang et al. 2010b, 2014; Ishibashi et al. 2011; Gondim et al. 2012, 2013; Hossain and Fujita, 2013). In rice plants (*Oryza sativa*) grown with salt or heat stress, Uchida et al. (2002) investigated the effects of H_2O_2 and nitric oxide (NO) pre-treatments on oxidative stress. The findings demonstrated that under salinity or heat stress, seedlings exposed to modest doses (10 M) of H_2O_2 or NO had healthier leaves and had more photosynthesis rate than the untreated control plants. Furthermore, Azevedo-Neto et al. (2005) discovered that adding H_2O_2 to the nutritional solution caused maize plants to become more tolerant to salinity by improving antioxidant activity and lowering lipid peroxidation in root and leaves. When seeds were immersed in H_2O_2 (1–120 M, 8 h) and then grew in salty environments, Wahid et al. (2007) observed that exogenous H_2O_2 increased salinity tolerance in *Triticum aestivum* (150 mM NaCl). The seedlings grown with the seed treated with H_2O_2 under saline conditions recorded low level of H_2O_2 and improved photosynthetic capacity than the seedling grown as normal control. It also indicated that the antioxidant system of seedling treated with H_2O_2 responded strongly than the normal seedlings grown as normal control. According to several experts, oxidative stress is brought on by drought stress by rising levels of singlet oxygen and H_2O_2 (De Carvalho 2013). But Jing et al. (2009) looked into how H_2O_2 priming could increase drought resistance in cucumber plants. Cucumber plants under drought conditions have rounded chloroplasts, fluffy chloroplast membranes and thylakoids as well. While H_2O_2 priming had no effect on chloroplast ultrastructure, the activity of the antioxidant enzymes GPOX, SOD, DHAR, CAT, APX and GR and the levels of AsA and GSH were increased due to H_2O_2 priming, and, consequently, MDA, H_2O_2 and O2 levels were reduced. The researchers of this work concluded that H_2O_2 priming mitigated some of the membrane damage detected in the chloroplasts of plants under drought conditions by enhancing antioxidant capacity and reducing the formation of ROS in treated plants. In a related study, Ishibashi et al. (2011) showed that soybean water-deficit signs might be reduced by foliar application of H_2O_2. Drought-stressed leaves in plants sprayed with H_2O_2 had higher photosynthetic rate, RWC content and stomatal conductance than the leaves of normal control plants. By examining ROS scavenging and MG metabolism, Hossain and Fujita (2013) investigated the probable biochemical pathways of H_2O_2 priming-induced drought resistance in mustard (*Brassica juncea* L.) seedlings. With re-treatment, their levels of endogenous H_2O_2 were noticeably higher than those of seedlings with H_2O_2 pre-treatment. In rebuttal to drought stress, lower CAT, APX and Gly II activities were recorded, while GPX, DHAR and Gly I activities considerably increased. Endogenously applied H_2O_2 has been shown to modulate cold-temperature stress tolerance in a beneficial way. According to Prasad et al. (1994a, b), the addition of H_2O_2 altered the sensitivity to freezing as a result of a brief surge in H_2O_2-activated adaptation mechanisms. Iseri et al. (2013) examined whether endogenously applied H_2O_2 may affect tomato plants' short-term chilling response and lead to adaptation. Before transplanting the seedlings to the soil, pre-treatments were carried out by submerging the roots for 1 h in a 1 mM H_2O_2 solution. When compared to plants that had not been stressed, the RWC of the control and non-acclimated groups was significantly lower under cold stress (3 °C for 16 h). Under stress, H_2O_2 encouraged the preservation of a higher RWC. The levels of anthocyanin in the leaves of cold-stressed acclimated plants were substantially higher than those of control and non-acclimated plants that were not under stress. High MDA levels showed that non-acclimated and control plants both suffered from oxidative damage brought on by cold temperatures. It was shown that the H_2O_2 acclimation process shielded the cells from cold-induced lipid peroxidation by the MDA levels in acclimated plants remaining similar to those of unstressed plants. Exogenous pre-treatments have been reported to boost the heat resistance of plants, and endogenous levels of H_2O_2 are found to rise in heat challenged plants at similar rates to other abiotic stresses (Hossain et al. 2013a, b; Mostofa et al. 2014b). In cucumber and tomato seedlings, H_2O_2 pre-treatments boosted the levels of APX and glucose-6-phosphate dehydrogenase (G6PDH) and induced resistance to heat stress (Kang et al. 2009). Exogenous H_2O_2 has been shown to improve turfgrass species' ability to withstand heat stress (Wang et al. 2014). Under thermal stress, seedlings that had been pre-treated with H_2O_2 had reduced levels of oxidative damage and H_2O_2, as well as higher levels of GST, APX, GR and GPX activity. These findings suggested that H_2O_2 may enhance antioxidant defence mechanisms, which in turn improves thermostability in turfgrass species. Plants have been found to produce excessive amounts of ROS, particularly H_2O_2, in response to exposure to heavy metals (Hossain et al. 2010; Mostofa and Fujita 2013; Mostofa et al. 2014a). It has also been discovered that H_2O_2 priming increases the plant's resistance to heavy metals. Table 12.2 summarises the roles and inherent potential of H_2O_2-driven plant defence systems against a variety of HMs, such as Cd, Cu and Ni. *Vigna radiata* plants were more resistant to Cu toxicity when exogenous sources of H_2O_2 were used (Fariduddin et al. 2014). The yield and growth characteristics, as well as photosynthetic pigments, stomatal movement, leaf water potential, antioxidant system and osmoprotectants against 100 mg kg1 Cu stress, were all significantly improved by root dipping treatment with H_2O_2 (Nazir et al. 2019). H_2O_2 protected endogenous cellular components in *Zea mays*, reducing the effects of Cu stress by boosting the amount of osmotic solutes in leaves and improving the minerals in the plant (Guzel and Terzi 2013). By increasing the metabolism of antioxidant enzymes like SOD, CAT and POX, root dipping treatment with 0.1 mM of H_2O_2 in *Solanum lycopersicum* increased Ni resilience, revived root development and improved photosynthetic efficiency and stomatal movement and growth characteristics and a variety of physiochemical traits (Nazir et al. 2019). H_2O_2 treatment ameliorates Ni stress by boosting PS II activity through improved photosynthetic nitrogen-use efficiency (NUE), sulphur-use efficiency (SUE) and glutathione (GSH)-lowered outcome alongside lowered ion leakage and lipid peroxidation in mustard plants (Khan et al. 2016). Depending on the number of species of plants and H_2O_2 concentrations used, different plant species respond differently to Cd stress (Hasanuzzaman et al. 2017). H_2O_2 may, at low concentrations, result in the production of antioxidants and the absorption of osmoprotectants, which will mitigate the negative effects of Cd poisoning (Wu et al. 2015). Additionally, 0.1 mM H_2O_2 restored crop growth and activated MAPKs (MPK1/2) by increasing the activities of CAT, APX, NADH peroxidase, AsA and GSH (Zhou et al. 2014, Table 3).

H_2O_2 has been shown to increase metallothionein activity and reduce the oxidative stress caused by Cd and Cu in rice (Zhang et al. 2017). Externally given H_2O_2 reduced lipid peroxidation, decreased the extended uptake of aluminium, preserved the nutritional make-up and finally reduced the toxicity of aluminium in wheat (Hossain et al. 2005). By increasing chlorophyll content, thiol content and antioxidant enzyme activities, activating the metallothionein protein (BnMP1) and reducing lipid peroxidation, H_2O_2 application also lowered Cr uptake and provided resistance to Cr in *Brassica napus* (Yıldız et al. 2013).

Table 12.2 Heavy metal stress in plants and its amelioration by H_2O_2 (Nazir et al. 2020)

Heavy metal stress	Mode of H_2O_2 treatment	Plant species	Response	References
Cu toxicity	Seed priming 0.5 mM	*Zea mays* *Zea mays*	Improves the dry matter production and the mineral ion distribution Enhancing the activities of some enzymes such as D1-pyrroline-5-carboxylatesynthetase (P5CS), glutamate dehydrogenase (GDH) and arginase and ornithine aminotransferase (OAT), involved in proline metabolism	Guzel and Terzi (2013) Wen et al. (2013)
	Foliar spray 2.5 mM Root dipping 0.1 or 0.5 mM	*Vigna radiata* *Solanum lycopersicum*	Increased activities of CAT, POX and SOD Significant increase in growth and yield attributes and also leaf water potential, photosynthetic pigments, stomatal movement, antioxidant system and osmoprotectant	Fariduddin et al. (2014) Nazir et al. (2019)
Ni toxicity	Foliar spray 50 mM	*Brassica juncea*	Enhanced PS II activity through enhanced photosynthetic nitrogen-use efficiency (NUE), sulphur-use efficiency (SUE) and glutathione (GSH)-reduced output along with reduced lipid peroxidation and electrolyte leakage	Khan et al. (2016)
	Root dipping 0.1 mM	*Solanum lycopersicum*	Decreased Zn toxicity by upregulation of GR, GS, DHAR and MDHAR genes	Nazir et al. (2019)
Cd toxicity	0.1 mM 100 m Mol L^{-1} Exogenous 50 mM	*Oryza sativa* *Oryza sativa*	Regained crop development and subsequent activation of MPK1/2 by enhancing CAT, APX and NADH peroxidase activities and AsA and GSH content Increased plant growth attributes and cysteine (Cys), glutathione (GSH) and phytochelatin mM *Brassica napus* Increased plant growth attributes and stimulated the activities of Gly I and Gly II and content of cysteine (Cys),	Zhou et al. (2014) Wu et al. (2015) Hasanuzzaman et al. (2017)

(continued)

Table 12.2 (continued)

Heavy metal stress	Mode of H_2O_2 treatment	Plant species	Response	References
			glutathione (GSH) and phytochelatin	
Al toxicity	Foliar spray 50 mM	*Triticum aestivum*	Stimulated AsA and GSH levels, as well as the ratios of AsA/(AsA+DHA) and GSH/GSSG	Xu et al. (2008)
Cr toxicity	Seed soaking 20 mM	*Brassica napus*	Improved thiol content, chlorophyll content, antioxidant enzyme activities, activation of metallothionein protein (BnMP1) mRNA and decreasing lipid peroxidation	Yıldız et al. (2013)
Pb toxicity	Foliar spray 5–100 mM	*Oryza sativa*	Increased the activities of CAT, SOD and APX	Jing et al. (2009)
B toxicity	Seed soaking 0.1–0.5 mM	*Cajanus cajan*	Regulate mineral nutrition, morpho-physiological parameters and ascorbate (AsA) and GSH metabolism	Chawla et al. (2010)

12.5 Conclusion and Future Prospects

Hydrogen peroxide (H_2O_2) plays a vital role in enhancing a variety of functions in plants and is a highly prospective chemical, but it is still difficult to come by. Furthermore, it is becoming increasingly clear that H_2O_2 functions in plants as a ubiquitous second messenger. A deeper understanding of how H_2O_2 significantly counteracts abiotic stress-induced alterations in plant growth and developmental processes including fruit growth and development, photosynthesis, pollen development and stomatal behaviour has been explored. The negative impacts of environmental stresses are mitigated by H_2O_2. Despite significant progress in recent years that has contributed to a basic understanding of how H_2O_2 influences physiology, many aspects of its signalling system remain unknown. Not only are these results fundamentally important, but they are also realistically important.

Plants have pathways that enable them to use ROS, notably H_2O_2, for signalling reasons that give approbatory stress resistance via modulating osmotic adjustment, ROS detoxification and photosynthetic carbon fixation. These processes are especially useful in stressful environmental conditions. Several investigations have hypothesised that H_2O_2 pre-priming of seeds or seedlings causes an inferential signal that aids in protecting plants from abiotic stresses by restoring cellular homeostasis; minimising oxidative stress to membranes, lipids and proteins; and attenuating stress signalling pathways; however, underlying mechanisms are not adequately understood. Future study on H_2O_2 priming should be productive and aid plant scientists in understanding the molecular underpinnings of abiotic stress resistance and advancing sustainable farming that is more environmentally friendly. Future research should focus on the specific roles that H_2O_2 signalling plays in the conflict between crops and ecological variables.

References

Al-Saikhan MS, Shalaby TA (2019) Effect of hydrogen peroxide (H_2O_2) treatment on physico-chemical characteristics of tomato fruits during post-harvest storage. Aust J Crop Sci 13:798–802

An Y, Feng X, Liu L, Xiong L, Wang L (2016) ALA-induced flavonols accumulation in guard cells is involved in scavenging H_2O_2 and inhibiting stomatal closure in Arabidopsis cotyledons. Front Plant Sci 7:1713

Antunes WC, de Menezes DD, Pinheiro DP, Williams TCR, Loureiro ME (2017) Guard cell-specific down-regulation of the sucrose transporter SUT1 leads to improved water use efficiency and reveals the interplay between carbohydrate metabolism and Kþ accumulation in the regulation of stomatal opening. Environ Exp Bot 135:73–85

Asada K (2006) Production and scavenging of reactive oxygen species in chloroplasts and their functions. Plant Physiol 141:391–396

Ashfaque F, Khan MI, Khan NA (2014) Exogenously applied H_2O_2 promotes proline accumulation. Water relations. Photosynthetic efficiency and growth of wheat (Triticum aestivum L.) under salt stress. Annu Res Rev Biol 105c:20

Azevedo-Neto AD, Prisco JT, Eneas-Filho J, Medeiros JVR, Filho E (2005) Hydrogen peroxide pre-treatment induces salt stress acclimation in maize plants. J Plant Physiol 162:1114–1122

Bagheri M, Gholami M, Baninasab B (2019) Hydrogen peroxide-induced salt tolerance in relation to antioxidant systems in pistachio seedlings. Sci Hort 243:207e213

Banerjee A, Roychoudhury A (2018) Abiotic stress, generation of reactive oxygen species, and their consequences: An overview. In: Singh VP, Singh S, Tripathi DK, Prasad SM, Chauhan DK (eds) Reactive oxygen species in plants: boon or bane-revisiting the role of ROS, 1st edn. Wiley, pp 23–50

Begara-Morales JC, Sánchez-Calvo B, Chaki M, Valderrama R, Mata-Pérez C, López-Jaramillo J, Padilla MN, Carreras A, Corpas FJ, Barroso BJ (2013) Dual regulation of cytosolic ascorbate peroxidase (APX)by tyrosine nitration and S-nitrosylation. J Exp Bot 65:527–538

Ben Rejeb K, Lefebvre-De Vos D, Le Disquet I, Leprince AS, Bordenave M, Maldiney R, Jdey A, Abdelly C, Savouré A (2015) Hydrogen peroxide produced by NADPH oxidases increases proline accumulation during salt or mannitols tress in Arabidopsis thaliana. New Phytol 208: 1138–1148

Bose J, Rodrigo-Moreno A, Shabala S (2014) ROS homeostasis in halophytes in the context of salinity stress tolerance. J Exp Bot 65(5):1241–1257

Brewer TF, Garcia FJ, Onak CS, Carroll KS, Chang CJ (2015) Chemical approaches to discovery and study of source sand targets of hydrogen peroxide redox signalling through NAPDH oxidase proteins. Annu Rev Biochem 84:765–790

Brychkova G, Yarmolinsky D, Fluhr R, Sagi M (2012) The determination of sulphite levels and its oxidation in plant leaves. Plant Sci 190:123–130

Cerny M, Habanova H, Berka M, Luklova M, Brzobohaty B (2018) Hydrogen peroxide, its role in plant biology and crosstalk with signalling networks. Int J Mol Sci 19:2812

Chang Q, Tang H (2014) Optical determination of glucose and hydrogen peroxide using a nanocomposite prepared from glucose oxidase and magnetite nanoparticles immobilized on graphene oxide. Microchim Acta 181:527–534

Chao Y-Y, Hsu YT, Kao CH (2009) Involvement of glutathione in heat shock-and hydrogen peroxide-induced cadmium tolerance of rice (Oryza sativa L.) seedlings. Plant Soil 318:37–45

Chawla S, Goyal SC, Angrish R, Rani C, Arora V, Datta KS, Madaan S, Devi S (2010) Acclimatory response to hydrogen peroxide and glutathione under salt boron stress through their impact on mineral nutrition and antioxidant defense system in pigeonpea (*Cajanus cajan* L.). Physiol Mol Biol Plants 16:295e304

Cona A, Rea G, Botta M, Corelli F, Federico R, Angelini R (2006) Flavin-containing polyamine oxidase is a hydrogen peroxide source in the oxidative response to the protein phosphatase inhibitor cantharidin in Zea mays L. J Exp Bot 57:2277–2289

Das K, Roychoudhury A (2014) Reactive oxygen species (ROS) and response of antioxidants as ROS-scavengers during environmental stress in plants. Front Environ Sci 2:53

Dat J, Vandenabeele S, Vranová E, Van Montagu M, Inzé D, Van Breusegem F (2000) Dual action of the active oxygen species during plant stress responses. Cell Mol Life Sci 57:779–795

De Carvalho MHC (2013) Drought stress and reactive oxygen species. Production, scavenging and signalling. Plant Signal Behav 3:156–165

Deng XP, Cheng YJ, Wu XB, Kwak SS, Chen W, Eneji AE (2012) Exogenous hydrogen peroxide positively influences root growth and exogenous hydrogen peroxide positively influences root growth and metabolism in leaves of sweet potato seedlings. Aust J Crop Sci 6:1572

Dickinson BC, Chang CJ (2011) Chemistry and bbiology of reactive oxygen species in signalling or stress responses. Nat Chem Biol 7:504–511

Dietz KJ, Mittler R, Noctor G (2016) Recent progress in understanding the role of reactive oxygen species in plant cell signaling. Plant Physiol 171:1535–1539

Dikilitas M, Karakas S, Ahmad P (2018) Predisposition of crop plants to stress is directly related to their DNA health. In: Egamberdieva D, Ahmad P (eds) Plant microbiome: stress response. Springer, Singapore, pp 233–254

Fan Y, Huang Y (2012) The effective peroxidase-like activity of chitosan-functionalized CoFe2O4 nanoparticles for chemiluminescence sensing of hydrogen peroxide and glucose. Analyst 137(5):1225–1231

Fariduddin Q, Khan TA, Yusuf M (2014) Hydrogen peroxide mediated tolerance to copper stress in the presence of 28-homobrassinolide in Vigna radiata. Acta Physiol Plant 36(10):2767–2778

Foyer CH, Noctor G (2003) Redox sensing and signalling associated with reactive oxygen in chloroplasts, peroxisomes and mitochondria. Physiol Plant 119: 355–364

Francoz E, Ranocha P, Nguyen-Kim H, Jamet E, Burlat V, Dunand C (2015) Roles of cell wall peroxidases in plant development. Phytochemistry 112:15–21

Geros H, Chaves M, Delrot S (2012) The biochemistry of the grape fruit: USA: Bentham Science Publishers. ISBN 9781608053605, 304

Gondim FA, Gomes-Filho E, Costa JH, Mendes Alencar NLM, Prisco JT (2012) Catalase plays a key role in salt stress acclimation induced by hydrogen peroxide pretreatment in maize. Plant Physiol Biochem 56:62–71

Gondim FA, Miranda RS, Gomes-Filho E, Prisco JT (2013) Enhanced salt tolerance in maize plants induced by H_2O_2 leaf spraying is associated with improved gas exchange rather than with non-enzymatic antioxidant system. Theor Exp Plant Physiol 25:251–260

Gong M, Chen B, Li ZG, Guo LH (2001) Heat-shock-induced cross adaptation to heat, chilling, drought and salt stress in maize seedlings and involvement of H_2O_2. J Plant Physiol 158:112–1130-00327

Grivennikova VG, Vinogradov AD (2013) Partitioning of superoxide and hydrogen peroxide production by mitochondrial respiratory complex I. Biochim Biophys Acta 1827:446–454

Guzel S, Terzi R (2013) Exogenous hydrogen peroxide increases dry matter production, mineral content and level of osmotic solutes in young maize leaves and alleviates deleterious effects of copper stress. Bot Stud 54:26

Hameed AM, Farooq SH, Iqbal NA, Arshad RU (2004) Influence of exogenous application of hydrogen peroxide on root and seedling growth on wheat (Triticum aestivum L.). Int J Agric Biol 6:366.e369

Hasanuzzaman M, Nahar K, Gill SS, Alharby HF, Razafindrabe BHN, Fujita M (2017) Hydrogen peroxide pretreatment mitigates cadmium-induced oxidative stress in Brassica napus L., an intrinsic study on antioxidant defense and glyoxalase systems Front Plant Sci 8:8.e115

He JM, Liu ZH, Xu H, She XP, Huang C (2006) The involvement of hydrogen peroxide in UV-B-inhibited pollen germination and tube growth of Paeonia suffruticosa and Paulownia tomentosa in vitro. Plant Growth Regul 49:199–208

Hossain MA, Fujita M (2013) Hydrogen peroxide priming stimulates drought tolerance in mustard (Brassica juncea L.). Plant Gene & Trait 4:109–123

Hossain MA, Hossain AZ, Kihara T, Koyama H, Hara T (2005) Aluminum induced lipid peroxidation and lignin deposition are associated with an increase in H_2O_2 generation in wheat seedlings. Soil Sci Plant Nutr 51:223.e230

Hossain MA, Hasanuzzaman M, Fujita M (2010) Up-regulation of antioxidant and glyoxalase systems by exogenous glycine betaine and proline in mungbean confer tolerance to cadmium stress. Physiol Mol Biol Plants 16:259–272

Hossain MA, Mostofa MG, Fujita M (2013a) Cross protection by cold-shock to salinity and drought stress-induced oxidative stress in mustard (Brassica campestris L.) seedlings. Mol Plant Breed 4: 50–70

Hossain MA, Mostofa MG, Fujita M (2013b) Heat-shock positively modulates oxidative protection of salt and drought-stressed mustard (Brassica campestris L.) seedlings. J Plant Sci Mol Breed 2: 1–14

Hu X, Bidney DL, Yalpani N, Duvick JP, Crasta O, Folkerts O, Lu G (2003) Overexpression of a gene encoding hydrogen peroxide-generating oxalate oxidase evokes defense responses in sunflower. Plant Physiol 133:170–181

Iseri OD, Körpe DA, Sahin FI, Haberal M (2013) Hydrogen peroxide pretreatment of roots enhanced oxidative stress response of tomato under cold stress. Acta Physiol Plant 35:1905–1913-1228-7

Ishibashi Y, Yamaguchi H, Yuasa T, Iwaya-Inoue M, Arima S, Zheng SH (2011) Hydrogen peroxide spraying alleviates drought stress in soybean plants. J Plant Physiol 168:1562–1567

Jahan AA, Anis M (2014) Changes in antioxidative enzymatic responses during acclimatization of in vitro raised plantlets of Cardiospermum halicacabum L. against oxidative stress. J Plant Physiol Pathol 4:2

Jing LZ, Kui GY, Hang LS, Gang BJ (2009) Effects of exogenous hydrogen peroxide on ultrastructure of chloroplasts and activities of antioxidant enzymes in greenhouse-ecotype cucumber under drought stress. Acta Hortic Sin 36:1140–1146

Kang NJ, Kang YI, Kang KH, Jeong BR (2009) Induction of thermotolerance and activation of antioxidant enzymes in H_2O_2 pre-applied leaves of cucumber and tomato seedlings. J Jpn Soc Hortic Sci 78:320–329

Kapoor D, Sharma R, Handa N, Kaur H, Rattan A, Yadav P, Gautam V, Kaur R, Bhardwaj R (2015) Redox homeostasis in plants under abiotic stress: role of electron carriers, energy metabolism mediators and proteinaceous thiols. Front Environ Sci 3:13

Khan MI, Khan NA, Masood A, Per TS, Asgher M (2016) Hydrogen peroxide alleviates nickel-inhibited photosynthetic responses through increase in use efficiency of nitrogen and sulfur, and glutathione production in mustard. Front Plant Sci 7:44

Khan TA, Yusuf M, Fariduddin Q (2018) Hydrogen peroxide in regulation of plant metabolism: Signalling and its effect under abiotic stress. Photosynthetica 56:1237–1248

Khandaker MM, Boyce AN, Osman N (2012) The influence of hydrogen peroxide on the growth, development and quality of wax apple (Syzygium samarangense, [Blume] Merrill & L.M. Perry var. jambu madu) fruits [Syzygium samarangense [Blume] Merrill & LM Perry var. jambu madu]. Plant Physiol Biochem 53:101–110

Kocyigit A, Guler EM, Dikilitas M (2017) Role of antioxidant phytochemicals in prevention, formation and treatment of cancer. In: Cristiana F, Elena A (eds) Reactive oxygen species (ROS) in living cells. Intech Open, pp 21–45

Krifka S, Hiller KA, Spagnuolo G, Jewett A, Schmalz G, Schweikl H (2012) The influence of glutathione on redox regulation by antioxidant proteins and apoptosis in macrophages exposed to 2-hydroxyethyl methacrylate (HEMA). Biomaterials 33:5177–5186

Kumari S, Verma VK (2019) Hydrogen peroxide mediates suberization, root thickness and stomatal movement in wheats. Int J Curr Microbiol App Sci 8:2448–2454

Li XP, Xu QQ, Liao WB, Ma ZJ, Xu XT, Wang M, Ren PJ, Niu LJ, Jin X, Zhu YC (2016) Hydrogen peroxide is involved in abscisic acid-induced adventitious rooting in cucumber (Cucumis sativus L.) under drought stress. J Plant Biol 59:536–548

Liu ZJ, Guo YK, Bai JG (2010) Exogenous hydrogen peroxide changes antioxidant enzyme activity and protects ultrastructure in leaves of two cucumber ecotypes under osmotic stress. J Plant Growth Regul 29(2):171–183

Martinez V, Nieves-Cordones M, Lopez-Delacalle M, Rodenas R, Mestre TC, Garcia-Sanchez F, Rubio F, Nortes PA, Mittler R, Rivero RM (2018) Tolerance to stress combination in tomato plants: new insights in the protective role of melatonin. Molecules 23(3):535

Mehler AH (1951) Studies on reactions of illuminated chloroplasts. II. Stimulation and inhibition of the reaction with molecular oxygen. Arch Biochem Biophys 33:339–351 90082–3, 65

Mittler R (2002) Oxidative stress, antioxidants and stress tolerance. Trends Plant Sci 7:405–410

Mostofa MG, Fujita M (2013) Salicylic acid alleviates copper toxicity in rice (Oryza sativa L.) seedlings by upregulating antioxidative and glyoxalase systems. Ecotoxicology 22:959–973. https://doi.org/10.1007/s10646-013-1073-x

Mostofa MG, Seraj ZI, Fujita M (2014a) Exogenous sodium nitroprusside and glutathione alleviate copper toxicity by reducing copper uptake and oxidative damage in rice (Oryza sativa L.) seedlings. Protoplasma 251:1373–1386

Mostofa MG, Yoshida N, Fujita M (2014b) Spermidine pretreatment enhances heat tolerance in rice seedlings through modulating antioxidative and glyoxalase systems. Plant Growth Regul 73:31–44

Munne-Bosch S, Pinto-Marijuan M (2017) Free radicals, oxidative stress and antioxidants. Encycl Appl Plant Sci 2:16–19

Navrot N, Rouhier N, Gelhaye E, Jacquot JP (2007) Reactive oxygen species generation and antioxidant systems in plant mitochondria. Physiol Plant 129:185–195

Nazir F, Hussain A, Fariduddin Q (2019) Hydrogen peroxide modulate photosynthesis and antioxidant systems in tomato (Solanum lycopersicum L.) plants under copper stress. Chemosphere 230:544–558

Nazir F, Fariduddin Q, Khan TA (2020) Hydrogen peroxide as a signalling molecule in plants and its crosstalk with other plant growth regulators under heavy metal stress. Chemosphere 252: 126486

Niu L, Liao W (2016) Hydrogen peroxide signaling in plant development and abiotic responses: crosstalk with nitric oxide and calcium. Front Plant Sci 7:230

Noctor G, Reichheld JP, Foyer CH (2018) ROS-related redox regulation and signaling in plants. Semin Cell Dev Biol 80:3–12. https://doi.org/10.1016/j.semcdb.2017.07.013

Nyathi Y, Baker A (2006) Plant peroxisomes as a source of signalling molecules. Biochim Biophys Acta 1763:1478–1495

Petrov VD, Van Breusegem F (2012) Hydrogen peroxide-a central hub for information flow in plant cells. AoB Plants 2012:pls014

Prasad TK, Anderson MD, Martin BA, Stewart CR (1994a) Evidence for chilling-induced oxidative stress in maize seedlings and a regulatory role for hydrogen-peroxide. Plant Cell 6:65–74

Prasad TK, Anderson MD, Stewart CR (1994b) Acclimation, hydrogen peroxide, and abscisic-acid protect mitochondria against irreversible chilling injury in maize seedlings. Plant Physiol 10: 619–627

Rani V, Deep G, Singh RK, Palle K, Yadav UC (2016) Oxidative stress and metabolic disorders: pathogenesis and therapeutic strategies. Life Sci 148:183–193

Remans T, Opdenakker K, Smeets K, Mathijsen D, Vangronsveld J, Cuypers A (2010) Metal-specific and NADPH oxidase dependent changes in lipoxygenase and NADPH oxidase gene expression in Arabidopsis thaliana exposed to cadmium or excess copper. Funct Plant Biol 37: 532–544

Sun Y, Zhang C, Qi A, Luo W (2009) Effects of hydrogen peroxide on the germination characteristics of cucumber seeds. In: International conference on environmental science and information application technology, pp 272.e275

Sun Y, Wang H, Liu S, Peng X (2016) Exogenous application of hydrogen peroxide alleviates drought stress in cucumber seedlings. S Afr J Bot 106:23–28

Uchida A, Jagendorf AT, Hibino T, Takabe T, Takabe T (2002) Effects of hydrogen peroxide and nitric oxide on both salt and heat stress tolerance in rice. Plant Sci 163:515–523. https://doi.org/10.1016/S0168-9452(02)00159-0

Wahid A, Perveen M, Gelani S, Basra SMA (2007) Pretreatment of seed with H_2O_2 improves salt tolerance of wheat seedlings by alleviation of oxidative damage and expression of stress proteins. J Plant Physiol 164:283–294

Wang P, Du Y, Li Y, Ren D, Song CP (2010a) Hydrogen peroxide mediated activation of MAP kinase 6 modulates nitric oxide biosynthesis and signal transduction in Arabidopsis. Plant Cell 22:2981–2998

Wang Y, Li J, Wang J, Li Z (2010b) Exogenous H_2O_2 improves the chilling tolerance of manila grass and Mascarene grass by activating the antioxidative system. Plant Growth Regul 61:195–204

Wang Y, Zhang J, Li JL, Ma XR (2014) Exogenous hydrogen peroxide enhanced the thermotolerance of Festuca arundinacea and Lolium perenne by increasing the antioxidative capacity. Acta Physiol Plant 36:2915–2924

Wen JF, Gong M, Liu Y, Hu JL, Deng MH (2013) Effect of hydrogen peroxide on growth and activity of some enzymes involved in proline metabolism of sweet corn seedlings under copper stress. Sci Hortic (Canterb) 164:366e371

Willekens H, Chamnongpol S, Davey M, Schraudner M, Langebartels C, Van Montagu M, Inzé D, Van Camp W (1997) Catalase is a sink for H_2O_2 and is indispensable for stress defence inC3 plants. EMBO J 16:4806–4816

Wu Z, Zhang C, Yan J, Yue Q, Ge Y (2015) Effects of sulfur supply and hydrogen peroxide pretreatment on the responses by rice under cadmium stress. Plant Growth Regul 77:299–306

Xiong J, Yang Y, Fu G, Tao L (2015) Novel roles of hydrogen peroxide (H_2O_2) in regulating pectin synthesis and dimethyl esterification in the cell wall of rice (Oryza sativa) root tips. New Phytol 206:118.e126

Xu PL, Guo YK, Bai JG, Shang L, Wang XJ (2008) Effects of long-term chilling on ultrastructure and antioxidant activity in leaves of two cucumber cultivar sunder lowlight. Physiol Plant 132: 467–478

Xu FJ, Jin CW, Liu WJ (2010) Pretreatment with H_2O_2 alleviates aluminum-induced oxidative stress in wheat seedlings. Plant Biol 54:44–53

Yıldız M, Terzi H, Bingül N (2013) Protective role of hydrogen peroxide pretreatment on defense systems and BnMP1 gene expression in Cr(VI)-stressed canola seedlings. Ecotoxicology 22: 1303–1312

You J, Chan Z (2015) ROS regulation during abiotic stress responses in crop plants. Front Plant Sci 6:1092

Zhang YH, She XP (2012) The involvement of nitric oxide and hydrogen peroxide in eATP-inhibited pollen germination and tube growth of Paulownia tomentosa in vitro. Adv Mater Res 343:347.e350

Zhang H, Lv S, Xu H, Hou D, Li Y, Wang F (2017) H_2O_2 is involved in the metallothionein-mediated rice tolerance to copper and cadmium toxicity. Int J Mol Sci 18:2083

Zhou J, Xia XJ, Zhou YH, Shi K, Chen Z, Yu JQ (2014) OH1-dependent H_2O_2 production and subsequent activation of MPK1/2 play an important role in acclimation-induced cross-tolerance in tomato. J Exp Bot 65:595.e607

Toward Sustainable Agriculture: Strategies Involving Phytoprotectants Against Reactive Oxygen Species

13

Priyanka Devi, Shipa Rani Dey, Lalit Saini, Prasann Kumar, Sonam Panigrahi, and Padmanabh Dwivedi

Abstract

A primary goal of this book chapter is to summarize knowledge on the effects of phytoprotectants of microbial origin (e.g., root growth-promoting rhizobacteria, root growth-promoting mycorrhiza fungi), phytohormones, osmoprotectants, antioxidants, as well as the metabolic compounds like melatonin which are related to reactive oxygen species. It is well-known that increasing environmental stresses have a detrimental effect on a plant's health as well as its productivity. The Food and Agriculture Organization (FAO) has in a recent report outlined the important task of devising strategies to deal with the impact of climate change. Increasing environmental stress and altered weather patterns can be considered as evidence of the effects of climate change on crop production. Among these serious issues are soil salinity, flooding, droughts, pollution, metal and metalloid toxicity, and extreme temperature changes. This is due to the fact that stress factors are more commonly present in plants because of the variability of these conditions combined with plant immobility. It is crucial to increase the adaptability of crop plants to these stresses in order to meet the increased food needs of the population. There are a number of mechanisms that can be used to improve plant tolerance to stress using exogenous phytoprotectants as a means of reducing these stresses. In order to enhance the resistance of plants to these stresses, a wide range

P. Devi · S. R. Dey · L. Saini · P. Kumar (✉)
Department of Agronomy, School of Agriculture, Lovely Professional University, Phagwara, India
e-mail: prasann.21784@lpu.co.in

S. Panigrahi
School of Life Sciences, Sambalpur University, Sambalpur, India

P. Dwivedi (✉)
Department of Plant Physiology, Institute of Agricultural Sciences, Banaras Hindu University, Varanasi, India

M. Faizan et al. (eds.), *Reactive Oxygen Species*,
https://doi.org/10.1007/978-981-19-9794-5_13

of phytoprotectants have been proved to be highly effective. It is the purpose of this chapter to discuss in detail a number of phytoprotectants including osmoprotectants, phytohormones, antioxidants, and nitric oxide.

Keywords

Agriculture · Biotic · Crops · Environment · Phytoprotectants · Salinity · Reactive oxygen species · Risk

Abbreviations

ADP	Adenosine diphosphate
ABA	Abscisic acid
AMF	Arbuscular mycorrhiza fungi
AsA	Ascorbate
Bt	*Bacillus thuringiensis*
Ca^{2+}	Calcium
CAT	Catalase
CO_2	Carbon dioxide
FAO	Food and Agriculture Organization
DHAR	Dehydroascorbate reductase
DNA	Deoxyribonucleic acid
GABA	Gamma-aminobutyric acid
GSH	Glutathione
GST	Glutathione S-transferases
GPX	Glutathione peroxidase
H_2O_2	Hydrogen peroxide
K^+	Potassium
MDHAR	Monodehydroascorbate reductase
NADPH	Nicotinamide adenine dinucleotide phosphate
Na^+	Sodium
NO	Nitric oxide
PA	Polyamines
PGPR	Plant growth-promoting rhizobacteria
PPPs	Pentose phosphate pathway
ROS	Reactive oxygen species
SOD	Superoxide dismutase
TFs	Transcription factors

13.1 Introduction

It is, unfortunately, the case that plants are increasingly at risk of being negatively affected by environmental stress in the modern era. In a recent report on climate change and agriculture, the FAO outlined the importance of devising strategies to combat the effects of the changes on agriculture (Dietz et al. 2016). Whenever environmental stresses increase, they affect plant development and threaten the sustainability of agriculture. Among them are waterlogging, droughts, salinity, high and low temperatures, as well as increased CO_2 levels. In recent years, environmental stresses have become a major issue. This is due to the concerns over the impacts of climate change on the world's plant resources, genetic diversity, and food safety. Scientific research has proven the fact that plants respond differently to multiple stress factors, as opposed to single stress factors. The mechanism of cellular, molecular, and genetic adjustments involved in these responses is highly intricate and involves alterations at many levels. Plants will be exposed to the negative effects of climate change, and this will have an effect on the sustainability of agriculture. Therefore, there is an urgent need to conduct further research into the developmental response of plants to these stress factors.

Plants that have long-term growth periods, such as perennial crops and woody plants, are more vulnerable to environmental stress than plants with short-term growth periods. Light stress, which is the result of excessive light, decreases the capacity of plants to photosynthesize and increases the light stresses, which in turn causes defects in plant growth and biomass production. The mechanisms which plants use to adjust their growth based on changing conditions outside the plant have evolved complex ways over time in many species. It is imperative that we understand the molecular mechanisms which determine the crop's water-use efficiency and crop quality as the ratio of biomass to water if we are to improve plants under drought stress conditions. As a result there are cellular mechanisms which contribute to the adaptive and developmental responses of plants to stress. A variety of environmental stressors are encountered by plants throughout their lives.

The presence of reactive oxygen species (ROS) is crucial to maintaining the health of plants as well as enhancing their tolerance to stresses. It is aimed at presenting an overview of recent progress in understanding ROS's role in apical meristem development, organogenesis, and response to abiotic stress in this review. This is accompanied by some novel findings from recent years. In addition, it is relevant to discuss the interaction between ROS and epigenetic modifications that play a role in regulating gene expression. Nevertheless, it is well-known that a coordinated growth of plant tissue and organs improves crop yield and productivity. In order to regulate plant morphogenesis, both genetic factors and environmental factors play an important role.

ROS can be found in several different cellular compartments as a byproduct of aerobic metabolism of plants, including chloroplasts (Dietz et al. 2016), mitochondria (Huang et al. 2013), and peroxisomes. It is imperative to acknowledge that ROS plays a crucial role in the regulation of normal plant growth and their response to stress. They are not just responsible for irreversible DNA damage and cell death. Miller et al. (2010) have found that ROS perform both pro- and antioxidative functions in the body, based on their reactivity, content, and potential for gliding across biological membranes.

13.2 Oxidative Stress Mechanism Is Affected by Phytoprotectants (ROS Production)

As plants find themselves in environments that are abundant in molecular oxygen, their cells are constantly exposed to various conditions such as environmental stresses and ultraviolet radiation. These conditions are triggered when some toxic chemical agents, such as reactive oxygen species, accumulate within them. Furthermore, in addition to oxygen radicals, there are superoxide anions and hydrogen peroxide (H_2O_2), which produce excess hydroxyl radicals, which cause proteins to be destroyed, deactivating enzymes and affecting gene functions that ultimately result in cell death (Choudhury et al. 2013; Mittler et al. 2004). It is thought that the ROS network genes, which contain almost 152 genes, act as chemical signals that help plants mitigate the effects of abiotic stress. The responses to ROS have evolved over time as plants have adapted to use them as chemical signals. As a result of the ROS network genes, which constitute almost 152 genes, plants are able to mitigate the effects of abiotic stress through a chemical signaling mechanism. Throughout the evolution process, plants have developed mechanisms to use ROS as chemical signals. Consequently, Choudhury et al. (2013) found that at any given moment, an active balance between ROS-producing pathways and ROS-scavenging pathways is conducive to cellular health. The plants are capable of expressing unique enzymes that scavenge and produce ROS in response to multiple abiotic stresses. Changes in the lipid profile have been observed as a result of altered levels of byproducts of lipid peroxidation; a rise in peroxidases, glutathione S-transferase, and catalase; and an accumulation of phytoprotectants which act as antioxidants, including ascorbate, phenolic compounds, carotenoids, alkaloids, sucrose, and trehalose (Choudhury et al. 2017). Soluble sugars, which are defined as mono- and disaccharides, play an essential role in phytoprotectant under abiotic stress conditions since they act as mediators of ROS generation like in mitochondrial respiration and antioxidants in oxidative pentose phosphate pathways (PPPs). Phytoprotectants like soluble sugars, which are involved in the production of ROS, can affect the oxidative mechanisms that damage plants (Couee et al. 2006). The sources of ROS in plants are varied, such as the NADPH oxidase in the cell membrane, the electron transport chain of the chloroplasts and mitochondria, the oxygenation of fatty acids, and the glycolate oxidase stage of photorespiration in peroxisomes and mitochondria, respectively (Doudican et al. 2005; Møller 2001). When plants increase their photosynthetic activity, soluble sugars lead to an increase in ROS production. A reduction in soluble sugars, on the other hand, suppresses the expression of photosynthetic genes, including those involved in metabolic processes such as the Calvin cycle, even under ordinary daylight conditions. It is now easier for us to understand the origin of this kind of simultaneous control of gene expression by soluble sugar and light since a link was found between light and sugar accumulation. The problem appears to be compounded under abiotic stress conditions such as chilling stress, where it is believed that sugar buildup acts as a cold protector (Ciereszko et al. 2001; Havaux and Kloppstech 2001). In the same way, the fluctuating glucose levels or carbohydrate deprivation that occurs during different developmental stages may stimulate the generation of ROS. There is also a hypothesis that starvation triggers lipid mobilization and alterations to the electron flow generation in peroxisomes due to the reduced ADP regeneration and the reduction in electron transport flow. As a result of this, acyl-CoA oxidase, protein synthesis, and mRNA synthesis will be increased. In a transcriptomic study, it is confirmed that sugar stress stimulates oxidative enzymes, including catalase, which results in ROS activation (Contento et al. 2004). As well, the involvement of soluble sugars and the interactions between several phytohormones (auxin, brassinosteroids, and ABA) have been reported in ROS generation. An important coupling between auxin signaling and oxidative stress is that auxins can elicit the production of reactive oxygen species (ROS) and regulate their equilibrium. It could be that auxins are responsible for triggering a Rho GTPase (RAC/ROP), which interacts with NADPH oxidases to produce apoplastic ROS (Duan et al. 2010). As a result of ROS production, oxidative signals are activated leading to the inhibition of auxin-dependent signaling pathways, while activating oxidative signaling pathways (Kovtun et al. 2000). Auxins induce changes in the cellular redox state in plants, which is due to the generation of reactive oxygen species (ROS) by auxin-induced changes in the redox state (Vivancos et al. 2011). The numerous other protective agents available to plant cells that assist in scavenging ROS produced and mitigating the effects of abiotic stresses may also enhance the understanding of ROS production-mediated effects imparted to plant cells during normal activities. Therefore, understanding ROS production-mediated effects imparted to plant cells during normal activities is also beneficial in understanding valuable ROS production-mediated effects imparted to plant cells during normal activities.

13.3 Phytoprotectants Affecting Oxidative Stress Mechanism (ROS Production)

It is well-known that plants' environment is abundant in molecular oxygen. The molecular oxygen accumulates in plant cells when they are stressed by external environmental stimuli or by exposure to ultraviolet radiation. Such established times in plants are called "reactive oxygen species accumulation." A reactive oxygen species can consist of a number of different molecules, including molecules such as hydroxyl radicals, superoxide anion, and hydrogen peroxide (H_2O_2); if they are produced in excess, they may contribute to widespread cell damage by degrading proteins, inactivating enzymes, or altering gene actions (Choudhury et al. 2013; Mittler et al. 2004).

Over time, plants developed mechanisms to use ROS as chemical signals to mitigate the effects of abiotic stress by manipulating ROS network genes. These genes represent a transcriptional network of almost 152 genes within each plant. There are ROS regulatory networks that involve redox-sensitive transcription factors (TFs) as well as other proteins, receptors, and the actions of ROS on phosphatases. Consequently, a balanced level of ROS-producing and ROS-scavenging pathways at any given moment is conducive to cellular well-being (Choudhury et al. 2013).

Under abiotic stress conditions, plants exhibit unique patterns of ROS-producing and scavenging enzymes, showing them to be ROS-scavenging enzymes. This change has been observed in several different forms, including altered levels of byproducts of lipid peroxidation, an increase in enzymes such as glutathione S-transferase and CAT, and an accumulation of phytoprotectants which act as antioxidants, such as ascorbate, phenolic compounds, carotenoids, alkaloids, sucrose, and trehalose (Choudhury et al. 2017).

There are a number of chemicals that are known to act as phytoprotectants in abiotic stress conditions. Among the numerous chemicals, soluble sugars which are monosaccharides and disaccharides have double roles, acting as mediators of ROS production, such as mitochondrial respiration, and as antioxidants in oxidative pentose phosphate pathways (PPPs). As this topic is currently being addressed, it is imperative to comprehend how phytoprotectants such as soluble sugars protect plants from ROS production by their ability to inhibit oxidative mechanisms present in plants (Couee et al. 2006). It can be observed that in plants ROS are accumulated from several different sources within the cells, such as the NADPH oxidase in the cell membrane, the electron transport chain between the chloroplast and mitochondria, the oxidation of nutrients, and the glycolate oxidizing stage of photosynthesis in the peroxisomes and metabolism in the mitochondria, respectively (Doudican et al. 2005; Møller 2001).

Plants that are exposed to soluble sugars will increase the production of ROS as a result of an increase in photosynthetic activity. The opposite effect is observed when dissolved sugar levels are reduced which negatively regulates the expression of photosynthetic genes, even under normal daylight conditions. This is true, especially for genes involved with Calvin cycle expression. It has been shown that the relationship between light and sugar accumulation has been elucidated. As a consequence of the findings, this method of regulating gene expression concurrently by soluble sugar and light is now better understood. It is also shown that the accumulation of sugar grains is an effective factor in cold-protective mechanisms during conditions of abiotic stress, such as chilling stress (Ciereszko et al. 2001; Havaux and Kloppstech 2001).

At specific developmental stages, a condition where carbohydrate levels fluctuate or carbohydrates are deficient may additionally result in an increase in ROS production. As a result of the electron transport involved in the mechanism of cytochrome c oxidase, the regeneration rate of ADP is significantly reduced, resulting in increased levels of ROS in mitochondria (Dutilleul et al. 2003). The process of sugar starvation is also thought to activate the mobilization of lipids and the activation of β-oxidation in peroxisomes. The activation of acyl-CoA oxidase, proteins, and mRNA is a result of improving the levels of acyl-CoA oxidase, proteins, and mRNA. There is also evidence for ROS activation due to sugar stress through a transcriptomic analysis where catalase enzymes are activated in response to sugar stress (Contento et al. 2004).

Additionally, soluble sugars have been observed to play an active role in ROS production through the interaction with phytohormones (auxin, brassinosteroids, and ABA). There appears to be a link between auxin signaling and oxidative stress because auxins can trigger the production of reactive oxygen species (ROS) and regulate ROS homeostasis. It is known that auxin activates Rho GTPase (RAC/ROP) that interacts with NADPH oxidases, resulting in apoplastic ROS production (Duan et al. 2010). On the other hand, ROS activate a MAPK signaling cascade that inhibits auxin-dependent signaling and triggers oxidative communication cascades to occur (Kovtun et al. 2000). Changes in the cellular redox state caused by auxin-induced ROS production, also known as auxin-induced redox-state changes, regulate the plant cell cycle (Vivancos et al. 2011). It is clear from the above that plant cells are equipped with numerous other protective factors that enable them to scavenge ROS produced and, as a result, mitigate the consequences of abiotic stresses. However, understanding the role that such protective compounds play as dual-role agents is also very helpful in understanding the critical ROS-mediated effects imparted to plant cells during normal activities.

13.4 Phytoprotectants as ROS Scavengers

Since the onset of oxidative stress, plants have learned how to mitigate the effects of ROS, by utilizing a variety of phytoprotectant that can either be nonenzymatic, such as carotenoids, ascorbate (AsA), tocopherol, and glutathione, or enzymatic, such as catalase (CAT), superoxide dismutase (SOD), and glutathione S-transferase. These redox buffers act as redox controllers and influence the expression of genes involved in abiotic stress (Foyer and Noctor 2005). We will now explain in more detail how such phytoprotectants can help plants to reduce oxidative stress by scavenging ROS in the presence of abiotic stresses. Additionally, it can reduce the stresses associated with oxidative stress.

All plant tissues contain ascorbic acid (AsA), a powerful antioxidant that plays a critical role in photosynthetic activity. Photosynthesizing is a key function of the plant's meristem. Among the numerous free radicals that work as a reductant are O_2^-, HO^-, and H_2O_2. Its ability to neutralize these radicals is the basis of its antioxidant properties. AsA occurs naturally in plants in sufficient quantities. The AsA-GSH cycle not only scavenges antioxidant components from the body but also preserves them as antioxidants. Plants are capable of coping with high temperatures and several possible abiotic stresses such as those caused by increased levels of monodehydroascorbate reductase (MDHAR) and dehydroascorbate reductase (DHAR). These enzymes are prominently in AsA metabolism and recycling (Meena et al. 2017). This recycling process ensures that sufficient quantities of AsA are maintained in order to provide plants with a tolerance to heat stress. The experiments carried out on *Arabidopsis* under high-temperature stress support the theory that overexpression of DHAR in cellular compartments is able to increase

AsA levels by 2 to 4.25 times, which reduces membrane damage and improves chlorophyll content to a significant extent than normal plants (Wang et al. 2010). It is also thought that glutathione (GSH) might be a potential antioxidant since it is abundant in almost all subcellular compartments such as mitochondria, endoplasmic reticulum, and cytosol, where it demonstrates considerable ability to scavenge ROS. As a ROS scavenger, it can act either directly by reacting with ROS such as HO^- and O_2^- or indirectly by regulating the levels of other antioxidants such as tocopherol and zeaxanthin resulting in reduced levels of ROS. It also acts as a substrate for a number of other enzymatic antioxidants, including glutathione S-transferases (GST) and glutathione peroxidase (GPX) (Hasanuzzaman et al. 2013). It is thought that the mechanism of action of GSH is such that the accumulation of GSH occurs upon the onset of abiotic stress. It is known that increased GSH concentrations counterbalance the stress-induced oxidation of GSH and can cause changes in gene expression either directly or indirectly through interaction with regulatory proteins and/or transcription factors. In addition to its importance in signal transduction and defense against ROS, this increase also involves a multilevel control mechanism, which involves the coordinated activation of genes coding for GSH biosynthetic enzymes and GR (Srivalli and Khanna-Chopra 2008). In that regard, GSH acts as a redox sensor of environmental cues, and the increase in GSH helps plants cope with oxidative stress. Tocopherols, as amphiphilic antioxidants and protective molecules, contribute to the scavenging of ROS in photosynthetic membranes. As they reduce free radicals such as lipid peroxyl (LOO^-) to their respective hydroperoxides, they limit the extent of lipid peroxidation (Maeda et al. 2005). They are also involved in numerous ROS-regulated signaling networks that are controlled by phytohormones and some antioxidants, making them ideally suited to take part in abiotic stress signaling. The majority of studies indicates that under optimal conditions, tocopherols do not play an influential role in plant survival. However, an adequate amount of redox-state tocopherols in chloroplasts can provide plants with the ability to deal with abiotic stresses. In a study involving the application of exogenous tocopherol to *H. annuus* seeds grown in salt stress, there were marked increases in the activities of antioxidant enzymes and a decrease in salt-induced leaf senescence (Hasanuzzaman et al. 2013). Overall, all cellular compartments are equipped with antioxidants to scavenge ROS at the site of production by local antioxidants. This serves to scavenge ROS molecules before they enter the cell. When the state of stress is severe or if the body's antioxidant capacity is not sufficient to deal with it, the free radicals (H_2O_2) can leak into the cytosol and move to other compartments within the cell. Cells are also equipped with mechanisms that allow them to combat the production of excessive H_2O_2 by transporting it in vacuoles for detoxification. During periods of abiotic stresses, plants' vacuoles become rich in protective molecules like flavonoids and ascorbate, which scavenge these substances and stabilize them within the plant (Gechev et al. 2006).

13.5 Physiological and Molecular Adaptations in Response to Phytoprotectants

Whenever a plant is faced with environmental stress, the most important thing for it to do is to allocate its energy in a way that will allow it to adapt to the environment in a better way but still maintain its growth and productivity. Plants need to adapt to these functions which results in changes in their physiological mechanisms, such as the activation of many metabolic reactions, the regulation of ion homeostasis, and the expression of stress-responsive genes (Ahmad and Wani 2013). A phytoprotectant functions as a chemical barrier to withstand environmental constraints and maintain the health and stability of plants. It is important to keep in mind that depending on the severity of the stress encountered, the physiological adaptation may not be robust enough to withstand the stress and the plant may be damaged. Our aim here is to explore different physiological and molecular adaptations that occur in plants as a result of phytoprotectant application and that improve the effects of phytoprotectant in producing stress-tolerant crops as a result of the application.

There are several physical and chemical changes that are caused by abiotic stresses, such as drought, cold, and salinity. As a result of these conditions, the plants' physiology is altered in many ways, such as low nutrient availability, toxic salt accumulation, reduced seed germination rate, early senescence, and even death. It is supposed that several potential low-molecular-weight organic compounds, such as compatible solutes, are produced by plants when they are exposed to osmotic stress and that they may play a role in the adaptation of the physiology of plants in order to cope with these conditions. A phytoprotectant known as an osmolyte is composed of carbohydrates (sucrose trehalose, polyamines), amino acids (propionyl), and sugar alcohols (sorbitol, mannitol), which are involved in maintaining cell turgor, reducing ionic toxicity, and protecting cell structures. Enhanced levels of sugar in drought- and salt-tolerant rice varieties suggest that these protective compounds can contribute to stress tolerance in rice (Ahmad and Wani 2013).

According to reports, osmolytes may play an extremely significant role in adapting plant physiology to environmental conditions by carrying out alternative protective functions. Under salt stress, rice plants accumulate soluble nitrogenous compounds such as polyamines (PA), betaines, imides, amino acids, and a number of other nitrogenous compounds. Researchers have shown that salt stress increases the amount of a polyamine called putrescene, which is involved in the process of developing tolerance to salt (Do et al. 2014). Interestingly, exogenous application of PA has also demonstrated the capability of overcoming the damaging effects of salinity in several other plant species (Mansour 2000). The physiological role of PA-induced adaptation in salt stress has been extensively studied. As a result of their polycationic nature, PA can directly interact with the surface of membranes or indirectly affect some membrane-binding enzymes while maintaining their structure. Besides serving as ROS scavengers and ammonia detoxifiers, they also serve as osmotic adjusters, but due to their low number, their influence is not as noticeable as that of other osmoprotectants (Mansour 2000; Kushad and Dumbroff 1991).

The amino acid proline (pro), which is elevated during conditions of abiotic stress such as salinity, drought, intense radiation, or oxidative stress, is most often an effective alleviator of these stresses. The antioxidative property of this substance allows it to maintain redox balance, stabilize enzymes, and protect cellular structures (Ahmad and Wani 2013). A study on transgenic *Arabidopsis* that contained 90% less proline than the wild type showed that they produced significantly more ROS and lipid peroxidation products compared to the wild type (Szekely et al. 2008). Furthermore, proline functions as a cellular osmotic adjuster when exposed to abiotic stresses, and this function is accomplished by lowering the cellular osmotic potential, which in turn allows reabsorption of water. The proline amino acid also plays a role in protecting cellular membranes against salt-induced injuries as well as stabilizing membrane structures. It can be noted, from a molecular perspective, that this membrane stabilization is accompanied by a decrease in the accumulation of Na^+ and Cl^- in shoots, and this, in turn, allows salt-affected plants to grow faster in response to proline treatment (Mansour 2000).

When plants are exposed to distinct abiotic stresses, they accumulate nonreducing sugars like trehalose in high concentrations. Stress increases these levels to a certain extent, which aids in stabilizing proteins and membranes (Ahmad and Wani 2013). The carbohydrate trehalose is also a precursor to glucose and is catabolized by the carbohydrate trehalose to produce glucose (Brodmann et al. 2002). It has been demonstrated that trehalose causes a significant increase in the transcripts of antioxidant enzyme genes such as superoxide dismutase, ascorbate peroxidase, and catalase in salt-stressed rice plants. Plants that have been treated with trehalose recover immediately compared to those that are left untreated (Nounjan et al. 2012).

13.6 Phytoprotectants Role in Crosstalk Mechanisms

It is known that plants produce unique metabolites when they are under stress, and these metabolites contain a variety of biochemicals. Stress is detected by certain receptors within plants that trigger response mechanisms such as signal transduction pathways that ensure the plant's survival and well-being when under stress. The signaling components that directly or indirectly participate in plant abiotic stress responses include oxidative burst; reactive oxygen species (ROS); ion efflux and influx, particularly Ca^{2+} through Ca^{2+} signaling; acidification or alkalization of the cytoplasm; nitric oxide; abscisic acid receptors; jasmonate activation; lipid communication; and cyclic nucleotides such as cyclic guanosine monophosphate. Plant welfare under stress conditions is dependent on the cross talk of different signaling pathways, and this is caused by transcription factors or other cellular components that integrate multiple signaling pathways. Cross talk between signaling pathways is one of the mechanisms that plants use to control the expression of a large number of genes in a spatiotemporal manner, allowing them to respond to abiotic stressors in an array of ways. There is a possibility of cross talk between signaling pathways at any stage of the process, including transcription, translation, and RNA splicing, editing, and posttranslational modification.

In many signaling pathways, transcription factors (TFs) are converging sites, and they can be synthesized and activated directly by signals from the stress response/ signaling pathways or indirectly through a feedback loop regulated by other transcription factors. All of these conditions leading to the association of multiple signaling pathways through TFs have been linked to drought, cold, injury, salinity, infections, and hormonal treatment.

As a result of oxidative stress, the generation of reactive oxygen species (ROS) alters a cell's redox status and contributes to the generation of reactive oxygen species (ROS). ROS has the ability to regulate a wide variety of genes that are involved in defense and antioxidant responses. A phytoprotectant interferes with ROS signaling through its ability to interfere with the antioxidant system. This interferes with the process of scavenging ROS by increasing the activity of enzymes such as ascorbate peroxidase and catalase. Plant growth and development are controlled by these pathways, which also function in partnership with nitrogen metabolism pathways, photosynthetic pathways, and hormonal pathways. It has been demonstrated that several phytoprotectants interact with various signaling pathways such as those mediated by nitric oxide (NO), lipid signaling, hormone signaling, and Ca^{2+} signaling through ROS signaling.

13.7 Phytoprotectants Involved in Signaling Pathway Engineering

There has been a struggle for food security among humans since the dawn of time. Global food security is under threat as a result of climate change and population expansion, resulting in resource depletion. It is often said that advances in genetic engineering have made it possible for modern man to overcome some of these risks (Koning 2017). We are still working on identifying new biotechnological approaches for enhancing crop types in an efficient and effective way. There is no doubt that the synthesis of *Bacillus thuringiensis* (Bt) bacterial proteins in plants was a decisive step forward when genetically modified crops first appeared.

Research has been conducted extensively on phytoprotectants in terms of their applicability in plants, as well as their role in the stress signaling network and the metabolic response to stress in plants. There is evidence that the phytoprotection of plants has been supported by a genetic transformation strategy under stress conditions, and improved crop performance has been reported even under suboptimal conditions. There is no doubt that these protective substances can be accumulated in the body either by activating or inhibiting the process of biosynthesis and catabolism. The ROS detoxification system of plants is activated by glycine betaine, an amine that has a protective effect against ROS. Genetically modified tomatoes produce 10–30% more than their non-modified counterparts, which suggests that GB could be a viable candidate for drought and cold resistance (Park et al. 2007). Besides polyols, simple sugars have also been targeted for genetic engineering due to their osmoprotective properties. As a type of molecular chaperone, polyols serve as a molecular chaperone. Thus, they help in scavenging ROS. A very comprehensive overview of the chemistry of straight-chain polyols has been reported for a number of plant species including *A. thaliana*, poplar, wheat, and tobacco. Straight-chain polyols like mannitol enhance salt and drought tolerance; however, sorbitol could damage plants because it interferes with carbon metabolism (Ahmad and Wani 2013). It has been demonstrated that transgenic maize that is drought and cold tolerant can serve as a viable target for genetic modification (He et al. 2013). There is evidence that GABA (gamma-aminobutyric acid) plays a role in scavenging reactive oxygen species (ROS) as well as contributing to the nitrogen-carbon pool (Liu et al. 2011). Trehalose, a nonreducing sugar that plays an important role in stress responses, plays a key role in these responses. As a result of genetic studies, GABA has been shown to be a good candidate as a phytoprotectant to be used in biosynthetic pathway engineering in order to produce transgenic plants that are resistant to salt and drought stress (Nounjan et al. 2012; Renault et al. 2010).

As a result, genes involved in the mannitol biosynthesis pathway hold promise as potential targets for pathway engineering. There are a growing number of studies being done to use biotechnological technologies to generate plant crops that are capable of tolerating stress. However, when it comes to filling the gaps in pathway engineering, there is a need for a better understanding of the process (Fig. 13.1).

13.8 Phytoprotectants of Microbial Origin

13.8.1 Mycorrhizal Fungi

The mycorrhiza term is used to refer to more than 6000 fungi that form symbiotic interactions between plants that have a wide range of species (Smith and Read 2010). As a helpful microorganism, arbuscular mycorrhiza fungi (AMF) act as a vital component of agriculture and natural ecosystems by enhancing the fertility of the soil. It consists of the direct interaction between soil and the roots of plants, which is an important component of the natural ecosystem (Khan 2005). A symbiosis between arbuscular mycorrhiza fungi and plants has several advantages as it enhances the absorption of nutrients, minerals, as well as water by plants. The fungus consumes more than 20% of net photosynthesis for its respiration and growth (Smith and Read 2010). As a result of the presence of AMF, plants become more resistant to abiotic and biotic stresses, allowing them to carry out other important functions. There are some AMF species that have been isolated in saline environments, and these species can survive under such conditions (Fernández and Juárez 2011). It is known that the presence of specific ions and salts that are toxic to living organisms can lead to osmotic effects limiting colonization and growth (Juniper and Abbott 2006; Sheng et al. 2008). Moreover, experiments have shown that symbiosis between AMF and various kinds of plants exhibits an increase in salinity tolerance, such as in maize, cucumber, tomato, citrus, acacia, and lettuce (Navarro et al. 2014; Rosendahl and Rosendahl 1991).

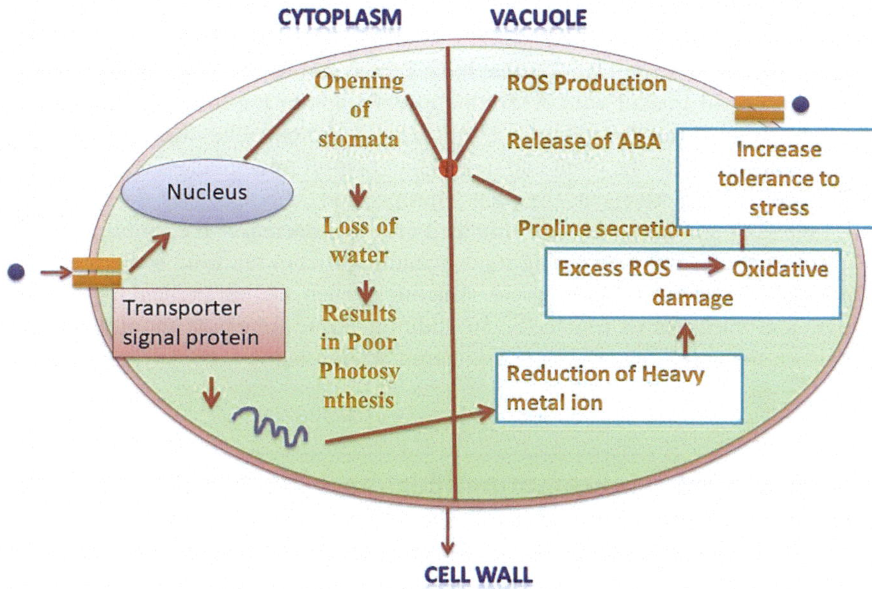

Fig. 13.1 Phytoprotectants including nitric oxide and polyamine interact with abiotic stress signaling to generate stress responses in stressed plants

In salt stress conditions, AMF boosts the uptake of soil water with the help of mycelium and increases the rate of plant mobilization; it also increases the soil contact with the roots and reduces the toxic effects of ions on both organelles of cell and plasma membrane. It changes the balance of phytohormones, boosts the defense of antioxidants (nonenzymatic as well as enzymatic), and increases gene expression in the plants. It has been observed that the application of AMF in the condition of low-water potential may enhance the hydraulic conductivity of the root (Kapoor et al. 2008), improve the architecture system of the roots by modifying them, and subsequently increase the conductance of the stomata by upgrading the capacity of gaseous exchange (Sheng et al. 2008; Navarro et al. 2014). A very recent discovery has discovered that inoculation of AMF into plants diminished chlorophyll production under saline-stressed conditions (Sheng et al. 2008). As a consequence, AMF can also be used as a phytoprotectant, causing an increase in plant growth and reducing the need for fertilizer and water. As a result, the selection of the correct AMF is crucial to the growth of fungi under stress conditions in order to ensure their effectiveness. In the initial stages of saline conditions, the symptoms that show up on the plants are due to water shortage as a result of reduced water intake, the processes by which AMF can improve plant tolerance against salinity or drought conditions. As a result of inoculating AMF under salt stress, the structure of the soil is improved (Zhang et al. 2017). The amount of water available is dependent on the availability of the water itself (Zou et al. 2015). Moreover, it has the advantage of enhancing the nutritional status of plants through the transportation of nutrients and minerals across the interface between the root and the soil, improving their resistance to osmotic stress conditions.

There have been several studies that have shown that when plants are exposed to osmotic stress and treated with AMF, the effect helps to improve their nutritional status. This mechanism functions by eliminating phytotoxic chemicals by allowing them to be absorbed by the plant and turned into helpful mineral elements which are then transported selectively in the plant (Juniper and Abbott 2006; Navarro et al. 2014; Lehmann and Rillig 2015). Activated charcoal decreases the mobilization of Na^{+} in tissues of plants preventing its concentration from reaching toxic levels by maintaining the capacity of ions in structures, such as intraradical mycelium, vesicles, and vacuoles of root cells. In addition, these fungi are thought to help break down the toxic ions and stop them from accumulating in plant tissues (Augé 2001).

13.9 Plant Growth-Promoting Rhizobacteria (PGPR)

Rhizobacteria that promote plant growth is known as plant growth-promoting rhizobacteria (PGPR) and is the association between a plant and a free-living bacterium in the soil (Antoun 2013). PGPR is a group of bacteria that makes up 1 to 2% of the total number of bacteria on earth. The *Bacillus* and *Pseudomonas* spp. groups are composed of a variety of genera (Podile and Kishore 2007). The ability of PGPR to colonize rhizospheres, radicular tissues, and root surfaces depends on the approach used to promote the growth of the plants (Gray and Smith 2005). There are both direct and indirect mechanisms that are responsible for the beneficial effects of the activities. It is thought that the growth of plants is indirectly influenced by competition against pathogens, the PGPR produces the hostile compound, or by inducing resistance to pathogens in the plants (Lugtenberg and Kamilova 2009; Beneduzi et al. 2012; Mhlongo et al. 2018). It may be of interest to study the production of antibiotics, proteases as lysing enzymes, lactonases, glucanases, cellulases, and siderophores, such as cyanide or 2-methylpentanoate, which are toxic compounds, or even dimethyl disulfide which are volatile compounds (Hernández-Calderón et al. 2018; Sharifi and Ryu 2018). By increasing the availability of nutrients (through the solubilization of K^{+} and P, as well as nitrogen fixation, among other nutrients), as well as by enhancing enzymatic synthesis and phytohormone concentrations, all of these factors directly contribute to growth promotion (Oke and Long 1999; Beattie 2007; Pii et al. 2015). Additionally, PGPRs have also been linked to ethylene regulation, enhanced solubilization of nutrients, and absorption of roots, as well as enhanced phytohormone synthesis (Glick et al. 2007). In a study conducted by Gomaa, Mohamed, and colleagues (2012), it was found that PGPR, under salt stress conditions, increased total chlorophyll in radish by increasing the proline level and the total free amino acid balance, which slowed down the degradation of chlorophyll. These mechanisms suggest that PGPR is beneficial to various plant species in terms of increasing their yields, while also contributing to the vegetative growth of the plants (Belimov et al. 2009; Bashan and de-Bashan 2015; Backer et al. 2018). Apart from these, the PGPR also reduces the use of pesticides and chemical fertilizers. The result shows that the PGPR has a greater capacity to absorb nutrients and provide protection for the plants. It has been documented that soil fertility, economic viability, and environmental sustainability improved as a result of these practices (Kang et al. 2014, Compant et al. 2016; Ilangumaran and Smith 2017). The PGPR formulation is already available for sale as a commercial product for the production of agriculture in a different formulation. It is possible that the ability for plants to develop their responses in response to specific PGRPs depends on the interaction between soil, plant, and bacteria conditions (Van Loon 2007; Pan et al. 2019). In subsequent efforts, there are wide ranges of rhizobacteria that are being evaluated around the world for their ability to confer resilience to different environmental stresses, including salt stress (Mayak et al. 2004; Tank and Saraf 2010; Chang et al. 2014; Fan et al. 2016). As a general rule, "induced systemic tolerance" refers to the process of regulating how plants react to abiotic stresses by inoculating PGPR into the plants to facilitate chemical and physiological changes. It was found that plants inoculated with PGPR and cultivated in salinity showed improved nutrient, water, and root balance as well as improved plant growth and development. In leaves of wheat that were inoculated with PGPR, the level of Na^+ concentration in the leaves was reduced. As well as this, the selectivity of Ca^{2+}, K^+, and Na^+ has been observed to be altered, resulting in an increase in the ratio of K^+/Na^+, as well as increased uptake of nitrogen, phosphorus, and potassium by numerous crops, including maize and wheat (Yang et al. 2009).

13.10 Conclusion and Future Perspective

The growth of plants depends on a number of factors, including climatic changes and climate-related abiotic stresses like heat, drought, and cold. There are physiological and molecular responses that plants coordinate in order to produce various immunological responses, which are important defense mechanisms. On the other hand, the process of adjusting to environmental conditions involves a lot of signal cascades operating simultaneously, which makes the adaptation process more complicated. In the face of a constantly changing environment, it is necessary to develop new techniques in order to overcome the risks of plant development. The use of phytoprotectants as immune boosters under stress conditions can increase photosynthetic rate, yield, and antioxidative capacity by enhancing photosynthetic activity. Plant-based phytoprotectants, however, are known to play an important role in interactions with physiological and metabolic processes in an obvious way, with metabolites and signaling molecules acting as signaling molecules. Activation of transcription factor genes associated with stress and scavenging of ROS as well as calcium-mediated resistance to environmental stresses has a high level of integration in response to environmental stress. There has been extensive research that has been done using strong molecular methods such as transcriptomes and proteomes to better understand plant abiotic stress signaling pathways that are mediated by phytoprotectants. In order to gain additional insights into the underlying molecular mechanism of the emergence of antibiotics, a genetic transformation technique can be used to enhance the endogenous production of protective compounds, resulting in more productive plants under suboptimal conditions, for other reasons. Additionally, it can be possible to generate new pathways in plants by transferring genes from other species into them. Through genetic engineering, more complex and high-throughput strategies can be used to strengthen the signaling pathways mediated by phytoprotectants in plants to strengthen their resistance to abiotic stress. This will help them better resist the challenges of the environment.

Acknowledgement All the authors acknowledge Lovely Professional University, India, and Banaras Hindu University, India, for providing possible assistance.

Conflict of Interest We wish to confirm that there are no known conflicts of interest associated with this publication that could have influenced its outcome.

References

Ahmad P, Wani MR (eds) (2013) Physiological mechanisms and adaptation strategies in plants under changing environment: vol 2. Springer Science & Business Media

Antoun H (2013) Plant-growth-promoting rhizobacteria. Brenner's encyclopedia of Genetics, 2nd edn, vol 5

Augé RM (2001) Water relations, drought and vesicular-arbuscular mycorrhizal symbiosis. Mycorrhiza 11(1):3–42

Backer R, Rokem JS, Ilangumaran G, Lamont J, Praslickova D, Ricci E, Smith DL (2018) Plant growth-promoting rhizobacteria: context, mechanisms of action, and roadmap to commercialization of biostimulants for sustainable agriculture. Front Plant Sci 1473

Bashan Y, de-Bashan LE (2015) Inoculant preparation and formulations for Azospirillum spp. In: Handbook for Azospirillum. Springer, Cham, pp 469–485

Beattie GA (2007) Plant-associated bacteria: survey, molecular phylogeny, genomics and recent advances. In: Plant-associated bacteria, pp 1–56

Belimov AA, Dodd IC, Hontzeas N, Theobald JC, Safronova VI, Davies WJ (2009) Rhizosphere bacteria containing 1-aminocyclopropane-1-carboxylate deaminase increase yield of plants grown in drying soil via both local and systemic hormone signalling. New Phytol 181(2): 413–423

Beneduzi A, Ambrosini A, Passaglia LM (2012) Plant growth-promoting rhizobacteria (PGPR): their potential as antagonists and biocontrol agents. Genet Mol Biol 35:1044–1051

Brodmann D, Schuller A, Ludwig-Müller J, Aeschbacher RA, Wiemken A, Boller T, Wingler A (2002) Induction of trehalose in Arabidopsis plants infected with the trehalose-producing pathogen Plasmodiophora brassicae. Mol Plant-Microbe Interact 15(7):693–700

Chang P, Gerhardt KE, Huang XD, Yu XM, Glick BR, Gerwing PD, Greenberg BM (2014) Plant growth-promoting bacteria facilitate the growth of barley and oats in salt-impacted soil: implications for phytoremediation of saline soils. Int J Phytoremediation 16(11):1133–1147

Choudhury S, Panda P, Sahoo L, Panda SK (2013) Reactive oxygen species signaling in plants under abiotic stress. Plant Signal Behav 8(4):e23681

Choudhury FK, Rivero RM, Blumwald E, Mittler R (2017) Reactive oxygen species, abiotic stress and stress combination. Plant J 90(5):856–867

Ciereszko I, Johansson H, Kleczkowski LA (2001) Sucrose and light regulation of a cold-inducible UDP-glucose pyrophosphorylase gene via a hexokinase-independent and abscisic acid-insensitive pathway in Arabidopsis. Biochem J 354(1):67–72

Compant S, Saikkonen K, Mitter B, Campisano A, Mercado-Blanco J (2016) Editorial special issue: soil, plants and endophytes. Plant Soil 405(1):1–11

Contento AL, Kim SJ, Bassham DC (2004) Transcriptome profiling of the response of Arabidopsis suspension culture cells to Suc starvation. Plant Physiol 135(4):2330–2347

Couee I, Sulmon C, Gouesbet G, El Amrani A (2006) Involvement of soluble sugars in reactive oxygen species balance and responses to oxidative stress in plants. J Exp Bot 57(3):449–459

Dietz KJ, Turkan I, Krieger-Liszkay A (2016) Redox-and reactive oxygen species-dependent signaling into and out of the photosynthesizing chloroplast. Plant Physiol 171(3):1541–1550

Do PT, Drechsel O, Heyer AG, Hincha DK, Zuther E (2014) Changes in free polyamine levels, expression of polyamine biosynthesis genes, and performance of rice cultivars under salt stress: a comparison with responses to drought. Front Plant Sci 5:182

Doudican NA, Song B, Shadel GS, Doetsch PW (2005) Oxidative DNA damage causes mitochondrial genomic instability in Saccharomyces cerevisiae. Mol Cell Biol 25(12):5196–5204

Duan Q, Kita D, Li C, Cheung AY, Wu HM (2010) FERONIA receptor-like kinase regulates RHO GTPase signaling of root hair development. Proc Natl Acad Sci 107(41):17821–17826

Dutilleul C, Garmier M, Noctor G, Mathieu C, Chétrit P, Foyer CH, de Paepe R (2003) Leaf mitochondria modulate whole cell redox homeostasis, set antioxidant capacity, and determine stress resistance through altered signaling and diurnal regulation. Plant Cell 15(5):1212–1226

Fan P, Chen D, He Y, Zhou Q, Tian Y, Gao L (2016) Alleviating salt stress in tomato seedlings using Arthrobacter and bacillus megaterium isolated from the rhizosphere of wild plants grown on saline–alkaline lands. Int J Phytoremediation 18(11):1113–1121

Fernández F, Juárez J (2011) Procedimiento de obtención de un agente micorrizógeno. Solicitud de Patente Invención 201130566:17

Foyer CH, Noctor G (2005) Redox homeostasis and antioxidant signaling: a metabolic interface between stress perception and physiological responses. Plant Cell 17(7):1866–1875

Gechev TS, Van Breusegem F, Stone JM, Denev I, Laloi C (2006) Reactive oxygen species as signals that modulate plant stress responses and programmed cell death. BioEssays 28(11): 1091–1101

Glick BR, Cheng Z, Czarny J, Duan J (2007) Promotion of plant growth by ACC deaminase-producing soil bacteria. In: New perspectives and approaches in plant growth-promoting Rhizobacteria research, pp 329–339

Gray EJ, Smith DL (2005) Intracellular and extracellular PGPR: commonalities and distinctions in the plant–bacterium signaling processes. Soil Biol Biochem 37(3):395–412

Hasanuzzaman M, Nahar K, Fujita M (2013) Plant response to salt stress and role of exogenous protectants to mitigate salt-induced damages. In: Ecophysiology and responses of plants under salt stress. Springer, New York, NY, pp 25–87

Havaux M, Kloppstech K (2001) The protective functions of carotenoid and flavonoid pigments against excess visible radiation at chilling temperature investigated in Arabidopsis npq and tt mutants. Planta 213(6):953–966

He C, He Y, Liu Q, Liu T, Liu C, Wang L, Zhang J (2013) Co-expression of genes ApGSMT2 and ApDMT2 for glycinebetaine synthesis in maize enhances the drought tolerance of plants. Mol Breed 31(3):559–573

Hernández-Calderón E, Aviles-Garcia ME, Castulo-Rubio DY, Macías-Rodríguez L, Ramírez VM, Santoyo G, Lopez-Bucio J, Valencia-Cantero E (2018) Volatile compounds from beneficial or pathogenic bacteria differentially regulate root exudation, transcription of iron transporters, and defense signaling pathways in Sorghum bicolor. Plant Mol Biol 96:291–304

Huang SS, Kirchoff BK, Liao JP (2013) Effect of heat shock on ultrastructure and calcium distribution in *Lavandula pinnata* L. glandular trichomes. Protoplasma 250(1):185–196

Ilangumaran G, Smith DL (2017) Plant growth promoting rhizobacteria in amelioration of salinity stress: a systems biology perspective. Front Plant Sci 8:1768

Juniper S, Abbott LK (2006) Soil salinity delays germination and limits growth of hyphae from propagules of arbuscular mycorrhizal fungi. Mycorrhiza 16(5):371–379

Kang SM, Radhakrishnan R, Khan AL, Kim MJ, Park JM, Kim BR, Lee IJ (2014) Gibberellin secreting rhizobacterium, pseudomonas putida H-2-3 modulates the hormonal and stress physiology of soybean to improve the plant growth under saline and drought conditions. Plant Physiol Biochem 84:115–124

Kapoor R, Sharma D, Bhatnagar AK (2008) Arbuscular mycorrhizae in micropropagation systems and their potential applications. Sci Hortic 116(3):227–239

Khan AG (2005) Role of soil microbes in the rhizospheres of plants growing on trace metal contaminated soils in phytoremediation. J Trace Elem Med Biol 18(4):355–364

Koning N (2017) Food security, agricultural policies and economic growth: Long-term dynamics in the past, present and future. Routledge

Kovtun Y, Chiu WL, Tena G, Sheen J (2000) Functional analysis of oxidative stress-activated mitogen-activated protein kinase cascade in plants. Proc Natl Acad Sci 97(6):2940–2945

Kushad MM, Dumbroff EB (1991) Metabolic and physiological relationships between the polyamine and ethylene biosynthetic pathways. Biochemistry and physiology of polyamines in plants, pp 77–92

Lehmann A, Rillig MC (2015) Arbuscular mycorrhizal contribution to copper, manganese and iron nutrient concentrations in crops–a meta-analysis. Soil Biol Biochem 81:147–158

Liu C, Zhao L, Yu G (2011) The dominant glutamic acid metabolic flux to produce γ-amino butyric acid over proline in Nicotiana tabacum leaves under water stress relates to its significant role in antioxidant activity. J Integr Plant Biol 53(8):608–618

Lugtenberg B, Kamilova F (2009) Plant-growth-promoting rhizobacteria. Annu Rev Microbiol 63(1):541–556

Maeda H, Sakuragi Y, Bryant DA, DellaPenna D (2005) Tocopherols protect Synechocystis sp. strain PCC 6803 from lipid peroxidation. Plant Physiol 138(3):1422–1435

Mansour MMF (2000) Nitrogen containing compounds and adaptation of plants to salinity stress. Biol Plant 43(4):491–500

Mayak S, Tirosh T, Glick BR (2004) Plant growth-promoting bacteria confer resistance in tomato plants to salt stress. Plant Physiol Biochem 42(6):565–572

Meena KK, Sorty AM, Bitla UM, Choudhary K, Gupta P, Pareek A, Minhas PS (2017) Abiotic stress responses and microbe-mediated mitigation in plants: the omics strategies. Front Plant Sci 8:172

Mhlongo MI, Piater LA, Madala NE, Labuschagne N, Dubery IA (2018) The chemistry of plant–microbe interactions in the rhizosphere and the potential for metabolomics to reveal signaling related to defense priming and induced systemic resistance. Front Plant Sci 9:112

Miller GAD, Suzuki N, Ciftci-Yilmaz SULTAN, Mittler RON (2010) Reactive oxygen species homeostasis and signalling during drought and salinity stresses. Plant Cell Environ 33(4): 453–467

Mittler R, Vanderauwera S, Gollery M, Van Breusegem F (2004) Reactive oxygen gene network of plants. Trends Plant Sci 9(10):490–498

Mohamed HI, Gomaa EZ (2012) Effect of plant growth promoting Bacillus subtilis and Pseudomonas fluorescens on growth and pigment composition of radish plants (Raphanus sativus) under NaCl stress. Photosynthetica 50(2):263–272

Møller IM (2001) Plant mitochondria and oxidative stress: electron transport, NADPH turnover, and metabolism of reactive oxygen species. Annu Rev Plant Biol 52(1):561–591

Navarro JM, Pérez-Tornero O, Morte A (2014) Alleviation of salt stress in citrus seedlings inoculated with arbuscular mycorrhizal fungi depends on the rootstock salt tolerance. J Plant Physiol 171(1):76–85

Nounjan N, Nghia PT, Theerakulpisut P (2012) Exogenous proline and trehalose promote recovery of rice seedlings from salt-stress and differentially modulate antioxidant enzymes and expression of related genes. J Plant Physiol 169(6):596–604

Oke V, Long SR (1999) Bacterial genes induced within the nodule during the rhizobium–legume symbiosis. Mol Microbiol 32(4):837–849

Pan J, Peng F, Xue X, You Q, Zhang W, Wang T, Huang C (2019) The growth promotion of two salt-tolerant plant groups with PGPR inoculation: a meta-analysis. Sustainability 11(2):378

Park EJ, Jeknić Z, Chen TH, Murata N (2007) The codA transgene for glycinebetaine synthesis increases the size of flowers and fruits in tomato. Plant Biotechnol J 5(3):422–430

Pii Y, Mimmo T, Tomasi N, Terzano R, Cesco S, Crecchio C (2015) Microbial interactions in the rhizosphere: beneficial influences of plant growth-promoting rhizobacteria on nutrient acquisition process. A review. Biol Fertil Soils 51(4):403–415

Podile AR, Kishore GK (2007) Plant growth-promoting rhizobacteria. In: Plant-associated bacteria. Springer, Dordrecht, pp 195–230

Renault H, Roussel V, El Amrani A, Arzel M, Renault D, Bouchereau A, Deleu C (2010) The Arabidopsis pop2-1mutant reveals the involvement of GABA transaminase in salt stress tolerance. BMC Plant Biol 10(1):1–16

Rosendahl CN, Rosendahl S (1991) Influence of vesicular-arbuscular mycorrhizal fungi (glomus spp.) on the response of cucumber (Cucumis sativus L.) to salt stress. Environ Exp Bot 31(3): 313–318

Sharifi R, Ryu CM (2018) Revisiting bacterial volatile-mediated plant growth promotion: lessons from the past and objectives for the future. Ann Bot 122(3):349–358

Sheng M, Tang M, Chen H, Yang B, Zhang F, Huang Y (2008) Influence of arbuscular mycorrhizae on photosynthesis and water status of maize plants under salt stress. Mycorrhiza 18(6):287–296

Smith SE, Read DJ (2010) Mycorrhizal symbiosis. Academic Press

Srivalli S, Khanna-Chopra R (2008) Role of glutathione in abiotic stress tolerance. In: Sulfur assimilation and abiotic stress in plants. Springer, Berlin, pp 207–225

Szekely G, Abraham E, Cseplo A, Rigy G, Zsigmond L, Csiszar J, Ayaydin F, Strizhov N, Jasik J, Schmelzer E, Koncz C, Szabados L (2008) Duplicated P5CS genes of Arabidopsis play distinct roles in stress regulation and developmental control of proline biosynthesis. Plant J 53:11–28

Tank N, Saraf M (2010) Salinity-resistant plant growth promoting rhizobacteria ameliorates sodium chloride stress on tomato plants. J Plant Interact 5(1):51–58

Van Loon LC (2007) Plant responses to plant growth-promoting rhizobacteria. New perspectives and approaches in plant growth-promoting Rhizobacteria research, pp 243–254

Vivancos PD, Driscoll SP, Bulman CA, Ying L, Emami K, Treumann A, Foyer CH (2011) Perturbations of amino acid metabolism associated with glyphosate-dependent inhibition of shikimic acid metabolism affect cellular redox homeostasis and alter the abundance of proteins involved in photosynthesis and photorespiration. Plant Physiol 157(1):256–268

Wang GP, Zhang XY, Li F, Luo Y, Wang W (2010) Overaccumulation of glycine betaine enhances tolerance to drought and heat stress in wheat leaves in the protection of photosynthesis. Photosynthetica 48(1):117–126

Yang J, Kloepper JW, Ryu CM (2009) Rhizosphere bacteria help plants tolerate abiotic stress. Trends Plant Sci 14(1):1–4

Zhang Z, Wang Q, Wang H, Nie S, Liang Z (2017) Effects of soil salinity on the content, composition, and ion binding capacity of glomalin-related soil protein (GRSP). Sci Total Environ 581:657–665

Zou YN, Srivastava AK, Ni QD, Wu QS (2015) Disruption of mycorrhizal extraradical mycelium and changes in leaf water status and soil aggregate stability in rootbox-grown trifoliate orange. Front Microbiol 6:203

Signaling Pathway of Reactive Oxygen Species in Crop Plants Under Abiotic Stress

14

Sumera Iqbal, Summera Jahan, Khajista Jabeen, and Noshin Ilyas

Abstract

Global climate change is linked with various abiotic stresses like drought, salt, temperature, and heavy metals, which can rigorously affect growth, development, and physiological functioning and ultimately cause yield reduction of crops. Environmental stresses pose oxidative damage in plants due to the denaturation of structural and functional proteins. Oxidative damage is usually triggered by reactive oxygen species (ROS), for example, singlet oxygen, hydrogen peroxide, superoxide ions, and hydroxyl radical. As a defense mechanism, parallel cellular responses and cell signaling pathways are activated in plants for controlling the level of these reactive oxygen species. This cellular signaling induces the development of new metabolic pathways by synthesizing low-weight metabolites and by modifications in phytohormonal levels. It is well established by now that the role of ROS under abiotic stresses is very significant for crop plants. More importantly, there is a potential of increasing stress tolerance of crop plants by manipulation of ROS. However, their responses can vary according to stress conditions and crop species. The current chapter enlightens the roles of ROS under abiotic stresses and their significance as major signals for maintaining cellular homeostasis and improving plant survival under harsh environmental conditions.

S. Iqbal (✉) · K. Jabeen
Department of Botany, Lahore College for Women University, Lahore, Pakistan

S. Jahan
Department of Botany, University of Gujrat, Gujrat, Pakistan

N. Ilyas
PMAS Arid Agriculture University, Rawalpindi, Pakistan

M. Faizan et al. (eds.), *Reactive Oxygen Species*,
https://doi.org/10.1007/978-981-19-9794-5_14

Keywords

Abiotic stress · MAPK cascade · Genetic signaling · Epigenetic signaling · Hormonal signaling

14.1 Introduction

Plants are reinforced by numerous adaptive approaches to survive in harsh conditions such as drought, salinity, cold, heat, and high light, ensuing in combination or separately (Bulgari et al. 2019; Raza et al. 2019). Several complex signaling networks are initiated in plants after the perception of stress by receptor molecules, including the overproduction of reactive oxygen species (ROS), modifications in phytohormones and gene expression, and a myriad of phosphatase/kinase signaling pathways (Devireddy et al. 2021; Noctor et al. 2018). Now it is generally accepted that ROS (e.g., superoxide anion, $O_2^{\bullet-}$, hydrogen peroxide, H_2O_2, singlet oxygen, $^1O_2$, hydroxyl radicle, $OH^{\bullet}$) are essential signaling molecules associated with plant acclimation responses to diverse abiotic stresses (Marcec and Tanaka 2021; Mittler 2017; Waszczak et al. 2018). Usually, stresses implement high-energy or electron transfer chains leading to the conversion of atmospheric oxygen to partially reduced or activated species of molecular oxygen (Choudhury et al. 2017). ROS like $O_2^{\bullet-}$ and H_2O_2 also partake in plant developmental processes and protect against stress conditions (Nimse and Pal 2015; Saini et al. 2018). Thus, ROS formation below the threshold levels positively influences developmental processes such as tracheary element production and cell wall lignification and even can ameliorate mild abiotic stress (Schutzendubel and Polle 2002; Zhou et al. 2022). However, the overproduction of ROS in different cellular organelles (such as chloroplast, peroxisomes, mitochondria, plasma membrane, and apoplast) under abiotic stress conditions disturbs ROS homeostasis (Singh et al. 2019). It triggers irreversible damages to cellular ingredients such as DNA, proteins, carbohydrates, and lipids and switches signal transduction cascade (Dietz et al. 2016; Singh et al. 2019). It initiates signaling in an incredibly synchronized way to cope with stress and stimulates defense genes, kinases, phytohormones, antioxidants, protein phosphorylation, and Ca^{2+} ions (Sachdev et al. 2021). The ROS originated under stress conditions trigger signaling by protein oxidation and causes the accumulation of peptides which serve as secondary messengers and in turn maintain signaling pathway (Kumar et al. 2017). Under stress, the signal transduction takes place at a speed of 8.4 cm min^{-1} (Miller et al. 2009; Pucciariello et al. 2012). For acclimation of plants to abiotic stress, the retrograde signals travel toward the nucleus and assist in the anterograde control (Woodson and Chory 2008), by ROS production in the cells induced by respiratory burst oxidase homolog (RBOH) proteins (Suzuki et al. 2011). Generally, the RBOH proteins are localized in the plasma membrane and trigger NADPH oxidation, and upon facing stress, these are processes either by phosphorylation or by calcium (Ca^{2+}) ion binding at the symplast and releasing superoxide ($O_2^{\bullet-}$) radicals toward the cell wall. The superoxide produced is then converted into

H_2O_2 by the action of superoxide dismutase (Chapman et al. 2019; Marcec and Tanaka 2021). However, the H_2O_2 is a comparatively more stable type of ROS, traveling plasma membranes by aquaporins and switching signaling process. Hence, H_2O_2 serves as a perfect secondary messenger for signal transduction process (Sewelam et al. 2016; Tian et al. 2016). Moreover, the H_2O_2 production in the apoplast also stimulates the neighboring cells to generate ROS, which leads to the generation of ROS wave (Kerchev and Van Breusegem 2022; Mittler et al. 2011). The abiotic stresses are occurring either separately or in combination triggering overproduction of ROS in plant cells, which becomes a substantial challenge for attaining optimal plant productivity and yield. Hence, disclosing ROS signaling pathways in plants through molecular and biochemical means is essential in mitigating stress and securing sustainable cultivation of economically important food crops under changing climatic scenarios.

14.2 ROS and MAPK Cascade

Signal transduction enables the information to transfer from the extracellular environment to different functional units located inside the cell, i.e., the external signal trigger modifications at the transcriptomic level. Activation of ROS sensors following abiotic stress could induce complex signaling cascades inside the cells to control the gene expression. Signaling cascade involves heterotrimeric G-proteins (Pfannschmidt et al. 2003), protein tyrosine phosphatases (PTPs), and different mitogen-activated protein kinases (MAPK) that regulate the protein phosphorylation (Kiffin et al. 2006; Foyer and Noctor 2005). MAPK cascades assist the cells to respond to diverse types of abiotic stress conditions as aroused by drought, salts, heavy metals, UV radiations, light, and temperature fluctuations (Sinha et al. 2011; Jagodzik et al. 2018). The MAPK belongs to the serine or threonine protein kinases essential for retrograde and anterograde signal transduction, in addition to stress signals. The MAPK cascade depends on the phosphorylation process as MAPK is activated by phosphorylation with MAPKK (MAPK kinase), which is further regulated by MAPKKK (MAPKK kinase). These kinases are linked together and known as extracellular receptor kinases. The MAPK signaling cascade elicits diverse stress, hormone, and cytokinesis responses (Xu et al. 2017). Under abiotic stresses, H_2O_2 regulates numerous MAPKs as in *Arabidopsis*; H_2O_2 induced the activation of the MAPK3 and MAPK6 via MAPKKK atrial natriuretic peptide 1 (ANP1) (Kovtun et al. 2000). Further, it was also studied that overexpression of ANP1 in transgenic plants induced tolerance to cold, heat, and osmotic stress (Kovtun et al. 2000). Later, it was also found that the influence of NDPK2 (*Arabidopsis* nucleotide diphosphate kinase) might be moderated by the MAPK3 and MAPK6 because NDPK2 can also activate and regulate the MAPKs. Another salinity- and cold stress-responsive MAP kinase gene (CbMAPK3) was isolated from *Chorispora bungeana* plant as its transcription levels are upregulated by stresses (Zhang et al. 2006). It can be inferred from information mentioned above that abiotic stresses trigger ROS generation,

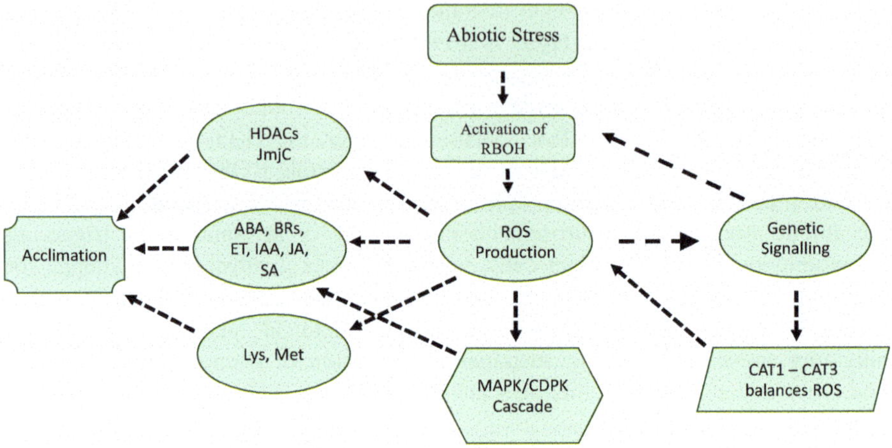

Fig. 14.1 Signaling pathway of ROS (reactive oxygen species) for acclimation responses in crop plants under abiotic stress conditions

which activates and regulates MAPK signaling pathway, suggesting that the MAPK kinase signaling pathway assists in mediating stresses in plants (Fig. 14.1).

14.3 ROS and Genetic Signaling

Massive pieces of evidence from previous literature have shown that environmental stresses such as cold, heat, drought, salinity, and heavy metal could promote ROS generation in plant cells (Ahammed et al. 2017; Kawarazaki et al. 2013; Lee et al. 2012; Wu et al. 2017; Zhou et al. 2018). Because of many types of interconvertible ROS, it is highly challenging to differentiate between a particular ROS's cytotoxic and signaling potential. Although ROS causes cell death, it is also crucial to confer resistance against abiotic stresses. But ROS should appear immediately after sensing stress to activate signal transduction and should be eliminated immediately with the disappearance of stress. Plants without an ATP-dependent protease in mitochondria, i.e., AtFtSH4, exhibited restricted growth at the shoot apical meristem and root apical meristem under heat stress (LD 31 °C). This was due to the overaccumulation of ROS and failure in mitochondrial functioning. Hence, mitochondrial function mainly depends on AtFtSH4, which is vital in preserving the cell division activity in shoot and root apical meristem and acclimation to heat stress (Dolzblasz et al. 2018).

The genes linked with NADPH oxidation, i.e., AtrbohF and AtrbohD, also trigger ROS production during osmotic stress that initiates signaling and regulates Na^+/K^+ ion balance in the cells. It was found that *Arabidopsis* mutants with missing *AtrbohF* genes were deficient in RBOH (the respiratory burst oxidase proteins) and were more sensitive to ion toxicity due to imbalanced Na^+/K^+ ions (Ma et al. 2012). Some latest findings suggest that under salinity stress, H_2O_2 may serve as a stress signal to regulate the Na^+/H^+ antiport system across the plasma membrane and maintain the

SOS1 mRNA in *Arabidopsis*, which is crucial for retaining K^+/Na^+ ion balance in the plant cells (Zhang et al. 2007a, b). The onset of hypoxia (low-oxygen stress) triggers the ROS production via RBOHs (Sasidharan et al. 2018).

It was found that *RBOH1* silencing affected the stomatal functioning in tomato (Yi et al. 2015). Previous literature indicated that genes associated with catalase (CAT) also respond to various abiotic stresses (Raza et al. 2021). In *the Arabidopsis thaliana*, the *CAT1* gene primarily responds to numerous stress conditions and regulates the H_2O_2 balance by scavenging it. Similarly, *CAT2* and *CAT3* also partake in H_2O_2 elimination and keep the ROS homeostasis in plant cells under light and dark cycles (Mhamdi et al. 2010). Under drought stress, ABA also maintains the balance of H_2O_2 by regulating the expression of CAT OsCATB and protecting cells from ROS-induced oxidative injuries (Ye et al. 2015). In *Cucumis sativus*, the expression of CAT1-CAT3 enhances in fruits, flowers, leaves, stem, and roots by ABA treatment, heat, drought, and osmotic stress (Hu et al. 2016).

In *Arabidopsis thaliana*, RRTF1 (Redox-Responsive Transcription Factor 1) is a transcription factor of AP2/ERF (APETALA2/ethylene response factor) and is a predominant part of the redox signaling pathway. Various ROS spawned by abiotic stress signals promptly stimulate its expression.

Increased RRTF1 levels in plants (facing stresses) induced ROS proliferation, indicating that RRTF1 amplifies ROS production. It seems as RRTF1 triggered ROS production for an instantaneous accumulation of ROS to arouse suitable downstream responses (Matsuo et al. 2015). The ROS burst triggered by high- or low-temperature stress also regulates the heat shock gene (HSG) upstream transcription (Pucciariello et al. 2012). The specific functions of HSPs during heat stress are to assist as molecular chaperones; prevent protein from denaturation, misfolding, and agglomeration; and also upgrade protein refolding (Ahuja et al. 2010; Pucciariello et al. 2012). Likewise, HSPs are also associated with other types of abiotic stress conditions as excess light stress causes the burst of H_2O_2 in the cells of *Arabidopsis thaliana* that is ameliorated by upstream regulation of HSPs, APX1 (Pnueli et al. 2003), moreover providing tolerance to the plants facing a combination of drought and heat stress conditions (Koussevitzky et al. 2008). The rapid ROS production following abiotic stress is a preserved signaling output for triggering immunity across the plant kingdom.

14.4 ROS and Epigenetic Signaling

Epigenetic changes are linked with the modifications in gene expression that are mitotically or meiotically conserved involving DNA methylation, histone modifications, histone variants, and chromatin remodeling in plants (Pikaard and Scheid 2014). Massive pieces of evidence indicated a connection between epigenetic regulation and ROS metabolism during different phases of plant development and environmental acclimatization. The removal of 5-methylcytosine from DNA is facilitated by four different types of DNA demethylases, ROS1 (REPRESSOR OF SILENCING 1), DME (DEMETER), DML2 (DME-like 2), and DML3 (DME-like 3)

(Duan et al. 2015). Previous reports indicated that ROS impacts the epigenetic mechanisms of gene regulation because DME and ROS1 can directly influence the Fe-S cluster in assembly machinery, which ROS can rapidly oxidize (RM et al. 2020). Hence, there exists a link between ROS metabolism and DNA methylation. The HDACs (histone deacetylases) operate in epigenetic modification with oxidative stress. Enhanced ROS causes an increase in modifications in different types of histones such as H3K79me3, H3K4me2/3, H3K9me2, and H3k27me3 due to arrest of HDAC activities (Chen et al. 2006; Niu et al. 2015).

The activities of HDA9 and HDA19 (histone deacetylases) are negatively influenced by cellular oxidation hence an enhancement in the expression of stress-responsive genes, and histone acetylation takes place in *Arabidopsis* (Liu et al. 2015). Further, increased ROS levels also trigger apparent epigenetic modification such as acylation; however, in rice leaves, the level of ROS is maintained by the functioning of ROS-related proteins (Zhou et al. 2018). The interaction between acylation and ROS might play essential roles in the posttranslational modifications in leaf proteins with vital metabolic operations. Two essential categories of enzymes catalyze the histone demethylation, i.e., the JmjC (Jumonji C) demethylases that are 2-oxoglutarate- and Fe^{2+}-dependent dioxygenases and LSD1 (lysine-specific demethylase 1), which is a FAD-dependent amino oxidase (Chen et al. 2011). After exposure to stress conditions, numerous JmjC proteins have the potential to regulate the stress-responsive genes, most probably by interacting with ROS produced under stress, hence establishing a complex system of defense responses (Shen et al. 2016). Consequently, the regulation of JmjC biosynthesis can control ROS levels and epigenetic modifications. However, further extensive investigation is required in this regard in plants. Further, the DDM1 (DNA methylation 1) is a critical SWI/SNF2 (switching defective 1/sucrose non-fermenting 2) chromatin remodeler that can mediate DNA methylation by MET1 (DNA methyltransferase 1), CMT2 (chromomethylase 2), and CMT3 (chromomethylase 3) and can also alter the nucleosome composition, particularly in heterochromatin regions of DNA in plants (Zemach et al. 2013). In *Arabidopsis thaliana*, a chromatin-linked factor, i.e., A subunit (AtTOP6A) of topo VI (topoisomerase VI), can control the $^1O_2$ signals from plastid to the nuclear region. Under overaccumulation of $^1O_2$, A subunit of Topo VI binds with the promoter regions of the $^1O_2$-responsive AAA-ATPase gene and other sets of $^1O_2$-responsive genes and instantly controls the transcription of these genes. The topo VI also controls the expression of H_2O_2-linked genes under high-light stress. However, AtTOP6A can differently modulate the transcription of H_2O_2- and $^1O_2$-linked genes, implying that topo VI can integrate multiple signal transduction pathways triggered in plants by ROS under stress conditions (Simkova et al. 2012).

14.5 ROS and Phytohormonal Cross Talk

After perceiving stress signals, complex signaling procedures, including phytohormonal modifications, operate in plants (Angelini et al. 2022; Noctor et al. 2018). Phytohormone and ROS signaling integrate multiple developmental and environmental signals to control the plant response under various types of abiotic stress. The perception of stress by receptors initiates signal transduction networks that further influence phytohormone and ROS, alter their levels, and control the plant growth responses to stress conditions (Mignolet-Spruyt et al. 2016; Kanojia and Dijkwel 2018).

The phytohormones-ROS interaction could also regulate the gene expression and introduce various physiological and biochemical modifications in plants in response to diverse abiotic stresses (Mittler and Blumwald 2015). For example, auxin can promote ROS accumulation to complete the cell cycle. The ROS, however, can impede auxin signaling and can regulate oxidative signaling by activating the MAPKs in response to stress conditions (Tognetti et al. 2012). Likewise, ROS can also upregulate or downregulate the SA (salicylic acid) signaling and stimulate CDPKs (Ca^{2+}-dependent protein kinases) for stomatal closure (Han et al. 2013; Prodhan et al. 2018). Besides, abiotic stresses can cause modifications in ROS levels by alteration in metabolic reactions (Choudhury et al. 2017); phytohormones can also induce additional modifications in ROS levels, mainly by triggering RBOHs (Lamers et al. 2020). For example, ROS-ABA cross talk induced by ABA (abscisic acid) can also regulate the functioning of RbohD and RbohF (Mittler and Blumwald 2015; Devireddy et al. 2018).

The predominant phytohormone that regulates stomatal functioning is ABA (Jiang and Song 2008; Qu et al. 2018). The ROS mediates ABA signaling in guard cells (Jammes et al. 2009). Previous studies revealing that ROS can activate MAPKs may imply that ABA and ROS signaling correspond at the MAPK level (Zhang et al. 2007a, b) and stomatal closure depends mainly on the interaction between these pathways. The H_2O_2 is another essential molecule that participates in signaling pathway for ABA-mediated stomatal closure (Li et al. 2017b; Rodrigues et al. 2017). Consequently, H_2O_2 generation triggered by ABA causes a reduction in stomatal aperture (Wang and Song 2008; Li et al. 2017a, b). In *Arabidopsis thaliana*, mutants lacking MAPKKK18 showed impaired stomatal closure by ABA. The MAPKKK18 might play a critical role in drought resistance by regulating stomatal closure upon perceiving the signals of drought stress (Li et al. 2017a). Further studies on the transcriptional control of the MAPKKK18 promotor indicated its increased promoter functioning after ABA signaling in guard cells of stomata (Mitula et al. 2015). The MAPKKK18 directly controls the two most critical proteins of the ABA signaling pathway, PP2C (protein phosphatase 2C) and phosphatase ABI1 (Abelson interactor 1) (Mitula et al. 2015), and kinase SnRK2.6 (Tajdel et al. 2016). In the absence of ABA signaling, ABI1 degrades MAPKKK18 by the UPS (ubiquitin-proteasome pathway) and also inhibits the MAPKKK18 kinase functionality by dephosphorylation (Ludwików 2015). However, MAPKKK18 degradation is inhibited by the binding of ABA with PYR/PYL receptors, and the stability of

MAPKKK18 enables it to stabilize allowing the kinase to activate downstream signaling modules (Mitula et al. 2015). Recent investigations also traced that ABA-regulated MAPKKK18 also plays a critical role in inducing drought stress resistance in plants (Li et al. 2017a).

ROS signaling can also trigger cross talks between different phytohormones (Czarnocka and Karpinski 2018). For example, ROS induced interaction between ABA and BRs (brassinosteroids) due to heat stress (Zhou et al. 2014). This cross talk enhanced the activity of NADPH oxidase and *RBOH1* gene transcription, resulting in elevated ABA and H_2O_2 levels in *Lycopersicum esculentum*, and improved its heat stress tolerance (Zhou et al. 2014). Similarly, ABA triggers the ROS wave in the plant cells in response to abiotic stresses. The ROS wave then switches on the expression of jasmonic acid, which enhances the stress resistance of plant tissues (Devireddy et al. 2018; Fichman and Mittler 2020).

In response to stress stimuli, ET (ethylene) also acts as an essential element in the signaling pathway such as salinity stress, dehydration, and iron deficiency (Ye et al. 2015; Khan et al. 2017). Salinity stress triggers ethylene (ET) biosynthesis inside the cell, which participates in internal signaling (Dong et al. 2011). Nevertheless, the external signals associated with salinity stress that trigger ET biosynthesis are unknown. Current investigations documented that in *Oryza sativa*, the RLK (receptor-like kinase) proteins induce osmotic tolerance by phosphorylating both MAPK3 and MAPK6 (Li et al. 2014).

The RLKs play an essential role in external signal transduction in the cell; most of the known RLKs from *Oryza sativa* and *Arabidopsis thaliana* are critical for plant growth (Osakabe et al. 2013) and acclimation to salinity and drought stress (Marshall et al. 2012; Vaid et al. 2013). The SIT1 (Salt Intolerance 1) is a type of RLK triggering salt and drought tolerance in *Oryza sativa*; *SIT1* shows its expression in epidermal cells of root immediately after sensing NaCl stimulus. Then it transfers the stress signal by activating MAPK3 and MAPK6 through phosphorylation (Li et al. 2014). Previous literature also reported that SIT1 also plays a critical role in antioxidant system activation (Jagodzik et al. 2018). Hence, during stress responses in plants, ROS production activates ERF1 (ethylene response factor 1) and initiates ET signaling (Vall-Llaura et al. 2022). The SIT1 induces ROS production, which activates ET signaling and activates MAPK3 and MAPK6 to proceed with the signal transduction cascade (Li et al. 2014). So, abiotic stresses modify the phytohormonal levels that in turn control RBOH activation and ROS formation by calcium binding, phosphorylation or dephosphorylation, nitrosylation, phosphatidic acid binding, and RBOH protein by different transcription factors (Jagodzik et al. 2018; Xia et al. 2015; Zandalinas et al. 2020; Zhang et al. 2022).

14.6 ROS and Amino Acid Signaling

After exposure to abiotic stress, plants accumulate different types of free amino acids (Ferreira Júnior et al. 2018; Huang and Jander 2017). Numerous amino acids serve as precursor molecules for signaling cascade and synthesizing secondary metabolites. For example, the amino acid Arg (arginine) participates in biosynthesis of polyamines (Alcázar et al. 2006), and Met (methionine) is used for ethylene synthesis (Amir 2010), and the conversion of Lys (lysine) to N-hydroxy pipecoline is needed for immune signaling after exposure to stress (Hartmann et al. 2018). After facing drought stress, massive amounts of amino acid get accumulated in tomato, maize, and cotton plants (Martinelli et al. 2007; Showler 2002; Zandalinas et al. 2022). Recently it was found that ABA-induced protein expression also increases the levels of free amino acids (Barros et al. 2017; Hirota et al. 2018). Further, under stress conditions, as the carbohydrate levels drop due to reduced photosynthetic rates, amino acids are used in mitochondrial respiration (Hildebrandt 2018; Hildebrandt et al. 2015). It was also reported that the degradation pathways of different types of amino acids Val (valine), Leu (leucine), and Ile (isoleucine) are critical for water stress tolerance in *Arabidopsis thaliana* plants (Pires et al. 2016).

14.7 Conclusion

This chapter depicts the ROS signaling pathway from the genetic level to the phytohormonal and amino acid level. Under abiotic stress, ROS production is under genetic control and triggered by RBOH proteins controlled at the genetic level. RBOH, in turn, activates the MAPK and CDPK cascade that causes modifications in the phytohormonal balance and assists in crop acclimation to abiotic stress. Hence, a complete understanding of ROS signaling pathways in plants through molecular and biochemical means is essential in mitigating stress and securing sustainable cultivation of economically important food crops under changing climatic scenarios.

References

Ahammed GJ, He BB, Qian XJ, Zhou YH, Shi K, Zhou J, Yu JQ, Xia XJ (2017) 24-Epibrassinolide alleviates organic pollutants-retarded root elongation by promoting redox homeostasis and secondary metabolism in *Cucumis sativus* L. Environ Pollut 229:922–931

Ahuja I, de Vos RC, Bones AM, Hall RD (2010) Plant molecular stress responses face climate change. Trends Plant Sci 15:664–674

Alcázar R, Marco F, Cuevas JC, Patron M, Ferrando A, Carrasco P, Tiburcio AF, Altabella T (2006) Involvement of polyamines in plant response to abiotic stress. Biotechnol Lett 28(23): 1867–1876

Amir R (2010) Current understanding of the factors regulating methionine content in vegetative tissues of higher plants. Amino Acids 39:917–931

Angelini J, Klassen R, Široká J, Novák O, Záruba K, Siegel J, Novotná Z, Valentová O (2022) Silver nanoparticles Alter microtubule arrangement. Dyn Stress Phytohormone Levels Plants 11(3):313

Barros JAS, Cavalcanti JHF, Medeiros DB, Nunes-Nesi A, Avin-Wittenberg T, Fernie AR, Araújo WL (2017) Autophagy deficiency compromises alternative pathways of respiration following energy deprivation in Arabidopsis thaliana. Plant Physiol 175:62–76

Bulgari R, Franzoni G, Ferrante A (2019) Biostimulants application in horticultural crops under abiotic stress conditions. Agron 9:306

Chapman JM, Muhlemann JK, Gayomba SR, Muday GK (2019) RBOH-dependent ROS synthesis and ROS scavenging by plant specialized metabolites to modulate plant development and stress responses. Chem Res Toxicol 32(3):370–396

Chen H, Ke Q, Kluz T, Yan Y, Costa M (2006) Nickel ions increase histone H3 lysine 9 dimethylation and induce transgene silencing. Mol Cell Biol 26:3728–3737

Chen X, Hu Y, Zhou DX (2011) Epigenetic gene regulation by plant Jumonji group of histone demethylase. Biochim Biophys Acta 1809:421–426

Choudhury FK, Rivero RM, Blumwald E Mittler R (2017) Reactive oxygen species, abiotic stress and stress combination. Plant J 90:856–867

Czarnocka W, Karpinski S (2018) Friend or foe? Reactive oxygen species production, scavenging and signaling in plant response to environmental stresses. Free Radic Biol Med 122:4–20

Devireddy AR, Zandalinas SI, Gomez-Cadenas A, Blumwald E, Mittler R (2018) Coordinating the overall stomatal response of plants: rapid leaf-to-leaf communication during light stress. Sci Signal 11(518):eaam9514

Devireddy AR, Zandalinas SI, Fichman Y, Mittler R (2021) Integration of reactive oxygen species and hormone signaling during abiotic stress. Plant J 105(2):459–476

Dietz KJ, Turkan I, Krieger-Liszkay A (2016) Redox- and reactive oxygen species-dependent signaling into and out of the photosynthesizing chloroplast. Plant Physiol 171:1541–1550

Dolzblasz A, Gola EM, Sokołowska K, Smakowska-Luzan E, Twardawska A, Janska H (2018) Impairment of meristem proliferation in plants lacking the mitochondrial protease AtFTSH4. Int J Mol Sci 19(3):853

Dong H, Zhen Z, Peng J, Chang L, Gong Q, Wang NN (2011) Loss of ACS7 confers abiotic stress tolerance by modulating ABA sensitivity and accumulation in Arabidopsis. J Exp Bot 62:4875–4887

Duan CG, Wang XG, Tang K, Zhang HM, Mangrauthia SK, Lei MG, Hsu CC, Hou YJ, Wang CG, Li Y, Tao WA (2015) MET18 connects the cytosolic iron-sulfur cluster assembly pathway to active DNA demethylation in Arabidopsis. PLoS Genet 11:e1005559

Ferreira Júnior DC, Gaion LA, Sousa Júnior GS, Santos DMM, Carvalho RF (2018) Drought-induced proline synthesis depends on root-to-shoot communication mediated by light perception. Acta Physiol Plant 40:363

Fichman Y, Mittler R (2020) Rapid systemic signaling during abiotic and biotic stresses: is the ROS wave master of all trades? Plant J 102:887–896

Foyer CH, Noctor G (2005) Redox homeostasis and antioxidant signaling: a metabolic interface between stress perception and physiological responses. Plant Cell 17:1866–1875

Han Y, Chaouch S, Mhamdi A, Queval G, Zechmann B, Noctor G (2013) Functional analysis of Arabidopsis mutants points to novel roles for glutathione in coupling H_2O_2 to activation of salicylic acid accumulation and signaling. Antioxid Redox Signal 18:2106–2121

Hartmann M, Zeier T, Bernsdorff F, Reichel-Deland V, Kim D, Hohmann M, Scholten N, Schuck S, Bräutigam A, Hölzel T, Ganter C (2018) Flavin monooxygenase-generated N-hydroxypipecolic acid is a critical element of plant systemic immunity. Cell 173(2):456–469

Hildebrandt TM (2018) Synthesis versus degradation: directions of amino acid metabolism during Arabidopsis abiotic stress response. Plant Mol Biol 98:121–135

Hildebrandt TM, Nunes Nesi A, Araújo WL, Braun HP (2015) Amino acid catabolism in plants. Mol Plant 8:1563–1579

Hirota T, Izumi M, Wada S, Makino A, Ishida H (2018) Vacuolar protein degradation via autophagy provides substrates to amino acid catabolic pathways as an adaptive response to sugar starvation in *Arabidopsis thaliana*. Plant Cell Physiol 59:1363–1376

Hu L, Yang Y, Jiang L, Liu S (2016) The catalase gene family in cucumber: genome-wide identification and organization. Genet Mol Biol 39:408–415

Huang T, Jander G (2017) Abscisic acid-regulated protein degradation causes osmotic stress-induced accumulation of branched-chain amino acids in *Arabidopsis thaliana*. Planta 246: 737–747

Jagodzik P, Tajdel-Zielinska M, Ciesla A, Marczak M, Ludwikow A (2018) Mitogen-activated protein kinase cascades in plant hormone signaling. Front Plant Sci 9:1387

Jammes F, Song C, Shin D, Munemasa S, Takeda K, Gu D, Cho D, Lee S, Giordo R, Sritubtim S, Leonhardt N (2009) MAP kinases MPK9 and MPK12 are preferentially expressed in guard cells and positively regulate ROS-mediated ABA signaling. Proc Natl Acad Sci 106(48): 20520–20525

Jiang J, Song CP (2008) MEK1/2 and p38-like MAP kinase successively mediate $H_{(2)}O_{(2)}$ signaling in Vicia guard cell. Plant Signal Behav 3:996–998

Kanojia A, Dijkwel PP (2018) Abiotic stress responses are governed by reactive oxygen species and age. Annu Plant Rev 1:295–326

Kawarazaki T, Kimura S, Iizuka A, Hanamata S, Nibori H, Michikawa M, Imai A, Abe M, Kaya H, Kuchitsu K (2013) A low temperature-inducible protein AtSRC2 enhances the ROS-producing activity of NADPH oxidase AtRbohF. Biochim Biophys Acta 1833(12):2775–2780

Kerchev PI, Van Breusegem F (2022) Improving oxidative stress resilience in plants. Plant J 109(2): 359–372

Khan NA, Khan MIR, Ferrante A, Poor P (2017) Ethylene: a key regulatory molecule in plants. Front Plant Sci 8:1782

Kiffin R, Bandyopadhyay U, Cuervo AM (2006) Oxidative stress and autophagy. Antioxidants Redox Sig 8:152–162

Koussevitzky S, Suzuki N, Huntington S, Armijo L, Sha W, Cortes D, Shulaev V, Mittler R (2008) Ascorbate peroxidase 1 plays a key role in the response of *Arabidopsis thaliana* to stress combination. J Biol Chem 283:34197–34203

Kovtun Y, Chiu WL, Tena G, Sheen J (2000) Functional analysis of oxidative stress-activated mitogen-activated protein kinase cascade in plants. Proc Natl Acad Sci U S A 97:2940–2945

Kumar V, Khare T, Sharma M, Wani SH (2017) ROS-induced signaling and gene expression in crops under salinity stress. In: Khan MIR, Khan NA (eds) Reactive oxygen species and antioxidant Systems in Plants: role and regulation under abiotic stress. Springer, Singapore, pp 159–184

Lamers J, van der Meer T, Testerink C (2020) How plants sense and respond to stressful environments. Plant Physiol 182:1624–1635

Lee S, Seo PJ, Lee HJ, Park CM (2012) A NAC transcription factor NTL4 promotes reactive oxygen species production during drought-induced leaf senescence in Arabidopsis. Plant J 70: 831–844

Li CH, Wang G, Zhao JL, Zhang LQ, Ai LF, Han YF, Sun DY, Zhang SW, Sun Y (2014) The receptor-like kinase SIT1 mediates salt sensitivity by activating MAPK3/6 and regulating ethylene homeostasis in rice. Plant Cell 26(6):2538–2553

Li Y, Cai H, Liu P, Wang C, Gao H, Wu C, Yan K, Zhang S, Huang J, Zheng C (2017a) Arabidopsis MAPKKK18 positively regulates drought stress resistance via downstream MAPKK3. Biochem Biophys Res Commun 484(2):292–297

Li Q, Wang Y, Liu C, Pei ZM, Shi WL (2017b) The crosstalk between ABA, nitric oxide, hydrogen peroxide, and calcium in stomatal closing of Arabidopsis thaliana. Biologia 72:1140–1146

Liu P, Zhang H, Yu B, Xiong L, Xia Y (2015) Proteomic identification of early salicylate- and flg22-responsive redox-sensitive proteins in Arabidopsis. Sci Rep 5:8625

Ludwików A (2015) Targeting proteins for proteasomal degradation-a new function of Arabidopsis ABI1 protein phosphatase 2C. Front Plant Sci 6:310

Ma L, Zhang H, Sun L, Jiao Y, Zhang G, Miao C, Hao F (2012) NADPH oxidase AtrbohD and AtrbohF function in ROS-dependent regulation of Na+/K+ homeostasis in Arabidopsis under salt stress. J Exp Bot 63(1):305–317

Marcec MJ, Tanaka K (2021) Crosstalk between calcium and ROS signaling during flg22-triggered immune response in Arabidopsis leaves. Plan Theory 11(1):14

Marshall A, Aalen RB, Audenaert D, Beeckman T, Broadley MR, Butenko MA, Caño-Delgado AI, de Vries S, Dresselhaus T, Felix G, Graham NS (2012) Tackling drought stress: receptor-like kinases present new approaches. Plant Cell 24(6):2262–2278

Martinelli T, Whittaker A, Bochicchio A, Vazzana C, Suzuki A, Masclaux-Daubresse C (2007) Amino acid pattern and glutamate metabolism during dehydration stress in the 'resurrection' plant Sporobolus stapfianus: a comparison between desiccation-sensitive and desiccation-tolerant leaves. J Exp Bot 58:3037–3046

Matsuo M, Johnson JM, Hieno A, Tokizawa M, Nomoto M, Tada Y, Godfrey R, Obokata J, Sherameti I, Yamamoto YY, Böhmer FD (2015) High redox responsive transcription factor1 levels result in accumulation of reactive oxygen species in *Arabidopsis thaliana* shoots and roots. Mol Plant 8(8):1253–1273

Mhamdi A, Queval G, Chaouch S, Vanderauwera S, Van Breusegem F, Noctor G (2010) Catalase function in plants: a focus on Arabidopsis mutants as stress-mimic models. J Exp Bot 61:4197–4220

Mignolet-Spruyt L, Xu E, Idänheimo N, Hoeberichts FA, Mühlenbock P, Brosche M, Van Breusegem F, Kangasjärvi J (2016) Spreading the news: subcellular and organellar reactive oxygen species production and signalling. J Exp Bot 67:3831–3844

Miller G, Schlauch K, Tam R, Cortes D, Torres MA, Shulaev V, Dangl JL, Mittler R (2009) The plant NADPH oxidase mediates rapid systemic signaling in response to diverse stimuli. Sci Signal 2:ra45

Mittler R (2017) ROS are good. Trends Plant Sci 22:11–19

Mittler R, Blumwald E (2015) The roles of ROS and ABA in systemic acquired acclimation. Plant Cell 27:64–70

Mittler R, Vanderauwera S, Suzuki N, Miller G, Tognetti VB, Vandepoele K, Gollery M, Shulaev V, Van Breusegem F (2011) ROS signaling: the new wave? Trends Plant Sci 16: 300–309

Mitula F, Tajdel M, Cieśla A, Kasprowicz-Maluśki A, Kulik A, Babula-Skowrońska D, Michalak M, Dobrowolska G, Sadowski J, Ludwików A (2015) Arabidopsis ABA-activated kinase MAPKKK18 is regulated by protein phosphatase 2C ABI1 and the ubiquitin–proteasome pathway. Plant Cell Physiol 56(12):2351–2367

Nimse SB, Pal D (2015) Free radicals, natural antioxidants, and their reaction mechanisms. RSC Adv 5:27986–28006

Niu Y, DesMarais TL, Tong Z, Yao Y, Costa M (2015) Oxidative stress alters global histone modification and DNA methylation. Free Radic Biol Med 82:22–28

Noctor G, Reichheld JP, Foyer CH (2018) ROS-related redox regulation and signaling in plants. Semin Cell Dev Biol 80:3–12

Osakabe Y, Yamaguchi-Shinozaki K, Shinozaki K, Tran LS (2013) Sensing the environment, key roles of membrane-localized kinases in plant perception and response to abiotic stress. J Exp Bot 64:445–458

Pfannschmidt T, Schutze K, Fey V, Sherameti I, Oelmuller R (2003) Chloroplast redox control of nuclear gene expression-a new class of plastid signals in interorganellar communication. Antioxidants and Redox Sig 5:95–101

Pikaard CS, Scheid OM (2014) Epigenetic regulation in plants. Cold Spring Harb Perspect Biol 6(12):019315

Pires MV, Pereira Júnior AA, Medeiros DB, Daloso DM, Pham PA, Barros KA, Engqvist MK, Florian A, Krahnert I, Maurino VG, Araújo WL (2016) The influence of alternative pathways of respiration that utilize branched-chain amino acids following water shortage in Arabidopsis. Plant Cell Environ 39(6):1304–1319

Pnueli L, Liang H, Rozenberg M, Mittler R (2003) Growth suppression, altered stomatal responses, and augmented induction of heat shock proteins in cytosolic ascorbate peroxidase (Apx1)-deficient Arabidopsis plants. Plant J 34:187–203

Prodhan MY, Munemasa S, Nahar MNEN, Nakamura Y, Murata Y (2018) Guard cell salicylic acid signaling is integrated into abscisic acid signaling via the Ca^{2+}/CPK-dependent pathway. Plant Physiol 178:441–450

Pucciariello C, Banti V, Perata P (2012) ROS signaling as common element in low oxygen and heat stresses. Plant Physiol Biochem 59:3–10

Qu Y, Song P, Hu Y, Jin X, Jia Q, Zhang X, Chen L, Zhang Q (2018) Regulation of stomatal movement by cortical microtubule organization in response to darkness and ABA signaling in Arabidopsis. Plant Growth Regul 84:467–479

Raza A, Razzaq A, Mehmood SS, Zou X, Zhang X, Lv Y, Xu J (2019) Impact of climate change on crops adaptation and strategies to tackle its outcome: a review. Plan Theory 8:34

Raza A, Su W, Gao A, Mehmood SS, Hussain MA, Nie W, Lv Y, Zou X, Zhang X (2021) Catalase (CAT) gene family in rapeseed (*Brassica napus* L.): genome-wide analysis, identification, and expression pattern in response to multiple hormones and abiotic stress conditions. Int J Mol Sci 22(8):4281

RM SK, Wang Y, Zhang X, Cheng H, Sun L, He S, Hao F (2020) Redox components: key regulators of epigenetic modifications in plants. Int J Mol Sci 21(4):1419

Rodrigues O, Reshetnyak G, Grondin A, Saijo Y, Leonhardt N, Maurel C, Verdoucq L (2017) Aquaporins facilitate hydrogen peroxide entry into guard cells to mediate ABA-and pathogen-triggered stomatal closure. Proc Natl Acad Sci 114(34):9200–9205

Sachdev S, Ansari SA, Ansari MI, Fujita M, Hasanuzzaman M (2021) Abiotic stress and reactive oxygen species: generation, signaling, and defense mechanisms. Antioxidants 10(2):277

Saini P, Gani M, Kaur JJ, Godara LC, Singh C, Chauhan SS, Francies RM, Bhardwaj A, Kumar NB, Ghosh MK (2018) Reactive oxygen species (ROS): a way to stress survival in plants. In: Zargar SM, Zargar MY (eds) Abiotic stress-mediated sensing and signaling in plants: an omics perspective. Springer, Singapore, pp 127–153

Sasidharan R, Hartman S, Liu Z, Martopawiro S, Sajeev N, van Veen H, Yeung E, Voesenek LA (2018) Signal dynamics and interactions during flooding stress. Plant Physiol 176:1106–1117

Schutzendubel A, Polle A (2002) Plant responses to abiotic stresses: heavy metal-induced oxidative stress and protection by mycorrhization. J Exp Bot 53:1351–1365

Sewelam N, Kazan K, Schenk PM (2016) Global plant stress signaling: reactive oxygen species at the cross-road. Front Plant Sci 7:187

Shen Y, Issakidis-Bourguet E, Zhou DX (2016) Perspectives on the interactions between metabolism, redox, and epigenetics in plants. J Exp Bot 67:5291–5300

Showler AT (2002) Effects of water deficit stress, shade, weed competition, and kaolin particle film on selected foliar free amino acid accumulations in cotton, *Gossypium hirsutum* (L.). J Chem Ecol 28:631–651

Simkova K, Moreau F, Pawlak P, Vriet C, Baruah A, Alexandre C, Hennig L, Apel K, Laloi C (2012) Integration of stress-related and reactive oxygen species-mediated signals by topoisomerase VI in *Arabidopsis thaliana*. Proc Natl Acad Sci 109:16360–16365

Singh A, Kumar A, Yadav S, Singh IK (2019) Reactive oxygen species-mediated signaling during abiotic stress. Plant Gene 18:100173

Sinha AK, Jaggi M, Raghuram B, Tuteja N (2011) Mitogen-activated protein kinase signaling in plants under abiotic stress. Plant Signal Behav 6:196–203

Suzuki N, Miller G, Morales J, Shulaev V, Torres MA, Mittler R (2011) Respiratory burst oxidases: the engines of ROS signaling. Curr Opin Plant Biol 14:691–699

Tajdel M, Mitula F, Ludwikow A (2016) Regulation of Arabidopsis MAPKKK18 by ABI1 and SnRK2, components of the ABA signaling pathway. Plant Signal Behav 11:e1139277

Tian S, Wang X, Li P, Wang H, Ji H, Xie J, Qiu Q, Shen D, Dong H (2016) Plant aquaporin AtPIP1;4 links apoplastic H_2O_2 induction to disease immunity pathways. Plant Physiol 171: 1635–1650

Tognetti VB, Mühlenbock P, Van Breusegem F (2012) Stress homeostasis the redox and auxin perspective. Plant Cell Environ 35:321–333

Vaid N, Macovei A, Tuteja N (2013) Knights in action, lectin receptor-like kinases in plant development and stress responses. Mol Plant 6:1405–1418

Vall-Llaura N, Fernández-Cancelo P, Nativitas-Lima I, Echeverria G, Teixidó N, Larrigaudière C, Torres R, Giné-Bordonaba J (2022) ROS-scavenging-associated transcriptional and biochemical shifts during nectarine fruit development and ripening. Plant Physiol Biochem 171:38–48

Wang P, Song CP (2008) Guard-cell signalling for hydrogen peroxide and abscisic acid. New Phytol 178:703–718
Waszczak C, Carmody M, Kangasjärvi J (2018) Reactive oxygen species in plant signaling. Annu Rev Plant Biol 69:209–236
Woodson JD, Chory J (2008) Coordination of gene expression between organellar and nuclear genomes. Nat Rev Genet 9:383–395
Wu Y, Yang Z, How J, Xu H, Chen L, Li K (2017) Overexpression of a peroxidase gene (AtPrx64) of *Arabidopsis thaliana* in tobacco improves plant's tolerance to aluminum stress. Plant Mol Biol 95:157–168
Xia XJ, Zhou YH, Shi K, Zhou J, Foyer CH, Yu JQ (2015) Interplay between reactive oxygen species and hormones in the control of plant development and stress tolerance. J Exp Bot 66: 2839–2856
Xu C, Liu R, Zhang Q, Chen X, Qian Y, Fang W (2017) The diversification of evolutionarily conserved MAPK cascades correlates with the evolution of fungal species and development of lifestyles. Genome Biol Evol 9:311–322
Ye L, Li L, Wang L, Wang S, Li S, Du J, Zhang S, Shou H (2015) MPK3/MPK6 are involved in iron deficiency-induced ethylene production in Arabidopsis. Front Plant Sci 6:953
Yi C, Yao K, Cai S, Li H, Zhou J, Xia X, Shi K, Yu J, Foyer CH, Zhou Y (2015) High atmospheric carbon dioxide-dependent alleviation of salt stress is linked to RESPIRATORY BURST OXIDASE 1 (RBOH1)-dependent H2O2 production in tomato (*Solanum lycopersicum*). J Exp Bot 66(22):7391–7404
Zandalinas SI, Fichman Y, Devireddy AR, Sengupta S, Azad RK, Mittler R (2020) Systemic signaling during abiotic stress combination in plants. Proc Natl Acad Sci 117:13810–13820
Zandalinas SI, Balfagón D, Gómez-Cadenas A, Mittler R (2022) Responses of plants to climate change: metabolic changes during abiotic stress combination in plants. J Exp Bot erac073
Zemach A, Kim MY, Hsieh PH, Coleman-Derr D, Eshed-Williams L, Thao K, Harmer SL, Zilberman D (2013) The Arabidopsis nucleosome remodeler DDM1 allows DNA methyltransferases to access H1-containing heterochromatin. Cell 153:193–205
Zhang A, Jiang M, Zhang J, Tan M, Hu X (2006) Mitogen-activated protein kinase is involved in abscisic acid-induced antioxidant defense and acts downstream of reactive oxygen species production in leaves of maize plants. Plant Physiol 141:475–487
Zhang A, Jiang M, Zhang J, Ding H, Xu S, Hu X, Tan M (2007a) Nitric oxide induced by hydrogen peroxide mediates abscisic acid-induced activation of the mitogen-activated protein kinase cascade involved in antioxidant defense in maize leaves. New Phytol 175:36–50
Zhang F, Wang Y, Yang Y, Wu HAO, Wang DI, Liu J (2007b) Involvement of hydrogen peroxide and nitric oxide in salt resistance in the calluses from Populus euphratica. Plant Cell Environ 30(7):775–785
Zhang Y, Zhang Y, Luo L, Lu C, Kong W, Cheng L, Xu X, Liu J (2022) Genome wide identification of respiratory burst oxidase homolog (Rboh) genes in Citrus sinensis and functional analysis of CsRbohD in cold tolerance. Int J Mol Sci 23(2):648
Zhou J, Wang J, Li X, Xia XJ, Zhou YH, Shi K, Chen Z, Yu JQ (2014) H_2O_2 mediates the crosstalk of brassinosteroid and abscisic acid in tomato responses to heat and oxidative stresses. J Exp Bot 65:4371–4383
Zhou H, Finkemeier I, Guan W, Tossounian MA, Wei B, Young D, Huang J, Messens J, Yang X, Zhu J, Wilson MH (2018) Oxidative stress-triggered interactions between the succinyl-and acetyl-proteomes of rice leaves. Plant Cell Environ 41(5):1139–1153
Zhou X, Joshi S, Patil S, Khare T, Kumar V (2022) Reactive oxygen, nitrogen, carbonyl and sulfur species and their roles in plant abiotic stress responses and tolerance. J Plant Growth Regul 41(1):119–142

Adverse Impact of ROS on Nutrient Accumulation and Distribution in Plants 15

Iqra Akhtar, Sumera Javad, Khajista Jabeen, Amina Tariq, Komal Nawaz, Anis Ali Shah, Ramish Nida, and Nimra Kousar

Abstract

Reactive oxygen species are produced as a normal product of plant cellular metabolism and play important roles in maintaining normal plant growth and improving their tolerance to stress. Reactive oxygen species (ROS) serve as cell signaling molecules for normal biologic processes. ROS play two divergent roles in plants such as in low concentrations they act as signaling molecules that mediate several plant responses in plant cells, including responses under stresses, whereas in higher concentrations, they cause exacerbating damage to cellular components. Enhanced level of ROS causes oxidative damage to lipid, protein, and DNA leading to altered intrinsic membrane properties like fluidity, ion transport, loss of enzyme activity, protein cross-linking, inhibition of protein synthesis, and DNA damage and ultimately resulting in cell death. The regulatory network comprising enzymatic and nonenzymatic antioxidant systems tends to keep the magnitude of ROS within plant cells to a non-damaging level. ROS are similar to a double-edged sword and, when present below the threshold level, mediate redox signaling pathways that actuate plant growth, development, and acclimatization against stresses. ROS plays vital role in the mobilization and remobilization of minerals and nutrients. The filling of grain with macro- and micronutrients is partly the result of a direct allocation from root uptake and remobilization from vegetative tissues. Nitrogen is always remobilized from leaves of all plant species, with different efficiencies, while nutrients such as K, S, P, Mg, Cu, Mo, Fe, and Zn can be mobilized to a certain extent when plants are

I. Akhtar · S. Javad (✉) · K. Jabeen · A. Tariq · R. Nida · N. Kousar
Department of Botany, Lahore College for Women University, Lahore, Pakistan

K. Nawaz · A. A. Shah
Department of Botany, Division of Science and Technology, University of Education, Lahore, Pakistan

M. Faizan et al. (eds.), *Reactive Oxygen Species*,
https://doi.org/10.1007/978-981-19-9794-5_15

facing deficiencies. On the opposite side, there are very few evidences for leaf mobilization of Ca, Mn, Ni, and B. The role of ROS in mechanisms related to the mobilization, remobilization, and partitioning of these mineral nutrients will be discussed in this chapter.

Keywords

Abiotic · Biotic · Hydrogen peroxide · Metabolism · Nutrients · Stress

15.1 Introduction

Plants obtain various chemicals from their environment for maintaining their life cycle, i.e., vegetative and reproductive cycle. They use these substances as their own energy source and also for making their own food and cellular components. All these chemicals which they require from their environment are known as "nutrients." And all the processes that are involved in the conversion of simpler nutrients into cellular components or use of these nutrients as an energy source are called metabolism of that plant. Therefore, nutrition and metabolism are considered very close to each other (Mengel et al. 2001). Even supplementation of plants with extra nutrients can help plants to survive from various stress conditions like drought stress (Seleiman 2019).

Some of the nutrients required by plants are essential, as plants are not able to live without them. These nutrients are involved in making primary metabolites of plants. There are 16 nutrients considered as essential for all types of plants. But these nutrients may be required in different quantities by plants. They may be macronutrients (required in larger quantities) or micronutrients (required in lesser quantities). Nutrients may be mobile or immobile nutrients (Pandey 2018). Sometimes nutrients may become immobile, and their absorption by roots is restricted due to some specific conditions like pH, i.e., if the pH of soil solution is higher, zinc, iron, boron, and manganese are stuck and are not absorbed easily by plants, whereas some nutrients are stuck when there is low pH like nitrogen, magnesium, calcium, and potassium.

Deficiency of mobile nutrients first appears in the older parts of plants, whereas the deficiency of immobile nutrients appears in younger leaves first (Saleem et al. 2020). Mobile nutrients can reach to all areas of active growth and in all directions via vascular tissue. Mobile nutrients are potassium, nitrogen, sulfur, phosphorous, and magnesium. For mobile nutrient both methods of applications, i.e., foliar and soil application, can be applied, whereas immobile nutrients can't be distributed again and are only once distributed through xylem tissue. These nutrients can't move to active growth areas easily. Immobile nutrients include zinc, iron, boron, iron, calcium, copper, etc. (Vashisth and Oswalt 2020), Fig. 15.1.

Fig. 15.1 Distribution of mobile and immobile nutrients

15.2 Nutrient Accumulation and Distribution in Plants

Autotrophic organisms like plants produce carbohydrates by using carbon dioxide (CO_2), the energy of solar radiation, nutrients, and water as essential components of their biomass (Demura and Ye 2010). Plant roots transport minerals mostly in the form of nutrients after absorbing them from the soil (López-Arredondo et al. 2013).

Organic matter and sedimentation are common sources of nutrient accumulation. A large portion of nutrients in peat-dominated wetlands are stored in microbial biomass, stabilized soil organic matter, and live and detrital plant tissue. Natural decomposition and die-off methods released nutrients from microbial biomass and vegetation (Chen et al. 2004). Nutrient storage in the vegetation of herbaceous wetlands is a short-time process (Wang et al. 2018), as compared to forested wetlands where nutrient storage in tissues of woody trees is long term (Ayub et al. 2021).

Forest fire can alter the soil biota, pH, soil texture, soil organic matter, color, and macro- and micronutrients (Zhang and Biswas 2017). Plant nutrient availability increased after the combustion of organic matter and litter with low-intensity fires. Forests affected by high-intensity fires can start their recovery after supplying plant nutrients. On the other hand, all forest species required different amounts of nutrients for growth; hence, rapid recovery of species is possible after knowing optimum concentrations of nutrients (Mittler 2017).

Nutrients have a different impact on the length of root, stem, several internodes, reduction tillering, number of shoots, number of nodes, primary and secondary root growth, root hair growth, photosynthesis, and reproductive growth in all types of plants (Kaznina and Titov 2014). If nutrients available to plants are not in good quantity or are not distributed to all plant parts due to any reason, aberration may arise.

For productive crop growth, optimum application of nutrients is preessential like macro- (N, P, K, Ca, Mg, S, and Na) and micronutrients (Fe, Mn, Zn, Cu, and B) for proper development and growth of plants. During the life cycle of a typical plant, deficiency of any nutrient can affect its growth significantly (Schippers et al. 2012).

15.3 Macronutrients

In plant metabolism, mineral nutrients play an important role in plant development and growth. Nitrogen, phosphorus, and potassium are vital minerals. Nitrogen is required in high concentrations as it is the main component of protein. Phosphorus is the key element of NADPH and ATP (Pandey 2018; Marschner 1995). Both of these molecules are part of energy cycles of cells through the process of oxidation and reduction.

15.3.1 Nitrogen

Green plants use carbon from the atmosphere for photosynthesis, while doing this, they are not only reducing carbon levels from the atmosphere but also providing energy in the form of food for all life forms. In the synthesis of living material like protein and nucleic acid, nitrogen plays a significant role because for the translation of genetic material nucleic acid (having nitrogen as backbone) is required, while protein (amino group in basic structure) is required as an enzyme or catalyst. Proteins of thylakoid and Calvin cycle in the photosynthesis of C_3 plants represent leaf nitrogen. So, in the leaves, nitrogen is the main component, while the absence of nitrogen in leaves shows chlorosis. Chlorosis appears in older leaves than in younger ones. In the absence of nitrogen, protein production is reduced, and carbohydrate production is increased due to which slender stems are shown by plants. Therefore, the plant starts to produce purple coloration and anthocyanin in leaves, petioles, and stems (Bonfim-Silva et al. 2015).

When nitric oxide synthetase acts on L-arginine, it is converted to L-citrulline with production of nitric oxide which is a highly diffusible gas and acts as a signaling molecule. NO is considered as the signaling gas molecule that is a subgroup of ROS. It provides resistance to plants against a number of stresses like diseases (Pandey 2018). It is also known to increase salt and drought tolerance in plants by modulation of ion homoeostasis (Yu et al. 2014; Domingos et al. 2015).

15.3.2 Phosphorus

Phosphorus is an essential component of lipids, nucleic acids, and all membrane proteins. Phosphorus not only stimulates flowering and root growth but also increases the cell division due to high metabolism in different parts of plants. It plays a major role in energy transfer and storage during photosynthesis (Huang et al. 2016). Phosphorus shortage delays plant maturity and causes the dark green color of leaves and the stunted growth of young plants. Necrotic spots on leaves appear due to the absence of phosphorus. Due to excessive production of anthocyanins, leaves appear slightly purple, but it is not associated with chlorosis. The appearance of a slender stem due to a lack of phosphorus is similar to nitrogen. Phosphorus plays a role as sugar phosphate intermediate and also in photosynthesis and respiration. For the reproductive growth of crops, phosphorus is an essential component in soil (Che et al. 2020; Denison and Kiers 2005), and to maintain soils for crop production, it is compulsory to add accessible forms of P (White and Brown 2010; Fixen 2005; Sims 2000).

15.3.3 Potassium

Potassium plays a vital role in enzyme activation and osmotic potential to activate enzymes involved in photosynthesis, respiration, and regulation of plant cells. It plays a role in enzyme activation involved in respiration and photosynthesis. Studies investigated that in plants translocation of photosynthate is promoted by K. It is a key element in controlling the stomatal movements throughout the day. Its shortage causes mostly necrosis at the tips of leaves and marginal chlorosis. Mostly deficiency symptoms appear in mature leaves than in young leaves due to potassium meta-mobilization. The necrotic lesion appears in monocots initially at the margin and tips of leaves as compared to the base of leaves. Chances of fungal attacks on roots increase in potassium-deficient plant (Malvi 2011).

15.3.4 Sulfur

Coenzymes and vitamin complexes that contain sulfur-containing amino acids are important for metabolism. Sulfur deficiency symptoms are almost similar to nitrogen deficiency symptoms like stunning growth and chlorosis. In younger leaves, sulfur is not easily remobilized and at the initial level shortage of sulfur appears in mature leaves (Hussain et al. 2013).

15.3.5 Calcium

In the middle lamellae synthesis of the cell wall, calcium plays an important role. Calcium also plays a crucial role in mitotic spindle synthesis for membrane functioning and acts as a second messenger for signaling processes. It also forms a protein that regulates metabolic processes known as calcium-calmodulin which is present in the cytosol. Its deficiency causes the death of the young meristematic region due to Ca involvement in the cell wall and also leads to necrosis in the tips of young leaves and roots. Calcium is required for elongation, cell division, and stability of cell walls by the formation of calcium pectate (Draycott and Christenson 2003).

15.3.6 Magnesium

Magnesium ions are very important for the activation of some enzymes which are playing their role in respiration, nucleic acid synthesis, and photosynthesis. Mg is part of the chlorophyll molecule, and its long-term deficiency leads toward the development of yellowish and white leaves, leaf abscission, and chlorosis. For reactions associated with energy (ATP) transfer, respiration, enzyme activation, protein synthesis, and photosynthesis, Mg is needed by plants (Draycott and Christenson 2003; Christenson and Draycott 2006).

15.4 Micronutrients

Micronutrients are those elements that are necessary for plants but required in lesser quantities for metabolic activities of plants, particularly for enzyme activation designed for catalysis reaction.

15.4.1 Boron

Still, the exact function of boron is not clear, but it is recommended that it plays a crucial role in the nucleic acid synthesis and cell elongation. For normal growth of some diatom species and plants, it is essential. Boron is absorbed by roots in the form of $B(OH)_3$ following the passive diffusion pathway as it is a small molecule that is uncharged. There is direct or indirect role of boron in many physiological processes of plants including amino acid synthesis, carbohydrate metabolism, hormone synthesis, nucleotide synthesis, phenol metabolism, pollen tube formation, protein metabolism, RNA synthesis, root growth, and transport of sugars (Uluisik et al.

2018). Boron is also involved in the stability of domains in plasma membrane as it has the ability to bind with glycoproteins, phosphoinositides, and glycolipids on the plasma membrane of the plants (Brown et al. 2002). Boron is also required for cell division and elongation as it is the main requirement of the meristematic regions of plants. Young meristem requires boron very actively. If boron is not supplied to plants in proper amounts, it can cause poor development of xylem vessels (Kohli et al. 2022; Wimmer and Eichert 2013). However, deficiency of boron leads to anatomical changes, physiology, and biochemistry of plant cells. But to find out the exact function of boron is difficult; maybe it takes part in membrane-level functions. Other imaginable roles of boron are the integrity of cell wall structure, sugar transport, respiration, lignification, phenol metabolism, and IAA metabolism. The reported concentration of boron is 20–200 mg B/Kg but it varies from soil to soil (Ahmad et al. 2012).

15.4.2 Zinc

For enzyme activation, an ionic form of zinc is needed which is involved in many metabolic activities like activation of DNA polymerases, DNA replication, hydrogenase, chlorophyll biosynthesis, synthesis of cytochrome, carbonic anhydrase, and stabilization of ribosomal fractions. Zinc deficiency appears in small leaves, reduction in internodal growth, and chlorosis of leaves which predicts the need for zinc for chlorophyll biosynthesis. Zinc-activated plants are involved in pollen formation and carbohydrate metabolism. Zinc is also required for tryptophan biosynthesis which is essential for the synthesis of auxins (Alloway 2004; Marschner 1995). Zinc is having a very critical role in mitigating various stresses from plants like drought, salinity, and heavy metal stress (Umair Hassan et al. 2020). Interaction of zinc with phospholipids and sulfhydryl group of membrane proteins helps in the maintenance of the membrane (Kabata and Pendias 2001; Dang et al. 2010; Alloway 2004). Crops respond positively to zinc application (Welch 2002). Petrochemical and geochemical weathering processes from rocks add zinc to soil presently. Availability of zinc in soil reduces due to high $CaCO_3$, high pH, phosphate, and clay, as these causes fixation of available zinc in soil. In sandy and acidic soil, zinc is commonly found in lower concentrations (Natasha et al. 2022; Welch 2002; Sillanpää 1990).

15.4.3 Manganese

The ionic form of manganese is required for the activation of enzymes in plants that are involved in electron transport chain and tricarboxylic acid cycle. During photosynthesis, it helps in the assimilation of carbon dioxide and release of oxygen molecules after water molecule breakdown and biosynthesis of chlorophyll (Marschner 1995). Its activity is needed in the formation of riboflavin, carotene, and ascorbic acid. A lesser concentration of manganese is necessary for a plant than a higher concentration, and it interferes with other mineral utilization such as Ca, Fe,

Mg, and P via few inhibitory effects on translocation and absorption. Major symptoms of manganese deficiency are intravenous chlorosis with subsequent formation of a necrotic spot in older and younger leaves of different plants (Rengel 2015). The oxidation state of manganese defines the availability of manganese such as Mn^{2+} is available for plants than Mn4+. This reduction procedure may be chemical and biological (Rengel 2000). In an alkaline medium, manganese chemistry is not defined, and its availability is reduced (Pan et al. 2014; Clark and Baligar 2000). Soil distribution, microorganisms, and responses of plants define the supply of manganese; therefore it is a complex mechanism, and mobilization of manganese around the root zone via root exudate is not clear (George et al. 2014; Mora et al. 2009; Gherardi and Rengel 2004).

15.4.4 Molybdenum

In plants for the activation of enzymatic reactions which include purine degradation, nitrogen assimilation, sulfite detoxification, and hormone synthesis, molybdenum is required which is a transitional metal. In the native state, this enzyme is inactive; however, when it forms a complex structure with organic pterin and molybdenum cofactor, it is in an active state (Bittner 2014). Most molybdenum enzymes need iron-containing redox groups, so molybdenum and iron have a close connection. Nitrogenase and nitrate reductase enzymes have these ion components. Molybdenum deficiency may cause nitrogen deficiency, chlorosis between older leaves and veins, and prevention of flower formation in plants. Crop production can increase after adding molybdenum to the soil in low concentrations. pH above 5.5 is good for availability of molybdate, but lower-pH conditions limited the assimilation of molybdenum, and as a result, growth and yield of plants are reduced (Srivastava et al., 2018). Excessive molybdenum can cause yellowish leaves (Kaiser et al. 2005) and a reduced seedling growth (Kumchai et al. 2013). In agriculture, soil concentration of molybdenum from 0.2 to 5.0 mg/Kg is required (Scheffer and Schachtschabel 2002). Molybdate ion concentration in plants and soil solutions is affected by Mn, Fe, clay minerals, Al oxides, and organic carbon, while soil pH releases ions into soil solution (Xu et al. 2013). Anions of molybdate are adsorbed on Mn, Al oxides, and Fe, and also on organic colloids and clay minerals during acidic conditions (Jiang et al. 2015). Due to leaching, lesser concentration of molybdenum is present in well-drained sandy soils, while high concentration is present in the wet soils. The concentration range of molybdenum in soil varies with soil type and regular phosphorus supply (Rutkowska et al. 2017).

15.4.5 Iron

As an enzyme component, iron plays an important role during redox reactions and in electron transfer during photosynthesis and respiration, while oxidizing reversibly from Fe^{2+} to Fe^{3+}. It is involved in the synthesis of chlorophyll and maintenance of chloroplast structure. It also helps to mitigate the effects of heavy metal stress. Iron deficiency symptom is intravenous chlorosis and chlorotic which appears due to prolonged iron deficiency, and the whole leaf may turn white. For chlorophyll-protein complex synthesis in leaves, iron is required. Phytoferritin is also observed in leaves and other parts of plants (Rout and Sahoo 2015; Oh et al. 1996).

15.4.6 Copper

Enzymes that are involved in redox reactions are associated with a copper element just like iron. One example of this is the plastocyanin enzyme which transfers electrons during the light reaction of photosynthesis. Copper is an essential part of the structure of various proteins with regulatory function and can play a number of essential roles for plants including cell wall metabolism, ethylene sensing, hormone signaling, mitochondrial respiration, photosynthetic electron transport, protein synthesis, and responses to oxidative stress. Enzymes like amino oxidase, Cu/Zn superoxide dismutase, cytochrome *c* oxidase, laccase, plastocyanin, and polyphenol oxidase have copper as their cofactor (Shabbir et al. 2020). Copper deficiency's initial symptoms are necrotic spots and dark green leaves, while under extreme deficiency of copper, leaves abscise prematurely (Ishka and Vatamaniuk 2020; Haehnel et al. 1989).

15.4.7 Nickel

In the earth's crust, nickel is the 22nd most abundant element and is also found in trace amounts in natural soils (Hussain et al. 2013). It is important for metabolic activities, for example, for some superoxide dismutases, glyoxalase-I, hydrogenases, methyl-coenzyme M reductase, and carbon monoxide dehydrogenase (Brown 2007). In higher plants only the urease enzyme is present which contains nickel. To reprocess hydrogen gas which is released during nitrogen fixation, nitrogen-fixing bacteria require nickel. Deficiency symptoms of nickel appear as urea accumulation and leaf tip necrosis (Polacco et al. 2013).

15.4.8 Chlorine

To split water complex during photosynthesis, plants require chlorine ions. In a cell division of roots and leaves, it plays a vital role. Due to chlorine deficiency, a bronze color appears in plant leaves and stunted and thickened root tips. For normal metabolic activities, some plants required high concentrations of chlorine (Clarke and Eaton-Rye 2000).

15.5 Reactive Oxygen Species (ROS)

Reactive oxygen is generated when the molecular oxygen (O_2) of an organism readily accepts electrons from other molecules. In plants, oxygen molecule is definitely a stable molecule that is not reactive, but it is a source of all ROS directly or indirectly. However, it may be changed into highly reactive oxygen species (ROS) in a number of organelles through various mechanisms that disrupt plant metabolism (Mittler 2017). Species that are in more reactive state than molecular oxygen are called ROS.

For example, superoxide (primary ROS) is formed when molecular oxygen is reduced by one electron Eq. (15.1). NADPH acts as a catalyst and supplies electrons Eq. (15.2). If oxygen is further reduced, hydrogen peroxide is produced. At low pH, it will be formed due to dismutation of superoxide Eq. (15.3).

$$O_2 + e^- \rightarrow O_2^- \text{ (superoxide)} \tag{15.1}$$

$$2O_2 + NADPH^+ \rightarrow 2O_2 + NADP^+ + H^+ \tag{15.2}$$

$$2O_2^- + 2H^+ \rightarrow H_2O_2 + O_2 \tag{15.3}$$

Superoxide dismutase (SOD) enzyme acts as a catalyst in this reaction. Through the Fenton or Haber-Weiss reaction in the presence of metal ion, further reduction of oxygen will lead to formation of (OH') hydroxyl radical. This radical is highly reactive and will react with the first molecule it encounters and has short half-life.

So it seems that ROS production will occur after the synthesis of superoxide anions. Some ROS are beneficial like hydrogen peroxide or nitrogen oxide which acts as a key signaling molecule, while some are harmful to biological systems (Domingos et al. 2015; Hancock et al. 2001). These may be useful up to a conc. limit o, above which they become harmful, even deadly.

Reactive oxygen species (ROS) in biological systems primarily comprise of the following:

1. Hydrogen peroxide (H_2O_2)
2. Hydroxyl radical (OH^-)
3. Singlet oxygen ($^1O_2$)
4. Superoxide anion radical (O_2^{-2})
5. Peroxyl radical (HOO-)
6. Hypochlorous acid (HOCl) (Rasool et al. 2013; Koyro et al. 2012; Ahmad et al. 2010a, 2010b, 2011, 2012; Ahmad and Umar 2011; Apel and Hirt 2004)

15.6 Sites of ROS Production

In plant cells, chloroplast, the cell membrane, peroxisomes, and mitochondria are sites where ROS are produced (Movafegh and Hoseini 2013). In both normal and stress condition, ROS is produced at different sites like cell wall, plasma membranes, chloroplasts, endoplasmic reticulum, mitochondria, and peroxisomes. The major sources of ROS production in the presence of light are chloroplasts and peroxisomes, while in dark conditions the major source of ROS production is mitochondria (Choudhury et al. 2013).

The sites of ROS production include the following.

(a) Chloroplast

In the thylakoids the major source of ROS production is the photosystems (PSI and PSII). The stressful conditions like salinity, drought, and temperature extremes cause water shortage and limit the concentration of CO_2 which will lead to formation of ROS ($O_2^{\bullet -}$) through Mehler reaction.

$$2O_2 + 2Fd_{red} \rightarrow 2O_2^{\bullet -} + 2Fd_{ox}$$

In PSI the $O_2^{\bullet -}$ is converted into H_2O_2 by a membrane-bound SOD (Cu/Zn) (Miller et al. 2010). The 2Fe-2S and 4Fe-4S clusters are involved in the electron leakage from PSI's ETC. In PSII, QA and QB electron acceptors are responsible for electron leakage and generation of $O_2^{\bullet -}$. At Fe-S center by Fenton reaction, $O_2^{\bullet -}$ is converting into more toxic ROS like $OH^{\bullet}$. PSII is involved in the generation of $1O_2$ by two ways which are as follows:

1. The balance between light harvesting and energy utilization is disturbed by environmental stress, and it leads to the formation of triplet Chl (3Chl$*$) which in turn reacts with dioxygen ($3O_2$) and liberates singlet oxygen ($1O_2$) (Karuppanapandian et al. 2011).
2. With LHC (light-harvesting complex) at PSII, $1O_2$ is generated when the ETC is over-reduced (Asada 2006). The major source of ROS production is chloroplast. The accumulation of singlet oxygen ($1O_2$) in chloroplast leads to peroxidation of membrane lipids and polyunsaturated fatty acids and damages membrane proteins due to which the reaction center of PSII is at risk (Triantaphylidès et al. 2008; Møller et al. 2007).

(b) Mitochondria

Harmful ROS (H_2O_2 and $O_2^{\bullet -}$) are generated in mitochondria on a smaller scale (Navrot et al. 2007). The primary offender is the mitochondrial ETC (mtETC), which has enough energized electrons to reduce O_2 and produce ROS. Complex I and complex III are two major components of mtETC (Møller et al. 2007; Noctor et al. 2007). In flavoprotein region, O_2 is reduced into $O_2^{\bullet -}$ due to NADH dehydrogenase or complex I. When electron flows from complex III to complex I due to lack of NAD^+-linked substrates, production of ROS is increased. ATP hydrolysis controls this electron flow (Turrens 2003). In complex III, reduced ubiquinone donates an electron to cytochrome c1 leaving behind an unstable ubisemiquinone semi-radical which causes seepage of electrons to O_2 and generates $O_2^{\bullet -}$ (Murphy 2009). Various enzymes in the mitochondrial matrix are also responsible for production of ROS directly like aconitase and indirectly like 1-galactono-γ-lactone (GAL) dehydrogenase (Rasmusson et al. 2008). Mn-SOD and the APX convert $O_2^{\bullet -}$ into H_2O_2 (Sharma et al. 2012). Out of the total O_2 consumption from mitochondria, it is estimated that 1–5% is utilized for production of H_2O_2. In normal condition, mitochondria produce ROS normally, but in stress condition, it is highly boosted (Pastore et al. 2007).

(c) Peroxisomes

The single-membrane-bounded spherical bodies in cytoplasm are called peroxisomes. Oxidative metabolism occurs in these bodies which results in the production of H_2O_2, and these bodies are considered as a site of ROS synthesis (Luis et al. 2006; Palma et al. 2009). During various metabolic processes like in chloroplast and mitochondria, these also produce $O_2^{\bullet -}$ at two different sites. The sites for ROS synthesis are as follows:

1. In the peroxisomal matrix, xanthine oxidase converts xanthine and hypoxanthine into uric acid and as a byproduct produces $O_2^{\bullet -}$.
2. In the peroxisomal membrane, an ETC consists of NADH, and cytochrome b utilizes O_2 as the electron acceptor and releases $O_2^{\bullet -}$ into the cytosol.

Three membrane proteins are also responsible for production of $O_2^{\bullet -}$; they form peroxisomal membrane polypeptides (PMPs) that have molecular mass 18, 29, and 132 kDa. For 18 and 32 KDa PMPs, NADH acts as the electron donor, while 29 KDa PMP utilizes NADPH as electron donor for reduction of cyt c. In water stress, the availability of water is low which results in closure of stomata, due to which balance between CO_2 and O_2 is disturbed, as a result of which rate of photorespiration is increased and leads to the formation of glycolate. Glycolate is considered as a major source of H_2O_2, as glycolate oxidase in peroxisome oxidizes glycolate and produces H_2O_2 (Noctor et al. 2002). Other processes that are involved in ROS production are oxidation of flavin oxidase and β-oxidation of fatty acids.

(d) Plasma Membrane

Plasma membrane is an important membrane of a cell that interacts with environmental conditions and is responsible for maintaining the activities of cell and its survival. In plasma membrane, NADPH-dependent oxidases are present and play an important function in gene expression during stress conditions (Apel and Hirt 2004). SOD catalyzes the transfer of electron from cytosolic NADPH to O_2 and produces

$O_2^{\bullet-}$ which further dismutates to H_2O_2. In plant defense mechanism against pathogens and to combat with abiotic stress, NADPH oxidase plays an important role (Kwak et al. 2003).

(e) Apoplast

Space around the plant cell membrane which converts CO_2 into a soluble and diffusible form that enters into cytosol and is utilized for the process of photosynthesis is called apoplast. Under stress conditions, apoplast is a site for production of H_2O_2 due to the combination of stress signals with abscisic acid (ABA) (Hu et al. 2006). ABA induced stomatal closure, and NADPH oxidases catalyzed the expression of AtRbohD and AtRbohF in the guard cells and the mesophyll cells of *Arabidopsis* which in turn results in the production of ROS (Kwak et al. 2003). Cell wall-linked oxidases, pH-dependent peroxidases (POXs), polyamine oxidases, and germin-like oxalate oxidases all are additional ROS-generating enzymes and produce H_2O_2.

(f) Cell Walls

ROS ($OH^{\bullet}$, $O_2^{\bullet-}$, H_2O_2, and $1O_2$) are generated in cell wall due to hyperoxidation of polyunsaturated fatty acids (PUFA) by lipoxygenase (LOX) enzyme. Other enzymes like diamine oxidases utilize polyamines or diamines to generate ROS. During abiotic stress, precursors of lignin undergo extensive cross-linking, through H_2O_2-mediated pathways to combine lignin with cell wall (Higuchi 2006).

(g) Endoplasmic Reticulum

In endoplasmic reticulum, ETC consisting of NADPH and Cyt P450 generates $O_2^{\bullet-}$ (Mittler 2002). Cyt P450 interacts with RH (organic molecule) and is reduced by flavoproteins and give rise to Cyt P450 R– (free radical). This free radical is oxidized by a triplet oxygen, and a resultant Cyt P450-ROO– (oxygenated complex) is formed. On decomposition of Cyt P450-Rh, $O_2^{\bullet-}$ is generated as a byproduct.

Here, electron transport and β-oxidation in mitochondria and chloroplast are the main processes by which ROS are generated in plants (Dutilleul et al. 2003). The fundamental pathway for the production of ROS is the photoreduction process in chloroplasts (Gill and Tuteja 2010). Particularly complex I and complex III cannot usually reach the terminal oxidase in mitochondria; instead, they immediately react with oxygen to create O_2^{-2} (Moller 2001). Under normal metabolic activities, ROS synthesis and removal are generally in dynamic balance and are regulated by both the redox potential and energy level (Pospísil 2012).

Interestingly, in plant defense, ROS function as signal transduction and also play a vital role in preventing pathogen penetration in plants (Barna et al. 2012; Torres et al. 2002, 2005). The signal pathway might primarily include the control of heat shock protein synthesis and accumulation, hence activating defensive systems (Piterkova et al. 2013).

Plants establish an extensive defense system and allow ROS scavenging for dynamic equilibrium, and antioxidant defense systems comprise both enzymatic and nonenzymatic mechanisms. Plant antioxidant molecules such as glutathione (GSH), ascorbic acid (ASC), alkaloids, phenolic compounds, carotenoids, tocopherol, and others are examples of nonenzymatic antioxidant defense mechanisms (Caverzan et al. 2012; Noctor et al. 2012). The enzymatic defense mechanism not only neutralizes ROS but also nullifies the harm that ROS caused (Moller 2001). Additionally, because proline accumulation could not quench singlet oxygen ($^1O_2$) in neutral aqueous solutions, proline is an ineffective $^1O_2$ scavenger (Signorelli et al. 2013). These antioxidant defense mechanisms may not be able to neutralize the excessive amount of ROS if the biotic and abiotic stressors are too large, which may still lead to cell death and necrosis development (Barna et al. 2012).

15.7 Reactive Oxygen Species in Life Cycle of Plants

Throughout their life cycle, plants are vulnerable to a variety of environmental challenges. Reactive oxygen species (ROS) are essential to sustaining normal plant development and increasing their capability to tolerate stresses. In addition to causing permanent DNA damage and cell death, ROS serve as important signaling molecules that control how plants develop normally and react to stress. Some biological components are oxidized and modified by ROS because they are reactive molecules, which prevents them from performing their specific functions (Mittler et al. 2011; Apel and Hirt 2004; Mittler et al. 2004). Plants produce a large number of ROS species that are involved in the control of several activities, including pathogen defense, programmed cell death (PCD), and stomatal behavior, under adverse conditions (Gill and Tuteja 2010; Schippers et al. 2012). ROS can be either in ionic or molecular states in plants. Each kind of ROS influences various physiological and biochemical processes in plants that are controlled by various genes and have a distinct oxidative capability (Huang et al. 2019).

However, in adverse circumstances, excess amounts of ROS are produced and disrupting the dynamic equilibrium. An excess of ROS can lead to breakdown of macromolecule and protoplast membrane oxidation, which can damage or even kill cells by reducing their capacity to execute their regular physiological functions (Ahmad et al. 2010a, 2010b, 2011; Ahmad and Umar 2011; Koyro et al. 2012; Rasool et al. 2013). ROS, if found in higher concentration, will lead toward incorrect programmed cell death and DNA damage which will affect plant badly (Xie et al. 2014).

15.8 Effects of ROS on Nutrient Distribution

ROS when produced in lesser concentrations are in harmony with plant metabolism, and they act as an indicator for plant to be prepared for stress. In fact, production of ROS shows the attempt of plant to alleviate the harmful effects of stress. Using ROS as a signaling molecule when under stress requires a precise balance between the ROS formation and scavenging systems. However, prolonged stress raises the level of ROS, which then causes oxidative stress, which inhibits vital cellular functions and cell viability. As a result, important elements of stress signals are defined such as antioxidant signaling, redox homeostasis, and continual generation/scavenging of ROS. It is also known that early mycorrhiza-plant symbiotic interactions generate ROS. But when these ROS exceed the limit, they begin to cause destruction of plant structure, starting from membrane damage to DNA and protein loss. There are many other factors that are interacting with ROS.

Terrestrial plants develop mycorrhizal association with fungi to efficiently utilize soil nutrients. Mycorrhizal association magically helps plants to improve nutrient uptake even in poor soil conditions. But ROS released by plants sometimes damage mycorrhizal fungi and ultimately disturb the nutrient uptake and vice versa. To understand the whole phenomenon of ROS effects, we need to study the mechanism of ROS regulation.

It is evident that ROS activity depends upon different levels of reactivity, potential to cross biological membranes, and site of production; thus ROS have a dual role in vivo (Miller et al. 2010). O_2 is a ROS, but it is nonreactive and stable; in organelles due to different processes that affect plant metabolism, it can be converted into high-energy ROS (Mittler 2017).

15.9 Distribution of Micronutrients and Macronutrients

Woody species have also been found to remobilize macronutrients according to seasonal cycles. As was previously demonstrated for N, deciduous trees store nutrients over the winter, which are remobilized from the trunk each spring to maintain leaf development (Millard and Grelet 2010). In comparison to immature trees, mature trees require more nutrients on the remobilization of nitrogen reserves for their development each spring. It has been demonstrated that macronutrients like phosphorus and nitrogen are remobilized from leaves in evergreen trees (Cherbuy et al. 2001). This remobilization takes place in the summer, following vegetative growth and associated with leaf shedding. Similar to how nitrogen and phosphorus remobilization happens in the middle of the summer, potassium remobilization can be linked to potassium resorption prior to leaf shedding or the fulfillment of nutritional needs when soil availability is low (Milla et al. 2005).

Remobilization of nitrogen is crucial for perennial plant viability. There are two steps of nitrogen remobilization in trees, which often develop in low-nitrogen environment. In the autumn, nitrogen is remobilized from senescent leaves and stored in tree trunks for the winter. Before root N absorption uptake as the primary mechanism to supply tree nitrogen demands, nitrogen is remobilized from trunks to developing organs in the spring. The importance of internal nitrogen cycling in the overall tree nitrogen budget increases as trees become mature. The development of nitrogen storage pools and effective nitrogen management, both of which are crucial for plant survival throughout time, is facilitated by both nitrogen removal from senescing leaves and root nitrogen absorption (Masclaux-Daubresse et al. 2010).

Because micronutrient concentrations in tissues are lower and stable isotope usage is presumably more limited than for macronutrients in crops and woody species, remobilization of micronutrients from plants has gotten far less study than for macronutrients. To fulfil the nutritional needs of animals or humans, however, crops, which are the primary sources of critical micronutrients, may not always contain adequate levels of these nutrients. In contrast of either rising or remaining constant soil concentrations during the past 60 years, the amount of micronutrients (mostly Fe, Mg, Zn, and Cu) in consumable items has decreased (Fan et al. 2008). Therefore, increasing the transfer of these micronutrients into edible components by remobilization from vegetative tissues may be a strategy to meet micronutrient requirements. Particularly, the remobilization of Cu, Fe, and Zn has been studied. It has been shown that during grain filling in wheat, Cu and Fe concentrations in all plant vegetative organs decrease with time, 77 and 40–62%, respectively, as a result of their remobilization (Garnett and Graham 2005). Wheat exhibits significant remobilization of Zn from the leaves to the grain (Kutman et al. 2012) and barley (Hegelund et al. 2012).

Kutman et al. (2011) reported that wheat grains derived more Fe (80%) than Zn (50%) from leaf remobilization. TaNAM (NAC-type transcription factor) RNAi line plants perform Fe and Zn remobilization from wheat leaves less efficiently than WT plants, and the amount of net Fe and Zn remobilization in both lines depends on the availability of mineral input (Waters et al. 2009). N deficit promoted nutrient remobilization in barley senescing leaves, which in turn increased phytosiderophore production and, thus, Fe solubility. However, they maintained that Fe remobilization from mature leaves was independent of N remobilization utilizing dark-induced leaf senescence. For the expansion and maturation of higher plants, molybdenum (Mo) is a crucial component. Wheat (*Triticum aestivum* L.) development is inhibited by deficiency of molybdenum, and the nutritious value of the grain is decreased. However, little is known regarding the distribution of macro- and micronutrients in wheat organs in both adequate and inadequate amounts. The mineral content of grain and other organs, as well as grain yield, is all controlled by the presence of molybdenum. Increased plant dry biomass, grain yield, and distribution of macronutrients (nitrogen, phosphorus, and potassium), as well as micronutrients (copper, iron, manganese, zinc, and molybdenum), among various wheat organs were all seen after Mo treatment (Moussa et al. 2021).

15.10 Roles of ROS in Plant Growth and Development

Organisms get an opportunity to utilize oxygen as an electron acceptor in aerobic environment, while enabling them to harness its reactive properties for metabolism and signaling (Schippers et al. 2012; Foyer and Noctor 2016). It is evident that, evolution in oxygenic environment necessarily promotes the oxidative activities, ROS signaling, and sensing during development process. In various environmental conditions, ROS are generated and removed dynamically during developmental and growth processes from seed germination to leaf senescence. Due to temporal and geographical variability of ROS generation and its interaction with plants, the impacts of ROS on plant growth and development are more complicated.

However, in adverse circumstances, excess amounts of ROS are produced and disrupt the dynamic equilibrium. An excess of ROS can lead to breakdown of macromolecule and protoplast membrane oxidation, which can damage or even kill cells by reducing their capacity to execute their regular physiological functions (Ahmad et al. 2010a, 2010b, 2011; Ahmad and Umar 2011; Koyro et al. 2012; Rasool et al. 2013).

15.10.1 ROS-Mediated Control of the Cell Cycle, Cell Division, Cell Expansion, and Cell Death

In plants, molecular factors of cell cycle that are influenced by ROS mechanism are not studied in detail. Reductive and oxidative signals are necessary for transitions between the phases of the cell cycle, and it is known that redox cycles are maintained within the cell cycle (De Simone et al., 2017; Diaz-Vivancos et al. 2015; Menon and Goswami 2007). Complex machinery of interacting cyclins (CYCs) and cyclin-dependent kinases (CDKs) governs phase to phase transition and progression, and alterations in redox reactions affect the activity and levels of CYCs and CDKs (Foyer et al. 2018; Féher et al., 2008). With the help of Teosinte Branched-Cycloidea-Proliferating Cell Factor1 (TCP) transcription factors, redox reactions affect the components of cell cycle (Kadota et al. 2005). TCPs feature a conserved redox-sensitive cysteine residue that is necessary for DNA binding and transcriptionally and potentially control the amounts of CYCs via interactions with CYC promoters. This shows that the interaction between a TCP transcription factor and its promoter may be disrupted under oxidizing circumstances as a result of the formation of disulfide bonds (Viola et al. 2013).

In plants, a number of reactive oxygen species (ROS) that are oxygen derived are present. The singlet oxygen (1 O_2) is produced by excitation of oxygen, and on reduction it generates superoxide radicals ($O_2^{\dagger}$), hydrogen peroxide (H_2O_2), and hydroxyl radicals ($OH^{\dagger}$) (Fig. 15.2).

In S1-to-M phase transition of the cell cycle, CYKs and CDKs play a functional role and are also associated with cell cycle arrest in *Arabidopsis* that occurs due to differential expression of glutathione-deficient rml1(root meristem-less) mutant (Vernoux et al., 2000; Schnaubelt et al. 2015). The most important redox buffer in plants during cell division is glutathione, and it plays an important role in growth defect phenotype of rml1 mutants. The amount of glutathione in nucleus and in cytosol is at equilibrium, but on treatment of cell with buthionine sulfoximine, glutathione is easily depleted from cytosol than nuclei (García-Giménez et al. 2013; Pellny et al. 2009). The amount of ROS fluctuates along with fluctuation of ascorbate and glutathione levels and results in the transition of one phase to another

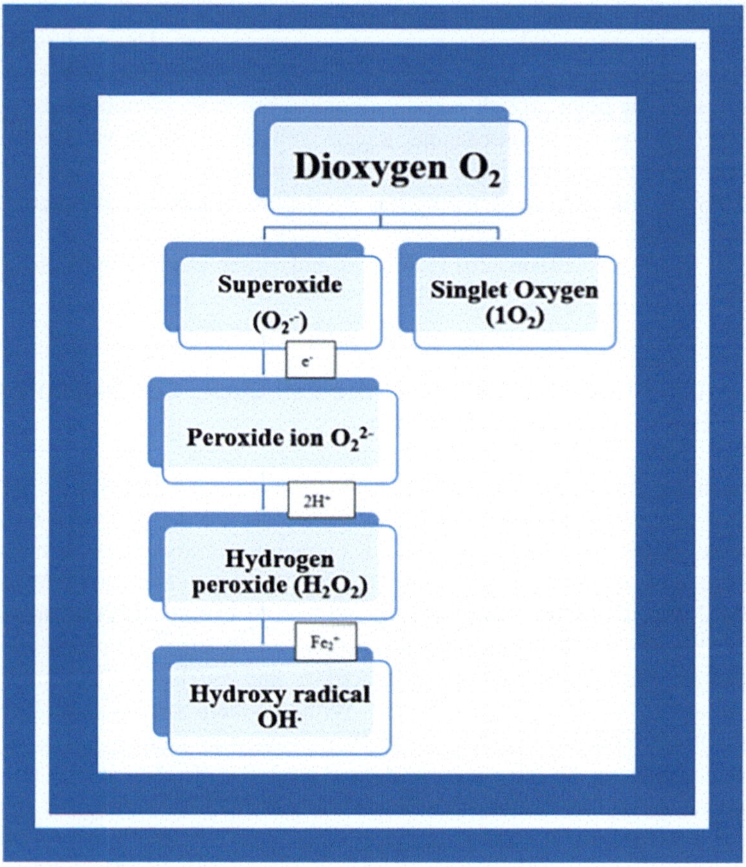

Fig. 15.2 Atmospheric oxygen-derived reactive oxygen species

during cell division (Diaz-Vivancos et al. 2010; Schnaubelt et al. 2015; Diaz-Vivancos et al. 2015; Tognetti et al. 2017).

For cytokinesis ROS and redox homeostasis are required. Cell wall stiffness, relaxation, and cell expansion rate are affected by apoplastic H_2O_2, hydroxyl radicals, and superoxides. On cell wall rigidity, different enzymes, like NADPH oxidases, amine, and oxalate oxidases, and activities like peroxidative and hydroxylation activities have antagonistic effects (Passardi et al. 2004; Schmidt et al. 2016). Peroxidative activities regulate the levels of H_2O_2 and contribute in the cross-linking of phenolics and extensions, which causes reduction in cell elongation and stiffness of cell wall. While on the other side, hydroxyl radical is involved in the cleavage of xyloglucans and pectins that lead toward the loosening of cell wall (Fry 1998; Passardi et al. 2004). It is evident that peroxidases are associated with both cell elongation and growth-restricting processes, like restricted leaf growth without affecting cell division (Lu et al. 2014; Raggi et al. 2015; Schmidt et al. 2016).

Table 15.1 List of enzymatic and nonenzymatic antioxidants along with their location and function

Sr#	Enzymatic antioxidants	Location	Reaction catalyzed
Nonenzymatic antioxidants			
01	Superoxide dismutase (SOD)	Peroxisomes, mitochondria, cytosol, and chloroplast	$O_2^{\bullet-}$ + 2 $O^{\bullet-}$ 2 + $2H^+ \rightarrow 2H_2O_2 + O_2$
02	Catalase (CAT)	Peroxisome and mitochondria	$2H_2O_2 \rightarrow O_2 + 2H_2O$
03	Ascorbate peroxidase (APX)	Peroxisomes, mitochondria, cytosol, and chloroplast	H_2O_2+ AA → $2H_2O$ + DHA
04	Monodehydroascorbate reductase (MDHAR)	Mitochondria, cytoplasm, and chloroplast	2MDHA + NADH → 2AA + NAD
05	Dehydroascorbate reductase (DHAR)	Mitochondria, cytoplasm, and chloroplast	DHA + 2GSH → AA + GSSG
06	Glutathione reductase (GR)	Mitochondria, cytoplasm, and chloroplast	GSSG + NADPH → 2GSH + NADP+
07	Guaiacol peroxidase (GPX)	Mitochondria, cytoplasm, chloroplast, and ER	H_2O_2 + DHA → $2H_2O$ + GSSG
Nonenzymatic antioxidants			
01	Ascorbic acid (AA)	Cytosol, chloroplast, mitochondria, peroxisome, vacuole, and apoplast	Detoxifies H_2O_2 via action of APX
02	Reduced glutathione (GSH	Cytosol, chloroplast, mitochondria, peroxisome, vacuole, and apoplast	Acts as a detoxifying co-substrate for enzymes like peroxidases, GR, and GST
03	α-Tocopherol	Mostly in membranes	Guards against and detoxifies products of membrane LPO
04	Carotenoids	Chloroplasts and other nongreen plastids	Quenches excess energy from the photosystems, LHCs
05	Flavonoids	Vacuole	Direct scavengers of H_2O_2 and $1O_2$ and $OH^{\bullet}$
06	Proline	Mitochondria, cytosol, and chloroplast	Efficient scavenger of $OH^{\bullet}$ and $1O_2$ and prevents damages due to LPO

There are various clocks and systems working in a specific plant that are controlling the level of ROS inside the cell and in the cell environment. These systems are consisting of enzymatic and nonenzymatic asntioxidants (Table 15.1). If ROS are not under control, they may start destroying cellular structure, proteins, DNA, and lipids of their own cells Caverzan et al. 2019.

15.10.2 Balance of ROS Production and ROS Scavenging

When soils are deficient in any nutrient like iron, zinc, nitrogen, sulfur, or phosphorous, they induce stress in plants as plants have to modify their physiology, cytology, metabolism, and even internal structure (e.g., lignification of cells) to cope with the stress. In such stress situation, ROS are produced that initially help plants to bear the stress. ROS help plants by activating many signaling pathways and by switching on alternative pathways. For example, at initial level, they may speed up influx of calcium and hydrogen peroxide at efflux of potassium at plasma membrane. These all change at membrane level and give rise to a signal that is meant for responding to nutrient deprivation. In such situation, auxins, sugars, and ROS are accumulated in the root cells. This may give rise to production of root hairs, thus causing more efficient absorption of least available nutrients (Iqbal et al. 2013; Wang et al. 2012). For example, when soils are deficient in potassium, inorganic phosphate, or nitrogen, there is overproduction of ROS that in turn causes an upstream regulation of calcium ion signaling. This results in the gene activation to control nutrient deficiency Neill et al. 2002).

In nutrient stress, ROS are produced and stored just behind the root elongation zone. This area of roots is considered as a very important and main region for nutrient distribution to aerial parts of the plants. ROS accumulated in this area are considered as the main agents causing the accumulation and translocation of potassium ions through affinity uptake (Shin and Schachtman 2004; Moritsugu et al. 1993). Various stresses like heavy metal stresses cause the depolarization of the plasma membranes, causing an increased potassium and sodium influx using various transporters in membranes (Vishwakarma et al. 2019).

The"expression of"genes encoding transport proteins in root cells is precisely regulated by the feeling of nutrient deprivation in plants, enabling them to boost nutrient uptake in nutrient-poor environments. However, despite the fact that numerous transporter proteins have been found to become active in response to food deprivation, it is still unclear how those genes are regulated. Recent scientific discoveries have revealed the complexity of plant transporter control, which involves both posttranscriptional regulation and translational regulation. Ion channels, membrane holes that are essential to plant physiology, are an illustration of these transporters. To meet the nutritional and osmotic needs of the cell and to establish an electric potential across the membranes, these channels were initially investigated as absorption mechanisms for minerals in their ionic state. These channels have recently been found to be crucial for a number of other essential cell processes, including the regulation of photosynthesis and the elongation of roots and pollen tubes, as well as cell senescence and programmed cell death (Checchetto et al. 2013; Herdean et al. 2016). The discovery that ion channels are involved in redox-dependent cell activities and have the ability to sense ROS increases during growth as a response to hormones under stressful situations is one of the most recent developments.

The first mention of ROS-activated ion channels was in relation to Cu2+ toxicity in algae. It is now known that these ROS-activated channels connect the production

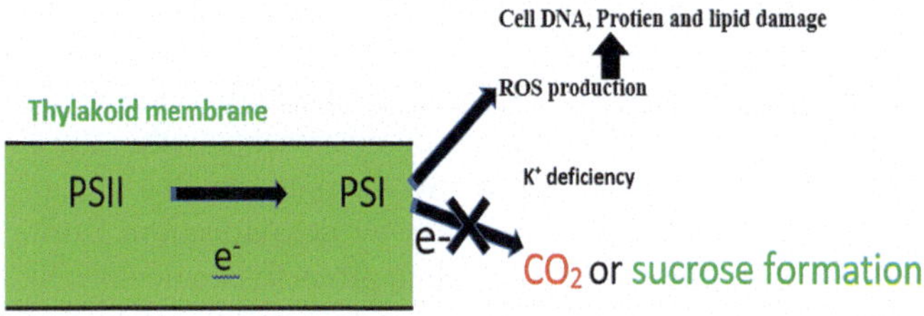

Fig. 15.3 Potassium deficiency and rerouting of electrons, ROS production, and destruction in cells

of ROS and other ions to Ca2+ and K+ transmembrane fluxes. Both events are typical and necessary for plant survival. The bulk of plant physiologic processes can be controlled when ROS generation, Ca2+, and K+ transport pathways interact. Lipid peroxidation results in elevated ROS levels, which weaken and increase the permeability of membranes (Zeng et al. 2019). Plants usually have complex methods to maintain ROS levels since high ROS concentrations can disrupt lipids, proteins, and DNA. As a result, it can be said that during nutrient deficiency, ROS are the primary signals that function to positively trade off distinct ions that are available for the ions that are not available using ethylene.

Potassium is necessary for many physiological functions, including photosynthesis, the movement of photosynthates into sink organs, the preservation of turgescence, the activation of enzymes, and the reduction of excessive ion uptake in flooded and salinized soils, such as Na and Fe.

Due to its close connection to fundamental metabolic functions, ROS generation is crucial in plant cells starting with ROS synthesis; the cell undergoes three states: oxidative stress, detoxification, and recovery (Czarnocka and Karpiński 2018). Damage imposed by oxidative stress that is caused by ROS could be lethal when the cycle of ROS generation and scavenging is disrupted.

Remobilization rates of nutrients are very influential for plant metabolism. It is required by plants when soil is deficient in nutrients and nutrients have to be remobilized to move toward actively growing young parts of plants or during reproductive growth of plants as reproductive parts become the sink for nutrients. In older leaves when remobilization of nutrients take place, ROS produced by photooxidation cause cell death and senescence. If this condition prolongs, the situation may become harsh for plant (Fig. 15.3).

Remobilization rates of nutrients are influenced by various factors:

- First, various types of crops and their accompanying seeds have extremely varying demands for each specific nutrient.
- Second, the nutritional condition of the vegetative component, where micronutrient remobilization is closely related to leaf concentration, with higher leaf concentration enabling faster rates of remobilization. Since they are not linked to structural components, highly mobile elements, including K and Na, are more

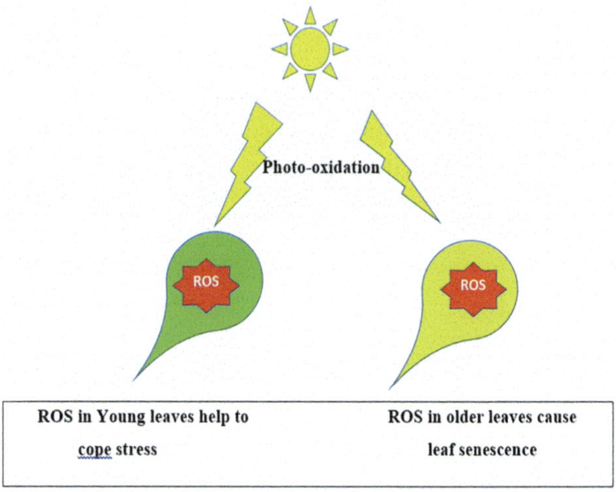

Fig. 15.4 Interaction of ROS and remobilization of nutrients in leaves during light stress

effectively remobilized in nutrient-deficient leaves (cell membranes, cell walls, enzymes) (White 2012).

- Third, while soil availability is high, direct nutrient absorption from roots to seed may occur, decreasing the need for remobilization throughout the reproductive phase (Liu et al. 2019).

Remobilization activities can also occur directly during reproductive development, when new sinks are developing and root activity and nutrient intake normally decline (Malagoli et al. 2005). Nutrient remobilization is typically linked to foliar senescence, which makes nutrients available for younger plant organs and increases nutrient usage efficiency (Himelblau and Amasino 2001; Avice and Etienne 2014). Phloem movement is largely needed for remobilization. N, S, P, Mg, K, and other macronutrients except Ca are known to be highly mobile in the phloem, whereas Fe, Cu, Zn, M“, Ni, B, Cl, and other micronutrients outside Mn exhibit at least moderate mobility (White 2012) (Fig. 15.4).

15.11 Conclusions and Perspectives

The management of micronutrients still remains unclear, despite substantial research on the storage and remobilization of macronutrients in particular under typical growth environments. Additionally, there hasn’t been much attention paid to the mechanisms underlying nutrient remobilization to the seeds in drought-like conditions. Under effect of ROS, the use of plant microbe interactions, which might improve micronutrient availability, plant absorption, storage, and remobilization to the seeds, may be combined with these efforts on the one hand with the screening of huge collections of genotypes on the other. During recent years, various sources, synthesis, and removal methods, as well as important antioxidant compounds and enzymes that scavenge ROS, have all been identified in plants. Firstly, since most ROS have short half-lives and are susceptible to chemical interactions, water or secondary ROS can be produced. It is still challenging to understand precisely how different ROS contribute to signaling and promote firmly confined and timely plant growth and development. Secondly, we don't fully comprehend how the spatiotemporal generation of distinct ROS and their activities interact. In some conditions, it can be exceedingly challenging to determine whether oxidative stress is the cause or a result of cellular damage. Thirdly, certain microRNA and proteins are oxidatively modified as a result of the aberrant concentration of ROS. The initiation of apoptosis, which ultimately results in cell death, may be influenced by the mismatch between degraded miRNAs and proteins. Oxidatively altered proteins or miRNA has a role in plant growth and development and would be characterized. Plant development and stress responses were significantly controlled by ROS levels and epigenetic changes, and both biotic and abiotic stimuli had a significant impact on plant growth and redox states. Understanding how ROS homeostasis, epigenetics, and plant adaptation and tolerance are interconnected will need an exploration of the interactions between ROS and epigenetic changes.

ROS play a number of roles in several biological processes. The involvement of ROS in hormone responses has been exposed by a number of innovative studies in recent years, and a number of new challenges have been raised as a result. Protein targets are altered by ROS to enable stimulus-specific cellular responses during signal transduction. Cellular mechanisms among those in charge of ROS formation in plant cells drive ROS production for particular signal transduction and how ROS producers and scavengers interact with one another to control cellular ROS level. How do plants control the NADPH oxidases' enzyme activity? There is undoubtedly still a lot to learn in this dynamic field.

References

Ahmad P, Umar S (2011) Reactive oxygen species, antioxidants and signaling in plants. J Plant Biol 51(3):167–173

Ahmad P, Jaleel CA, Salem MA, Nabi G, Sharma S (2010a) Roles of enzymatic and non-enzymatic antioxidants in plants during abiotic stress. Crit Rev Biotechnol 30:161175

Ahmad P, Nabi G, Jaleel CA, Umar S (2011) Free radical production, oxidative damage and antioxidant defense mechanisms in plants under abiotic stress. In: Ahmad P, Umar S (eds) Oxidative stress: role of antioxidants in plants. Studium Press Pvt. Ltd, New Delhi, p 1953

Ahmad P, Umar S, Sharma S (2010b) Mechanism of free radical scavenging and role of phytohormones during abiotic stress in plants. In: Ashraf M, Ozturk M, Ahmad MSA (eds) Plant adaptation and phytoremediation. Springer, Dordrecht, pp 99–108

Ahmad W, Zia MH, Malhi SS, Niaz A, Saifullah (2012) Boron deficiency in soils and crops: a review, crop plant. In: Goyal A (ed). INTECH. ISBN: 978-953-51-0527-5

Alloway BJ (2004) In zinc in soil and crop nutrition, vol 2. International Zinc Association, Brussels, pp 32–34

Apel K, Hirt H (2004) Reactive oxygen species: metabolism, oxidative stress, and signal transduction. Annu Rev Plant Biol 55:373–399. https://doi.org/10.1146/annurev.arplant.55.031903.141701

Asada K (2006) Production and scavenging of reactive oxygen species in chloroplasts and their functions. Plant Physiol 141(2):391–396

Avice JC, Etienne P (2014) Leaf senescence and nitrogen remobilization efficiency in oilseed rape (*Brassica napus* L.). J Exp Bot 65(14):3813–3382

Ayub MA, Ahmad Z, Umar W, Waris AA, Fatima H, Nadeem M, Iftikhar I (2021) Accumulation, partitioning, and bioavailability of micronutrients in plants and their crosstalk with phytohormones. In: Plant growth regulators. Springer, Cham, pp 39–73

Barna B, Fodor J, Harrach BD, Pogany M, Kiraly Z (2012) The Janus face of reactive oxygen species in resistance and susceptibility of plants to necrotrophic and biotrophic pathogens. Plant Physiol Biochem 59:37–43

Bittner F (2014) Molybdenum metabolism in plants and crosstalk to iron. Department of Plant Biology, Braunschweig University of Technology, Braunschweig

Bonfim-Silva EM, de Anicésio ECA, de Oliveira JR, de Freitas Sousa HH, da Silva TJA (2015) Soil water availabilities in the content and accumulation of nitrogen and chlorophyll index in the safflower. Am J Plant Sci 6(14):2280

Brown PH (2007) Nickel. In: Barkerand AV, Pilbean DJ (eds) Handbook of plant nutrition. CRC Taylor and Francis, NewYork, pp 395–402

Brown PH, Bellaloui N, Wimmer MA, Bassil ES, Ruiz J, Hu H, Pfeffer H, Dannel F, Römheld V (2002) Boron in plant biology. Plant Biol 4(02):205–223

Caverzan A, Passaia G, Rosa SB, Ribeiro CW, Lazzarotto F, Margis-Pinheiro M (2012) Plant responses to stresses: role of ascorbate peroxidase in the antioxidant protection. Genet Mol Biol 35:1011–1019

Caverzan A, Piasecki C, Chavarria G, Stewart JCN, Vargas L (2019) Defenses against ROS in crops and weeds: the effects of interference and herbicides. Int J Mol Sci 20(5):1086

Che J, Yamaji N, Miyaji T, Mitani-Ueno N, Kato Y, Shen RF, Ma JF (2020) Node-localized transporters of phosphorus essential for seed development in rice. Plant Cell Physiol 61(8): 1387–1398

Checchetto V, Teardo E, Carraretto L, Formentin E, Bergantino E, Giacometti GM, Szabo I (2013) Regulation of photosynthesis by ion channels in cyanobacteria and higher plants. Biophys Chem 182:51–57

Chen CR, Condron LM, Davis MR, Sherlock RR (2004) Effects of plant species on microbial biomass phosphorus and phosphatase activity in a range of grassland soils. Biol Fertil Soils 40(5):313–322

Cherbuy B, Joffre R, Gillon D, Rambal S (2001) Internal remobilization of carbohydrates, lipids, nitrogen and phosphorus in the Mediterranean evergreen oak *Quercus ilex*. Tree Physiol 21(1): 9–17

Choudhury S, Panda P, Sahoo L, Panda SK (2013) Reactive oxygen species signaling in plants under abiotic stress. Plant Signal Behav 8:23681. https://doi.org/10.4161/psb.23681

Christenson DR, Draycott AP (2006) Nutrition-phosphorus, sulphur, potassium, sodium, calcium, magnesium and micronutrients-limiting and nutrient deficiencies. Sugar Beet:185–220

Clark RB, Baligar VC (2000) Acidic and alkaline soil constraints on plant mineral nutrition. In: Wilkinson RE (ed) Plant-environment interactions. Marcel Dekker Inc, New York, pp 133–177. cel Dekker, New York, pp. 57–87

Clarke S, Eaton-Rye JJ (2000) Amino acid deletions in loop C of the chlorophyll a-binding protein CP47 alter the chloride requirement and/or prevent the assembly of photosystem II. Plant Mol Biol 44:591–601

Czarnocka W, Karpiński S (2018) Friend or foe? Reactive oxygen species production, scavenging and signaling in plant response to environmental stresses. Free Radic Biol Med 122:4–20

Dang HR, Li Y, Sun X, Zhang LY (2010) Absorption, accumulation and distribution of zinc in highly-yielding winter wheat. Agric Sci China 9(7):965–973

Demura T, Ye ZH (2010) Regulation of plant biomass production. Curr Opin Plant Biol 13(3): 298–303

Denison RF, Kiers ET (2005) Evolutionary stability of rhizobium mutualism depends on legume host sanctions. In: Biological nitrogen fixation, sustainable agriculture and the environment. Springer, Dordrecht, pp 221–224

De Simone A, Hubbard R, De La Torre NV, Velappan Y, Wilson M, Considine MJ, Soppe WJ, Foyer CH (2017) Redox changes during the cell cycle in the embryonic root meristem of Arabidopsis thaliana. Antioxid Redox Signal 27(18):1505–1519

Diaz-Vivancos P, Wolff T, Markovic J, Pallardo FV, Foyer CH (2010) A nuclear glutathione cycle within the cell cycle. Biochem J 431(2):169–178

Diaz-Vivancos P, de Simone A, Kiddle G, Foyer CH (2015) Glutathione–linking cell proliferation to oxidative stress. Free Radic Biol Med 89:1154–1164

Domingos P, Margarida AP, Wong A, Gehring C, Feijo AJ (2015) Nitric oxide: a multitasked signaling gas in plants. Mol Plant 8(4):506–520

Draycott AP, Christenson DR (2003) Nutrients for sugar beet production: soil-plant relationships, 7–32, 123–138, 180–181. CABI Publishing, Wallingford

Dutilleul C, Garmier M, Noctor G, Mathieu C, Chetrit P, Foyer CH (2003) Leaf mitochondria modulate whole cell redox homeostasis, set antioxidant capacity, and determine stress resistance through altered signaling and diurnal regulation. Plant Cell 15:1212–1226

Fan X, White IM, Shopova SI, Zhu H, Suter JD, Sun Y (2008) Sensitive optical biosensors for unlabeled targets: a review. Anal Chim Acta 620(1–2):8–26

Féher A, Ötvös K, Pasternak TP, Pettkó-Szandtner A (2008) The involvement of reactive oxygen species (ROS) in the cell cycle activation (G0-to-G1 transition) of plant cells. Plant Signal Behav 3(10):823–826

Fixen PE (2005) Decision support systems in integrated crop nutrient management. In: Proceedings no. 569. International Fertilizer Society, York

Foyer CH, Noctor G (2016) Stress-triggered redox signalling: what's in prospect. Plant Cell Environ 39:951–964. https://doi.org/10.1111/pce.12621

Foyer CH, Wilson MH, Wright MH (2018) Redox regulation of cell proliferation: bioinformatics and redox proteomics approaches to identify redox-sensitive cell cycle regulators. Free Radic Biol Med 122:137–149

Fry PS (1998) The development of personal meaning and wisdom in adolescence: a reexamination of moderating and consolidating factors and influences. Lawrence Erlbaum Associates Publishers

García-Giménez JL, Markovic J, Dasí F, Queval G, Schnaubelt D, Foyer CH, Pallardó FV (2013) Nuclear glutathione. Biochimica et Biophysica Acta (BBA)-General Subjects 1830(5): 3304–3316

Garnett TP, Graham RD (2005) Distribution and remobilization of iron and copper in wheat. Ann Bot 95(5):817–826

George TS, French AS, Brown LK, Karley AJ, White PJ, Ramsay L, Daniell TJ (2014) Genotypic variation in the ability of landraces and commercial cereal varieties to avoid manganese deficiency in soils with limited manganese availability: is there a role for root-exuded phytases. Physiol Plant 151:243–256

Gherardi MJ, Rengel Z (2004) The effect of manganese supply on exudation of carboxylates by roots of lucerne (*Medicago sativa*). Plant Soil 260:271–282

Gill SS, Tuteja N (2010) Reactive oxygen species and antioxidant machinery in abiotic stress tolerance in crop plants. Plant Physiol Biochem 48(12):909–930

Haehnel W, Ratajczak R, Robenek H (1989) Lateral distribution and diffusion of plastocyanin in chloroplast thylakoids. J Cell Biol 108(4):1397–1405

Hancock JT, Desikan R, Neill SJ (2001) Role of reactive oxygen species in cell signalling pathways. Biochem Soc Trans 29(2):345–349

Hegelund JN, Pedas P, Husted S, Schiller M, Schjoerring JK (2012) Zinc fluxes into developing barley grains: use of stable Zn isotopes to separate root uptake from remobilization in plants with contrasting Zn status. Plant Soil 361(1):241–250

Herdean A, Teardo E, Nilsson AK, Pfeil BE, Johansson ON, Ünnep R, Nagy G, Zsiros O, Dana S, Solymosi K, Garab G (2016) A voltage-dependent chloride channel fine-tunes photosynthesis in plants. Nat Commun 7(1):1–11

Higuchi ET (2006) Look back over the studies of lignin biochemistry. J Wood Sci 52:2–8. https://doi.org/10.1007/s10086-005-0790-z

Himelblau E, Amasino RM (2001) Nutrients mobilized from leaves of *Arabidopsis thaliana* during leaf senescence. J Plant Physiol 158(10):1317–1323

Hu X, Zhang A, Zhang J, Jiang M (2006) Abscisic acid is a key inducer of hydrogen peroxide production in leaves of maize plants exposed to water stress. Plant Cell Physiol 47:1484–1495. https://doi.org/10.1093/pcp/pcl014

Huang H, Ullah F, Zhou DX, Yi M, Zhao Y (2019) Mechanisms of ROS regulation of plant development and stress responses. Front Plant Sci 10:800

Huang J, Hu B, Qi K, Chen W, Pang X, Bao W, Tian G (2016) Effects of phosphorus addition on soil microbial biomass and community composition in a subalpine spruce plantation. Eur J Soil Biol 72:35–41

Hussain MB, Azam AS, Hina S, Farooq MA, Ali B (2013) Morphological, physiological and biochemical responses of plants to nickel stress: a review. Afr J Agric Res 8:1596–1602

Iqbal N, Trivellini A, Masood A, Ferrante A, Khan NA (2013) Current understanding on ethylene signaling in plants: the influence of nutrient availability. Plant Physiol Biochem 73:128–138

Ishka RM, Vatamaniuk OK (2020) Copper deficiency alters shoot architecture and reduces fertility of both gynoecium and androecium in *Arabidopsis thaliana*. Plant Direct 4(11):e00288

Jiang W, Yang ZF, Yu T, Hou QY, Zhong C, Zheng GD, Yang ZQ, Li J (2015) Evaluation of the potential effects of soil properties on molybdenum availability in soil and its risk estimation in paddy rice. J Soils Sediments 15:1520–1530

Kabata PA, Pendias H (2001) Trace elements in soils and plants. CRC Press, Boca Raton

Kadota M, Nakao T, Taniguchi N, Takata E, Mimura M, Nishiyama K, Komura T (2005) Surface acoustic wave duplexer for US personal communication services with good temperature characteristics. Jpn J Appl Phys 44(6S):4527

Kaiser BN, Gridley KL, Ngaire BJ, Phillips T, Tyerman SD (2005) The role of molybdenum in agricultural plant production. Ann Bot 96:745–754

Karuppanapandian T, Moon JC, Kim C, Manoharan K, Kim W (2011) Reactive oxygen species in plants: their generation, signal transduction, and scavenging mechanisms. Aust J Crop Sci 5:709–725

Kaznina NM, Titov AF (2014) The influence of cadmium on physiological processes and productivity of Poaceae plants. Biol Bull Rev 4(4):335–348

Kohli SK, Kaur H, Khanna K, Handa N, Bhardwaj R, Rinklebe J, Ahmad P (2022) Boron in plants: uptake, deficiency and biological potential. Plant Growth Regul:1–16

Koyro HW, Ahmad P, Geissler N (2012) Abiotic stress responses in plants: an overview. In: Ahmad P, Prasad MNV (eds) Environmental adaptations and stress tolerance of plants in the era of climate change. Springer, science1business Media, New York, pp 1–28

Kumchai J, Huang JZ, Lee CY, Chen FC, Chin SW (2013) Proline partially overcomes excess molybdenum toxicity in cabbage seedlings grown *in vitro*. Genet Mol Res 12:5589–5601

Kutman UB, Kutman BY, Ceylan Y, Ova EA, Cakmak I (2012) Contributions of root uptake and remobilization to grain zinc accumulation in wheat depending on post-anthesis zinc availability and nitrogen nutrition. Plant Soil 361(1):177–187

Kutman UB, Yildiz B, Cakmak I (2011) Improved nitrogen status enhances zinc and iron concentrations both in the whole grain and the endosperm fraction of wheat. J Cereal Sci 53(1):118–125

Kwak JM, Mori IC, Pei ZM, Leonhardt N, Torres MA, Dangl JL (2003) NADPH oxidase AtrbohD and AtrbohF genes function in ROS−dependent ABA signaling in Arabidopsis. EMBO J 22: 2623–2633

Liu DY, Liu YM, Zhang W, Chen XP, Zou CQ (2019) Zinc uptake, translocation, and remobilization in winter wheat as affected by soil application of Zn fertilizer. Front Plant Sci 10:426

Vishwakarma K, Mishra M, Patil G, Mulkey S, Ramawat N, Pratap Singh V, Deshmukh R, Kumar Tripathi D, Nguyen HT, Sharma S (2019) Avenues of the membrane transport system in adaptation of plants to abiotic stresses. Crit Rev Biotechnol 39(7):861–883

López-Arredondo DL, Leyva-González MA, Alatorre-Cobos F, Herrera-Estrella L (2013) Biotechnology of nutrient uptake and assimilation in plants. Int J Dev Biol 57(6–7-8):595–610

Lu L, Joannopoulos JD, Soljačić M (2014) Topological photonics. Nat Photonics 8(11):821–829

Luis A, Sandalio LM, Corpas FJ, Palma JM, Barroso JB (2006) Reactive oxygen species and reactive nitrogen species in peroxisomes. Production, scavenging, and role in cell signaling. Plant Physiol 141:330–335. https://doi.org/10.1104/pp.106.078204

Malagoli P, Laine P, Rossato L, Ourry A (2005) Dynamics of nitrogen uptake and mobilization in field-grown winter oilseed rape (*Brassica napus*) from stem extension to harvest: I. global N flows between vegetative and reproductive tissues in relation to leaf fall and their residual N. Ann Bot 95(5):853–861

Malvi UR (2011) Interaction of micronutrients with major nutrients with special reference to potassium. Karnataka J Agric Sci 24(1):106–109

Marschner H (1995) Mineral nutrition of higher plants, 2nd edn. Academic Press, Dordrecht

Masclaux-Daubresse C, Daniel-Vedele F, Dechorgnat J, Chardon F, Gaufichon L, Suzuki A (2010) Nitrogen uptake, assimilation and remobilization in plants: challenges for sustainable and productive agriculture. Ann Bot 105(7):1141–1157

Mengel K, Kirkby EA, Kosegarten H, Appel T (2001) Plant nutrients. In: Principles of plant nutrition. Springer, Dordrecht, pp 1–13

Menon SG, Goswami PC (2007) A redox cycle within the cell cycle: ring in the old with the new. Oncogene 26(8):1101–1109

Milla R, Castro-Díez P, Maestro-Martínez M, Montserrat-Martí G (2005) Relationships between phenology and the remobilization of nitrogen, phosphorus and potassium in branches of eight Mediterranean evergreens. New Phytol 168(1):167–178

Millard P, Grelet GA (2010) Nitrogen storage and remobilization by trees: ecophysiological relevance in a changing world. Tree Physiol 30(9):1083–1095

Miller GAD, Suzuki N, Ciftci-Yilmaz S, Mittler RON (2010) Reactive oxygen species homeostasis and signalling during drought and salinity stresses. Plant, Cell Environ 33(4):453–467

Mittler R (2002) Oxidative stress, antioxidants and stress tolerance. Trends Plant Sci 7:405–410. https://doi.org/10.1016/S1360-1385(02)02312-9

Mittler R (2017) ROS are good. Trends Plant Sci 22(1):11–19

Mittler R, Vanderauwera S, Gollery M, Van Breusegem F (2004) Reactive oxygen gene network of plants. Trends Plant Sci 9(10):490–498

Mittler R, Vanderauwera S, Suzuki N, Miller GAD, Tognetti VB, Vandepoele K, Van Breusegem F (2011) ROS signaling: the new wave? Trends Plant Sci 16(6):300–309

Møller IM, Jensen PE, Hansson A (2007) Oxidative modifications to cellular components in plants. Ann Revenue Plant Biol 58:459–481. https://doi.org/10.1146/annurev.arplant.58.032806.103946

Moller IM (2001) Plant mitochondria and oxidative stress: electron transport, NADPH turnover, and metabolism of reactive oxygen species. Ann Rev Plant Physiol Plant Mol Biol 52:561–591

Mora ML, Rosas A, Ribera A, Rengel Z (2009) Differential tolerance to Mn toxicity in perennial ryegrass genotypes: involvement of antioxidative enzymes and root exudation of carboxylates. Plant Soil 320:79–89

Moritsugu M, Shibasaka M, Kawasaki T (1993) Where is the most important and efficient site for absorption and translocation of cations in excised barley roots? Soil Sci Plant Nutr 39(2): 299–307

Moussa MG, Sun X, Ismael MA, Elyamine AM, Rana MS, Syaifudin M, Hu C (2021) Molybdenum-induced effects on grain yield, macro–micro-nutrient uptake, and allocation in Mo-inefficient winter wheat. J Plant Growth Regul 41:1–16

Movafegh S, Hoseini SM (2013) ROS in plant cell: a review. Int J Farm Allied Sci 2(2)

Murphy M (2009) How mitochondria produce reactive oxygen species. Biochem J 417:1–13. https://doi.org/10.1042/BJ20081386

Natasha N, Shahid M, Bibi I, Iqbal J, Khalid S, Murtaza B, Bakhat HF, Farooq ABU, Amjad M, Hammad HM, Niazi NK (2022) Zinc in soil-plant-human system: a data-analysis review. Sci Total Environ 808:152024

Navrot N, Rouhier N, Gelhaye E, Jacquot JP (2007) Reactive oxygen species generation and antioxidant systems in plant mitochondria. Physiol Plant 129:185–195. https://doi.org/10.1111/j.1399-3054.2006.00777.x

Neill SO, Gould KS, Kilmartin PA, Mitchell KA, Markham KR (2002) Antioxidant activities of red versus green leaves in *Elatostema rugosum*. Plant Cell Environ 25(4):539–547

Noctor G, De Paepe R, Foyer CH (2007) Mitochondrial redox biology and homeostasis in plants. Trends Plant Sci 12:125–134. https://doi.org/10.1016/j.tplants.2007.01.005

Noctor G, Mhamdi A, Chaouch S, Han YI, Neukermans J, Marquez-Garcia BELEN, Foyer CH (2012) Glutathione in plants: an integrated overview. Plant Cell Environ 35(2):454–484

Noctor G, Veljovic-Jovanovic S, Driscoll S, Novitskaya L, Foyer CH (2002) Drought and oxidative load in the leaves of C3 plants: a predominant role for photorespiration. Ann J Bot 89:841–850. https://doi.org/10.1093/aob/mcf096

Oh SH, Cho SW, Kwon TH, Yang MS (1996) Purification and characterization of phytoferritin. J Biochem Mol Biol 29:540–544

Palma JM, Corpas FJ, del Río LA (2009) Proteome of plant peroxisomes: new perspectives on the role of these organelles in cell biology. Proteomics 9:2301–2312. https://doi.org/10.1002/pmic.200700732

Pan Y, Koopmans GF, Bonten LTC, Song J, Luo Y, Temminghoff EJM, Comans RNJ (2014) Influence of pH on the redox chemistry of metal hydroxides and organic matter in paddy soils. J Soils Sedim 14:1713–1726

Pandey N (2018) Role of plant nutrients in plant growth and physiology. In: Plant nutrients and abiotic stress tolerance. Springer, Singapore, pp 51–93

Passardi F, Penel C, Dunand C (2004) Performing the paradoxical: how plant peroxidases modify the cell wall. Trends Plant Sci 9(11):534–540

Pastore D, Trono D, Laus MN, Di Fonzo N, Flagella Z (2007) Possible plant mitochondria involvement in cell adaptation to drought stress a case study: durum wheat mitochondria. J Exp Bot 58:195–210. https://doi.org/10.1093/jxb/erl273

Pellny TK, Locato V, Vivancos PD, Markovic J, De Gara L, Pallardó FV, Foyer CH (2009) Pyridine nucleotide cycling and control of intracellular redox state in relation to poly (ADP-ribose) polymerase activity and nuclear localization of glutathione during exponential growth of Arabidopsis cells in culture. Mol Plant 2(3):442–456

Piterkova J, Luhova L, Mieslerova B, Lebeda A, Petrivalsky M (2013) Nitric oxide and reactive oxygen species regulate the accumulation of heat shock proteins in tomato leaves in response to heat shock and pathogen infection. Plant Sci 207:57–65

Polacco JC, Mazzafera P, Tezotto T (2013) Nickel and urease in plants: still many knowledge gaps. Plant Sci 19:79–90

Pospísil P (2012) Molecular mechanisms of production and scavenging of reactive oxygen species by photosystem II. Biochim Biophys Acta 1817:218–231

Raggi A, Tasca D, Panerai S, Neri W, Ferri R (2015) The burden of distress and related coping processes in family caregivers of patients with Alzheimer's disease living in the community. J Neurol Sci 358(1–2):77–81

Rasmusson AG, Geisler DA, Møller IM (2008) The multiplicity of dehydrogenases in the electron transport chain of plant mitochondria. Mitochondrion 8:47–60. https://doi.org/10.1016/j.mito.2007.10.004

Rasool S, Ahmad A, Siddiqi TO, Ahmad P (2013) Changes in growth, lipid peroxidation and some key antioxidant enzymes in chickpea genotypes under salt stress. Acta Physiol Plant 35(4): 1039–1050

Rengel Z (2000) Uptake and transport of manganese in plants. In: Sigel A, Sigel H (eds) Metal ions in biological systems. March

Rengel Z (2015) Availability of Mn, Zn and Fe in the rhizosphere. J Soil Sci Plant Nutr 15(2): 397–409

Rout GR, Sahoo S (2015) Role of iron in plant growth and metabolism. Rev Agric Sci 3:1–24

Rutkowska B, Szulc W, Spychaj-Fabisiak E, Pior N (2017) Prediction of molybdenum availability to plants in differentiated soil conditions. Plant Soil Environ 63:491–497

Saleem H, Khan AR, Jilani TA (2020) Expert system for diagnosing mobile and immobile nutrients deficiency of plants. Int J Acad Inform Syst Res (IJAISR) 4(8):10–15

Scheffer F, Schachtschabel P (2002) Textbook of soil science, vol 15. Auflage Spektrum Akademischer Verlag, Heidelberg, p 593

Schippers JH, Nguyen HM, Lu D, Schmidt R, Mueller-Roeber B (2012) ROS homeostasis during development: an evolutionary conserved strategy. Cell Mol Life Sci 69(19):3245–3257

Schmidt JJ, Gagnon GA, Jamieson RC (2016) Microalgae growth and phosphorus uptake in wastewater under simulated cold region conditions. Ecol Eng 95:588–593

Schnaubelt D, Queval G, Dong Y, Diaz-Vivancos PEDRO, Makgopa ME, Howell G, Foyer CH (2015) Low glutathione regulates gene expression and the redox potentials of the nucleus and cytosol in *Arabidopsis thaliana*. Plant Cell Environ 38(2):266–279

Seleiman MF (2019) Use of plant nutrients in improving abiotic stress tolerance in wheat. In: Wheat production in changing environments. Springer, Singapore, pp 481–495

Shabbir Z, Sardar A, Shabbir A, Abbas G, Shamshad S, Khalid S, Murtaza G, Dumat C, Shahid M (2020) Copper uptake, essentiality, toxicity, detoxification and risk assessment in soil-plant environment. Chemosphere 259:127436

Sharma P, Jha AB, Dubey RS, Pessarakli M (2012) Reactive oxygen species, oxidative damage, and antioxidative defense mechanism in plants under stressful conditions. J Bot 2012:217037. https://doi.org/10.1155/2012/217037

Shin R, Schachtman DP (2004) Hydrogen peroxide mediates plant root cell response to nutrient deprivation. Proc Natl Acad Sci 101(23):8827–8832

Signorelli S, Arellano JB, Melo TB, Borsani O, Monza J (2013) Proline does not quench singlet oxygen: evidence to reconsider its protective role in plants. Plant Physiol Biochem 64:80–83

Sims T (2000) Soil fertility evaluation. In: Sumner ME (ed) Hand book of soil science. CRC Press, Boca Raton, London, New York, Washington DC

Sillanpää M (1990) Micronutrient assessment at the country level: an international study (no. 63)

Srivastava S, Briggs BR, Dong H (2018) Abundance and taxonomic affiliation of molybdenum transport and utilization genes in Tengchong hot springs, China. Environ Microbiol 20(7):2397–2409

Tognetti VB, Bielach A, Hrtyan M (2017) Redox regulation at the site of primary growth: auxin, cytokinin and ROS crosstalk. Plant, Cell and Environment 40(11):2586–2605

Torres MA, Dangl JL, Jones JD (2002) Arabidopsis gp91phox homologues AtrbohD and AtrbohF are required for accumulation of reactive oxygen intermediates in the plant defense response. Proc Natl Acad Sci 99:517–522

Torres MA, Jones JD, Dangl JL (2005) Pathogen-induced, NADPH oxidase-derived reactive oxygen intermediates suppress spread of cell death in *Arabidopsis thaliana*. Nat Genet 37: 130–134

Triantaphylidès C, Krischke M, Hoeberichts FA, Ksas B, Gresser G, Havaux M (2008) Singlet oxygen is the major reactive oxygen species involved in photooxidative damage to plants. Plant Physiol 148:960–968. https://doi.org/10.1104/pp.108.125690

Turrens JF (2003) Mitochondrial formation of reactive oxygen species. J Physiol 552:335–344. https://doi.org/10.1113/jphysiol.2003.049478

Uluisik I, Karakaya HC, Koc A (2018) The importance of boron in biological systems. J Trace Elem Med Biol 45:156–162

Umair Hassan M, Aamer M, Umer CM, Haiying T, Shahzad B, Barbanti L, Nawaz M, Rasheed A, Afzal A, Liu Y, Guoqin H (2020) The critical role of zinc in plants facing the drought stress. Agriculture 10(9):396

Vashisth T, Oswalt C (2020) Movement of plant nutrients: HS1373. EDIS 2020(6)

Vernoux T, Wilson RC, Seeley KA, Reichheld JP, Muroy S, Brown S, Maughan SC, Cobbett CS, Van Montagu M, Inzé D, May MJ, Sung ZR (2000) The root MERISTEMLESS1/CADMIUM SENSITIVE2 gene defines a glutathione-dependent pathway involved in initiation and maintenance of cell division during postembryonic root development. Plant Cell 12(1):97–110

Viola E, Tornabene F, Ferretti E, Fantuzzi N (2013) Soft core plane state structures under static loads using GDQFEM and cell method. CMES 94(4):301–329

Wang W, Wang C, Sardans J, Tong C, Ouyang L, Asensio D, Gargallo-Garriga A, Peñuelas J (2018) Storage and release of nutrients during litter decomposition for native and invasive species under different flooding intensities in a Chinese wetland. Aquat Bot 149:5–16

Wang YY, Hsu PK, Tsay YF (2012) Uptake, allocation and signaling of nitrate. Trends Plant Sci 17(8):458–467

Waters RD, Burnett E, Lamm A, Lucas J (2009) Engaging stakeholders through social networking: how nonprofit organizations are using Facebook. Public Relat Rev 35(2):102–106

Welch RM (2002) The impact of mineral nutrients in food crops on global human health. Plant Soil 247:83–90

White PJ (2012) Long-distance transport in the xylem and phloem. In: Marschner's mineral nutrition of higher plants. Academic Press, pp 49–70

White PJ, Brown PH (2010) Plant nutrition for sustainable development and global health. Ann Bot 105:1

Wimmer MA, Eichert T (2013) Mechanisms for boron deficiency-mediated changes in plant water relations. Plant Sci 203:25–32

Xie HT, Wan ZY, Li S, Zhang Y (2014) Spatiotemporal production of reactive oxygen species by NADPH oxidase is critical for tapetal programmed cell death and pollen development in Arabidopsis. Plant Cell 26(5):2007–2023

Xu N, Braida W, Christodoulatos C, Chen JP (2013) A review of molybdenum adsorption in soils/ bed sediments: speciation, mechanism, and model applications. Soil Sediment Contamination Int J 22:912–929

Yu Q, An L, Li W (2014) The CBL–CIPK network mediates different signaling pathways in plants. Plant Cell Rep 33(2):203–214

Zeng Q, Ling Q, Wu J, Yang Z, Liu R, Qi Y (2019) Excess copper-induced changes in antioxidative enzyme activity, mineral nutrient uptake and translocation in sugarcane seedlings. Bull Environ Contam Toxicol 103(6):834–840

Zhang Y, Biswas A (2017) The effects of forest fire on soil organic matter and nutrients in boreal forests of North America: a review. In: Adaptive soil management: from theory to practices, pp 465–476

Index

M. Faizan et al. (eds.), *Reactive Oxygen Species*,
https://doi.org/10.1007/978-981-19-9794-5

The manufacturer's authorised representative in the EU is Springer Nature Customer Service Centre GmbH, Europaplatz 3, 69115 Heidelberg, Germany. If you have any concerns regarding our products, please contact ProductSafety@springernature.com

Printed and bound by CPI Group (UK) Ltd, Croydon, CR0 4YY
30/09/2026
02225241-0003